STUBBORN LIGHT

STU

The Best of *The Sun*, Volume III

BBORN
LIGHT

A collection of writings from the second decade
of *The Sun*, published in Chapel Hill, North Carolina

Edited by Sy Safransky

Published by The Sun Publishing Company
Chapel Hill, North Carolina

The contents of this book first appeared in *The Sun*.

Published by:
The Sun Publishing Company
107 North Roberson Street
Chapel Hill, North Carolina 27516
(919) 942-5282
www.thesunmagazine.org

Cover photograph: William Carter
Cover and book design: Julie Burke

Ordering information:
To order additional copies directly from the publisher, send $22.95 (postpaid) for each book to The Sun Publishing Company, 107 North Roberson Street, Chapel Hill, North Carolina 27516.

Library of Congress Catalog Card Number:
00 131483

ISBN 1-883235-14-6

Manufactured in the United States of America.

10 9 8 7 6 5 4 3 2

A NOTE ON
THE COVER

When we first published William Carter's photograph on the cover of our April 1998 issue, many subscribers wrote to ask what was making this elderly couple laugh. We asked the photographer to tell us what happened that day.

My wife and I spend Septembers in a small village in northern Italy. The couple in the picture (she has since died) lived in an even smaller village nearby. I met them through a friend and made an appointment to come back and photograph them the next day.

When I returned, they had forgotten the appointment. I found the woman on a path, and she said her husband was off picking figs and would be along soon. While we waited for him, I sat her on a nearby stone bench and set up my tripod. The man appeared shortly and insisted on giving me two or three figs, which I accepted.

The husband sat down beside his wife and, as I worked behind the camera, he asked where I was from and what I did. When he found out I was a professional photographer, he said, "Then I should pay you for taking my picture."

"No," I said in my rudimentary Italian, "you pay me with figs."

Now, the Italian word for "figs" is fici. But I said, "Paga con fice." Fice is Italian slang for "vaginas." I took their picture as they laughed.

I didn't find out myself what they were laughing at until later, when my wife and I were chatting with a multilingual Italian monk from a nearby monastery. He, of all people, explained my mistake.

CONTENTS

A Note On The Cover V

Introduction
    Sy Safransky XIII

Progress
    Fiction by Gillian Kendall 3

On Seeing A Sex Surrogate
    Mark O'Brien 25

A Priest Calls On Sunday Night
    Poetry by Edwin Romond 43

The Heart Of Compassion:
An Interview With Ram Dass
    Sy Safransky 45

Christmas In Seattle
    Fred Hill 73

Sonderkommando
    Fiction by Ivor S. Irwin 79

Even Hitler
    Poetry by David C. Childers 103

Judaism's Mystical Heart:
An Interview With Dovid Din
    Howard Jay Rubin 105

A Clouded Visit With Rolling Thunder
    Pat LittleDog 131

The White Man's Vision-Quest Journal
    Fiction by Gloria Dyc 145

The End Of A Sixties Dream?
An Interview With Stephen Gaskin
    Michael Thurman 161

When Thieves Break In
    Stephen T. Butterfield 179

Notes
    Poetry by Chris Bursk 191

Aliens In The Garden
    Fiction by James Carlos Blake 203

Eating Head
    Lorenzo W. Milam 223

All The Panamas In The World And Herb's
    Fiction by T.L. Toma 231

Blood For Oil
Poetry by Chris Bursk 243

A Soccer Hooligan In America
Fiction by Carl-Michal Krawczyk 247

Why Schools Don't Educate
John Taylor Gatto 255

Giving Away Gardens
Dan Barker 267

Scavenger's Run
David Grant 277

Amazing Conversations
John Rosenthal 285

Born Too Young: Diary Of A Pilgrimage
Sparrow 297

The Myth Of Therapy:
An Interview With James Hillman
Sy Safransky 355

Miracle At Canyon De Chelly
Deena Metzger 367

Somewhere Along The Line
Poetry by Antler 377

Cowards
Dan Howell 379

Villanelle
Poetry by Richard Hoffman 387

Of Lineage And Love
Stephen T. Butterfield 389

Flesh Of My Flesh
Poetry by Ona Siporin 405

Heart Too Big
Fiction by John C. Richards 407

Jealousy
Poetry by Lou Lipsitz 429

Last Year's Poverty Was Not Enough
Fiction by Ashley Walker 431

My Mother Is Still Alive
Poetry by Kathleen Lake 441

Uniting The Opposites:
An Interview With M.C. Richards
Sy Safransky 443

Night Of Dying

Maureen Stanton 453

This Life, This Word Unsaid

Poetry by John Hodgen 461

What's Eating Me: A Memoir

David Guy 465

Finding Out About Your Heart

Fiction by Candace Perry 479

Letter To Maxim

Alison Luterman 485

She Said, Can't We Just Be Friends?

Poetry by R.T. Smith 497

What It's Like

Fiction by Dana Branscum 499

Trying To Quit

Eleanor Glaze 509

On Being Unable To Breathe

Stephen T. Butterfield 519

The Cure

Poetry by Kathleen Lake 535

The Unhealed Life
Yaël Bethiem 537

Combing
Poetry by Veronica Patterson 543

Why I Like Dead People
Sallie Tisdale 545

The Evidence Of Miracles
Poetry by Jaimes Alsop 553

Instrument Of The Immortals
Jake Gaskins 555

The Prayer Of The Body:
An Interview With Stephen R. Schwartz
Sy Safransky 565

The Body Knew
Poetry by Tim Seibles 589

Tully
Fiction by D. Patrick Miller 591

The Ward
Bruce Mitchell 605

Contributors
.... 611

INTRODUCTION

S omeone called recently to ask my advice about starting a magazine. I told her the most important advice I could give her was not to follow anyone's advice.

If I'd followed the advice I was given in 1974, I never would have started *The Sun*. You need a million dollars to start a magazine, people told me. That was exactly a million dollars more than I had. To print the first issue, I borrowed fifty dollars. I sold the magazine on the street. For more than ten years, there were fewer than a thousand subscribers and never enough money to pay the bills. Unfortunately, this was the plight of many small magazines in those years; few survived.

Last year, when *The Sun* celebrated its twenty-fifth anniversary, our assistant editor joked that, in independent-journal years, this made *The Sun* about two hundred years old. I'm grateful *The Sun* has evolved from a stapled stack of xeroxed pages to a handsome monthly magazine with nearly fifty thousand subscribers nationwide. I'm grateful, too — at a time when a handful of media conglomerates control most of what we read in books, newspapers, and magazines — that *The Sun* survives without corporate backing and without carrying any advertising. One of the few reader-supported magazines in the country, its very existence makes a statement.

The Sun's literary reputation has grown as well. The magazine has received four Alternative Press Awards from the *Utne Reader*. Works from *The Sun* have won the Pushcart Prize, been featured on National Public Radio, and appeared in *Best American Short Stories* and *Best American Essays*. But these accolades wouldn't mean a thing if *The Sun* hadn't stayed true to itself: a magazine that honors life in all its complicated glory; that avoids literary pretentiousness and spiritual posturing; that celebrates beauty without ignoring the destructive forces around (and within) us. Sometimes sorrowful, sometimes sensual, sometimes angry, sometimes mischievous, *The Sun* is a magazine whose politics are personal and whose God isn't way up in the sky. People who write for *The Sun* aren't afraid to look at what they've lost; nor are they afraid to insist on joy, to look joy in the eye no matter how hard they're crying. Their struggle, our struggle: where do we draw the line? *The Sun* reminds us there are no lines.

In the mideighties, we published *A Bell Ringing in the Empty Sky*, a two-volume anthology of the best essays, stories, poems, and interviews from the magazine's first ten years. *Stubborn Light*, the book you're holding in your hands, is volume three in the "Best of *The Sun*" series, covering the magazine's second decade.

Choosing what to include was a challenge. As I worked my way through the 120 issues we published from 1984 to 1993, my list of "definite" yeses grew longer and longer. I winced, knowing what lay ahead. Mind you, I'm accustomed to making tough decisions; we receive nearly a thousand submissions a month at *The Sun*. But this was different. These weren't unsolicited manuscripts. This was writing I'd published, writing I'd fallen in love with. Now I'd fallen in love again.

In the end, as I struggled to pare down my choices, I felt like a hiker who'd been told to lighten his pack until it contained nothing that wasn't absolutely essential — and then to get rid of half of that.

I consoled myself that no anthology, no matter how inclusive, could do justice to a decade of *The Sun*. Some of what it leaves out: All those heated exchanges on the Correspondence page. People in Readers Write revealing some secret shame or secret happiness. The amazing photographs. The page of quotations that seals each issue with a kiss. The best of *The Sun*? It's the smart, dedicated people who work beside me, and it's all the devoted men and women who have come and gone. It's our old house in Chapel Hill with the creaky wooden floors and a ghost I've never seen. It's a subscriber telling me he can't imagine life without *The Sun*. As if I could.

Many people worked long and hard on *Stubborn Light*. Special thanks go to *The Sun*'s art director, Julie Burke, for carefully overseeing the project and giving the book its quiet, elegant look; assistant editor Andrew Snee for his incisive suggestions and sensitive editing; copy editor Seth Mirsky for guarding the gates of grammar and usage; proofreader Lynda Malone for standing beside him; manuscript editor Colleen Donfield for helping me decide what to leave in and what to take out, and whose comment sheets are themselves works of art; and editorial assistant Rachel Elliott for doggedly typing the text, and doggedly entering the corrections, and doggedly keeping the rest of us moving ahead.

None of my own essays are included in *Stubborn Light*, since many of them have already been reprinted in my book *Four in the Morning*. Nor did we include excerpts from the Readers Write section; we're planning a best of Readers Write, to appear in the near future.

You can open *Stubborn Light* and begin reading anywhere. But we've arranged the book, as we do each issue of *The Sun*, to be read from front to back. If you read it this way, you may find that *Stubborn Light* is more than the sum of its parts.

Publication dates are noted at the end of each piece. When reading the interviews, it's important to remember that people change; their ideas may not be the same as they were ten or fifteen years ago. Three people we interviewed — Rabbi Dovid Din, M.C. Richards, and Stephen R. Schwartz — have died. In 1997, Ram Dass suffered a debilitating stroke, which has limited his ability to speak.

It also grieves me to note the death of two authors whose work appears in *Stubborn Light*. In 1996, on a blustery winter night in Shrewsbury, Vermont, Stephen T. Butterfield's heart gave out while he was walking his dog; he was fifty-three. Just the day before, Butterfield had spent hours playing guitar and bouzouki with his Celtic band, When the Wind Shakes the Barley. According to his friends, he had left the session filled with the joy of the music and the communal spirit that comes from making music together.

Mark O'Brien died in July 1999 at the age of forty-nine, in his apartment in Berkeley, California. Almost completely paralyzed by polio in childhood, O'Brien insisted on living on his own, with minimal assistance, most of his adult life. He wrote — with passion, humor, and breathtaking honesty — using a mouth-stick to press the keys of his computer. He died alone, in the middle of the night, in his iron lung.

If O'Brien isn't dancing in heaven now to one of Butterfield's lilting tunes, heaven's the worse for it.

Sy Safransky

STUBBORN LIGHT

What is to give light must endure burning.

VICTOR FRANKL

Progress

Gillian Kendall

for Mark O'Brien

The first time we had Joe over, one spring evening some years ago, he lay on his gurney with his face positioned toward us. Because we were sitting, we could see his face at eye level. When we spoke, we looked directly into his eyes, not acknowledging the gurney, the sheets and straps, the tubes, and the hard metal rails.

Only I knew the rest of Joe's body: the underdeveloped, paralyzed lower half, with its baby white flab and deep red creases. He allowed pictures of himself to be taken only from the neck up, and in them he looked like an attractive nineteen-year-old,

---

F i c t i o n

---

though, like us, he was in his midtwenties. He had contracted polio just before the vaccine became widely available, and it had left him dependent on an iron lung, unable to use his back, arms, or legs. His genitals, I'd noticed with relief for his ego, looked perfectly average.

I knew intimate details about him — things only a mother or lover should know, things we would not speak of over wine and cheese this evening — because it was my job, four days a week, to take care of Joe. About to graduate from San Francisco State with no desire for a nine-to-five, I had cast about for the least-robotic job I could find. So, for six dollars an hour, I fed and talked to Joe and washed and dressed him. Being a masseuse, I occasionally threw in a brisk back rub. (I couldn't support myself through massage in Berkeley; the competition was vicious.)

Tilting a glass toward Joe's open mouth, I moved to direct the straw, but Joe pulled it toward him with his strong tongue. He drank, his eyes smiling at me.

At the far end of our living room, with its cathedral ceiling, its fading velvet couch, and its tired leather armchairs, one of my housemates was reading Virginia Woolf out loud to us. East Coast and anorexic, Susan was ablaze, consumed by her graduate study of suicidal writers. Although she claimed to have overcome her eating disorder, she still measured out three minute meals a day with scientific precision, allowed herself no sweets but the occasional secret raisin, and stole bits of food from her boyfriend's plate when he glanced elsewhere. The slightest deviation in household routine — our forgetting to put out the recycling, a phone call after her bedtime — could cause her to skip meals for days, hoping we would notice and pretending to hope we would not. I noticed and pretended not to.

Over the rim of the book I could see her hair, short and straight like a boy's, and her owl-like glasses. She looked thin and serious, but this was joy for her.

Our plan to entertain Joe was succeeding. Not only was it giving him pleasure, but it also allowed us to feel kind, since Joe required more emotional effort than other guests. For one thing, there was the gurney to ignore, all six feet of it. And there were so many topics to avoid in the presence of a handicapped person. But having Joe in our home assured us that we were open-minded.

We accepted him because, like us, he wanted to be alternative and intellectual. Joe liked classic books and classical music, and he had a degree from Berkeley in psychology. He had gotten it in six years, mostly through a lot of independent studies. He envied Susan's graduate program. He envied my hands. We liked him.

We all liked Susan's Monday-evening poetry sessions, to which a small troop of smart young people regularly carpooled, bringing with their signed first editions small gifts of Irish soda bread, nitrate-free wine, and coffee from El Salvador that cost ten dollars a pound. On those nights, Susan polished the silver and made me go over the floorboards with linseed oil. Sometimes our friends brought their friends, and after they had gone we would discuss the newcomers. Generally they were not quite right, and Susan would arrange that they not be encouraged to return.

But this was only Friday — an informal Joe-rehearsal to see if he'd fit in on Mondays. Susan thrust her brittle chin outward as she half read, half recited the Woolf with a militant, hypno- tized inflection. After she had finished, she set the book down with a little pat, but kept her eyes on the cover. Virginia Woolf's doglike visage stared back.

In the silent interval that followed, Susan's boyfriend, Loyton, brushed back his bangs, gazing at the fire. He'd grown his wheat- colored mane six inches since coming to California. When Su- san found him in the Cafe Roma, he'd been wearing plaid pants, but under her coaching he'd made some changes, started buying camping equipment from L.L. Bean. On weekends they went hiking to hot springs and shot many rolls of film. He was turn- ing out all right.

Far more all right than my own boyfriend, I felt. Timmy's tubular body was bent into a grotesque pretzel, legs in a denim lotus, arms entwined like fighting snakes, and he was studying his new shape with a blank, puzzled expression. He was waiting for someone to ask him what he was doing, so he could untwist, shrug, and say, "I don't know. It's fun."

The silence went beyond a pause. It seemed likely to extend

indefinitely. I was impressed that no one needed to break it, that we seemed so comfortable with each other. Maybe we were. The four of us had been sharing a house for almost a year, and Joe was beginning to blend in. Eventually, though, it began to seem as if we were all demonstrating how we didn't need conversation. Loyton squinted determinedly into the fire. Susan kept up soulful eye contact with Virginia Woolf. Timmy stared slack-jawed at nothing. I was afraid to fidget.

"Could I have some more wine, please?" Joe's question was low and unobtrusive, but an opener if anyone wanted it.

"Of course," I muttered. As he drank, I ventured, "I like this zinfandel, Tim. Good job."

"Huh? Oh, yeah. It had a fancy-schmancy label." Tim swirled his ginger ale and pretended to sniff the bouquet. He hated alcohol.

"Speaking of wines," Susan perked up, "we're headed up the coast for July Fourth, aren't we, Loyt? My friend told me that we must stop at a few of the little vineyards on the way. Has anyone been there?"

She looked brightly around the group. It was a good effort, but it flopped. She knew Tim hadn't been there, nor I, and the only one left was Joe. In progressive northern California, the wineries would boast wheelchair ramps and outsized toilet stalls, but I doubted they'd be iron-lung accessible. Joe could leave his lung only if he had oxygen. The cylinder lay strapped under the mattress, and from its rubber hose he occasionally needed to suck a rejuvenating hit. That was how he went to church and how he was able to be at our house, but even so, he could stay out only about four hours — not long enough to tour the Napa Valley.

"I hear it's really beautiful up there," Joe said, the way someone else might speak of the Himalayas. "I'd love to go one day."

Susan drew breath to invite him, but held the invitation in; her pale cheeks warmed a little. I knew she was thinking of the gurney, Loyton's Toyota, the oxygen tank and pee bags. "We'll have to put on a slide show," she covered. "With samples."

"That would be wonderful," Joe said, as if he'd been promised something valuable.

No one spoke. I feared another silence.

"I hear," Joe continued, "they're trying to make that German frost wine up there now. They wait till the first freeze and press the grapes when all the water inside is frozen. It's sweeter that way."

"Where'd you learn that?" said Tim, rocking slightly.

"TV," said Joe. "It's how I find out a lot of things."

The phone rang and Timmy untwisted himself, then clambered up off the couch to get it, his heavy, spraddle-legged walk rattling the glasses. I heard Joe laughing and looked at him.

"Don't look so pissed off," he whispered.

At the time, I thought about Joe's disability this way: I regretted it had happened to such a nice guy — as if polio would've been OK for, say, Phyllis Schlafly or Al Haig — but I wondered if he was so nice partly because of what he'd been through. Was he made patient by hospitals, or humbled by pain? No, I thought, not everyone's character improves under stress: Susan's eating disorder didn't mellow her snobbery; Tim's awkwardness didn't make him more eager to please; my own vague loneliness didn't lessen my critical nature. If Joe were deparalyzed, I thought, he'd still speak quietly, sit straight but relaxed, remember birthdays, laugh at all jokes, return phone calls promptly, and visit the sick. He would remain reliable and sweet, better than my friends and I.

He was so polite that, although his drivers had gotten him to our driveway that night at 6:30, he had waited until 7, the time we'd mentioned, to come in.

"You mean to say," said Susan when she heard, "that you just stayed outside for half an hour?"

"It was no problem," Joe said gently. "A different ceiling to look at." After every sentence his lips stayed slightly parted, and he searched your eyes.

I said the drivers could have taken him out of the van so he could look up at the moon and tree branches, but of course they hadn't thought of that. Underpaid, poorly educated, and ultimately desensitized to suffering, the drivers hauled Joe and his gurney around like luggage. That night the driver was Manny, according to the name on his shirt. Manny rang the doorbell at seven and through his soda-flecked mustache grumbled, "CPS delivering Mr. Joseph Beardsley. CPS accepts no responsibility for client's welfare inside your home. Will you sign here?" He held a chipped clipboard a few inches from my face, like a gun. I signed an *x* in protest.

"OK," Manny said, sweating. "We're going to proceed inside." He and another man negotiated the gurney, Joe aboard, up the two flights of steps from street level and then through our narrow porch. The angles weren't all that steep, but to Joe, who had once been thrown off and broken a wrist, it was a roller coaster. He closed his eyes.

They labored up the first steps one at a time, but then, accelerating at the curve, they clipped the edge of the gurney.

"Hey!" Joe yelped, eyes wrinkling tighter.

They held him in midair. "What?"

"Take it easy, OK?"

"Shit." Manny jolted his end back up. Joe was lying at about a sixty-degree angle, pointing headfirst toward the ground. He slipped a little down the mattress.

"Slow down," he said. "I mean it!"

They ignored him.

The porch was too shallow for the length of the gurney, so to get him in the front door they had to tilt him up and sideways, like a banking jet. The pillow tumbled off, and Joe fell hard against his restraining strap. "God damn!" he yelled. "What are you trying to do to me?" He entered our house angry and afraid. But as soon as the door closed behind the drivers, he sighed and his forehead cleared. "I brought cookies," he said. "Mrs. Field's."

AFTER WE'D FINISHED two bottles of wine that first night, Joe asked me to retrieve some notes from his backpack. As I withdrew the wrinkled printout, Susan and Tim stared at the typed pages. "Joe writes with a mouth-stick," I told them. "He taps out words on a keyboard mounted over his head." I didn't add that I was always afraid it would fall on him.

"I average about fifteen words a minute," he said, rolling his eyes. "I can only hold down one key at a time. At first everything I wrote looked like e.e. cummings, till I figured out how to use CAPS LOCK."

The printout, he explained, contained notes for his talk at a medical conference at San Francisco State. The doctors hadn't invited him to speak; he'd volunteered after learning that they were studying a prototype mover called Progress — basically a bed with a small engine attached — that could go anywhere an electric wheelchair could. Joe's body couldn't be supported in a wheelchair, but in the mover, he had heard, he could lie flat and drive with a mouth-stick. He was going to the conference to campaign for a test drive.

He wanted to try out his talk — part poem, part speech — on us. I set aside my plate and held the crumpled paper up to his eyes.

It was hard to understand him. Joe couldn't speak for long without the respirator, and every few lines he had to suck in oxygen. Also, because his lungs held little air, his voice didn't carry. Since I was next to him and was used to his low, nasal voice, I could hear, but the others were straining and lost. Tim sat up straight and squinted. Susan and Loyton glanced at each other with polite, secret alarm.

Even I could make out only parts of the poem: the last day before polio, six-year-old, sun-freckled Joe had rushed ahead of his siblings into an outdoor pool. Pools to him since then, the poem said, were indoor, grayish whirlpool baths feebly pumping tepid water. Pools since he was six meant someone holding him, keeping him afloat, helping him to breathe. At six he'd dared the high board once, had lost or won countless races. Now he slumped on plastic rafts, drifting wherever the jets propelled him, chlorine covering the smell of urine. Now he didn't race for anything.

I hated it. He would be addressing doctors, not social workers, and they'd want details of his physical pain or relief, information on the side effects of his drugs. But when he stopped I said, "It's great."

Susan and Loyton echoed me and offered other encouragements. All of us, except Tim, praised Joe's talk because its pain was true, and we had never felt pain like that.

SEVERAL HOURS AFTER our party, someone called for Tim's word-processing services. A private detective would pay double for a report typed before eight. Such rush jobs didn't worry Tim, who was most coherent between midnight and dawn anyway, but the sleep interruptions irritated the rest of us, especially me. All the next day, calls for Tim poured into the house, and customers kept coming to pick up work. It felt like living in an office.

Later that day, Tim abruptly decided to get his own business line. He lumbered into our bedroom, where I was reorganizing our tape collection, and said, "If I get any more calls at night, Susan's going to starve to death."

"That's not funny."

"Oh, sure it is," he said. "You have no sense of humor about anything anymore."

"Well, fuck you," I said cleverly. I strained to think of something lighthearted to follow it.

"So, anyway," Tim said, "I need a reference for the phone company."

"Well?" I snapped. "I'll give you one."

"No, someone with a *phone* has to." He rubbed at the lobe of his ear nervously. "Susan's in class."

"Loyton?" I filed a Tom Waits tape under *W*, then moved it to *T*. "He has good credit."

"Joe." Tim reached for my phone book. "Joe's always at home. They can call him."

His using Joe, his assurance that Joe would be lying by the phone, annoyed me. I took his *Best of Gershwin* and hid it deeply among the Grateful Dead.

After the phone company and Tim had finished with him, Joe called to speak to me. "Thank you for including me yesterday," he said. "I tried to call you earlier, but it was busy. It was one of the best evenings I've had."

I waited for him to add "in months" or "for a long time," but he didn't. He invited me to dinner. "I'm sorry I can't ask Susan and Loyton and Tim, too," he said firmly, "but you know how small my place is."

So, on a day when I was not working for him, I went to Joe's place as his guest. As his attendant, I usually wore jeans; to be his guest I put on a long skirt and a shell necklace. I announced to Tim, "I'm going over to Joe's for dinner."

Tim barely looked up from his typing. "I'll eat a potpie."

"OK. You don't have to kiss me goodbye."

We kissed with lips puckered, like children. He looked back at the screen, but I put my arms around his shoulders from behind. "Hey," I breathed into his neck. "Maybe later." But I didn't much feel like it.

JOE HAD THE best-located apartment of anyone I knew. One block off the happening end of Telegraph, Dana Street was quiet and bushy, with houses in eccentric colors. Wind chimes dangled over doorways, windsocks sailed from verandas, and cats stalked through the flower beds, hunting songbirds.

Joe's apartment was in a long redwood building. On such a street, even with rent control, it could have cost six hundred dollars a month, but with his government subsidy his share was only fifty. He shared the apartment with an invisible female roommate who had the upstairs kitchen, bathroom, and bedrooms. Joe lived in his lung in the front room, through which she entered and exited at odd hours.

Sunk a few feet below street level, the place was terminally dim. Joe was parked near the only window so he could read and hear people passing, but the rest of the room tapered into darkness unless he turned on a lot of lights.

That night he had candles, which the last attendant must have set up. They glimmered on the window ledge, inches from his hair, and on the TV table and the radio. Two flames were reflected in the rounded side of his lung, which looked like a water tank lying on its side with Joe's head and legs sticking out. In the candlelight, his pale skin looked healthy. He started chattering as soon as I came in. "Hope you're hungry; I have a lot of food. Go ahead if you want to change the CD. I was playing *Madame Butterfly*, but I could handle something less dramatic. Would you like some wine? It's been breathing heavily for an hour."

I admired the candles, rolled a chair near his head, and began to act social, although I felt I should be opening the lung to straighten the pads Joe lay on, or getting out his toothbrush to clean his teeth. He said again how he'd enjoyed that Friday, and how nice it was to hear literature read aloud, although he found Woolf too alienating. And Tim, he said, was charming, not at all as I'd described him.

"He gets on my nerves," I said. "Obviously he's a good person, and smart, and all the etceteras, or I wouldn't be with him." This was a lie; the truth was that I was afraid to break up with Tim in case no one else materialized. "I just wish he were more mature. He acts like a little boy sometimes."

"Like when he piled up all the plates and cups into a tower?"

"Exactly. Loyton's talking about fission, and my boyfriend can't even keep his hands still."

"Well, isn't that appealing, though? Don't we like people who are free with their emotions?" Joe turned his face toward me more fully. "At least he isn't all hung up or a workaholic."

"I guess." I hadn't expected a challenge to my complaining. "But he isn't really emotional, you know? I thought he would be. When I met him at that Reiki-massage workshop, I thought he was really playful and open. He liked to touch everyone. Later he told me he'd gone there just to meet women."

"Looks like it worked."

"Yeah, I guess it did, for him."

"Spare me the self-pity," Joe said. "You told me he gives great acupressure."

"He also works on my car."

Despite my disappointment over not being sincerely in love with Tim, I doubted I'd be any less critical of anyone else. And when I looked around, impatient with Tim's refusal to get a real job, to save money, to attend to world events, or to think beyond next week, I didn't see any other available men.

A spicy aroma was making me hungry. "You had someone cooking?" I asked.

"Go see."

In the kitchen I found a tray, with china and daisies perfectly arranged on a blue cloth. The oven held warm ceramic bowls filled with vegetarian entrees and exotic side dishes from the Good Earth restaurant. He'd spent more than I made in a day. I'd gathered that he had wealthy, if absent, parents, as well as income from the state. I called out my appreciation: "Hot damn!"

The candlelight was gold, the wine was wine colored, Joe's cheeks looked pink, and I suppose mine glowed, too. We shared one plate, which I held on my knees, and we ate and ate, nearly moaning with greedy delight. The rice tasted of saffron and fennel, and the mushroom casserole was made with herbs I couldn't guess. I said, "You'll love this," and scooped a big mouthful for him.

He chewed it slowly, with his eyes shut.

"Do you know how they make this?" I said.

"Of course not."

"I mean, do you think it's sautéed first, or do you think they —"

" 'Sautéed'?" he said. "Is that like fried? Is that what the Galloping Gourmet does with onions?"

"We ought to cook together sometime." I slathered butter on a dark rye roll. "You could supervise."

"Sure," he said. "We'd end up with dog food. Don't eat all the stuffed artichokes."

"This is fun," I said around a hunk of bread. "You ought to feed me more often."

"Good," he said. "I get the feeling you could use more fun."

That stopped me. Did I come off as dull? Me, with my functioning legs, poetry readings, and bright Berkeley housemates? "I have fun." I tried to produce an example. "Sometimes Tim and I go to the rose garden and watch the sun set. And we play tennis. That's fun."

"OK, maybe I'm wrong."

"You're right. I don't even like tennis." I held on to the sunsets, though.

It was late when I rose to go. "I told Tim I'd be home early," I said.

"You did," Joe answered flatly.

"You didn't have to go to so much trouble." I folded up the linen napkin.

"No, I didn't, but I wanted to."

"Oh, I forgot to tell you. Susan wanted me to invite you to come over one Monday. She has a few people in to read poems and things."

Nothing.

"So," I said, "shall I brush your teeth, or . . ."

"Ha!" He glared at the ceiling.

"What?"

"Is that what you usually do after you have dinner with a man? Brush his teeth?"

"I don't usually have dinner with anyone but Tim," I said lamely.

"And? Do you brush his teeth afterward?"

"No," I said. "After we eat, we do the dishes." I quickly kissed the top of Joe's head and left.

Weaving down Telegraph, dodging the lousy street musicians and aggressive panhandlers, I wondered if Joe had interpreted my complaints about Tim as a request for a replacement. Well, I wondered, what if we became lovers? Could I bathe, make love to, then bathe again that soft, still body? No, I could not make love to just a head. I did not always want to be on top.

"I think Joe wanted me to kiss him good night," I announced the second I got home. Susan was leaning against the stove with the kettle on high, and Tim was sitting on the wooden table, kicking his heels and eating a carton of ice cream with a serving spoon. I settled on the bar stool.

"Well," said Susan, watching wisps of steam emerge from the kettle, "did you?"

"Of course not."

We decided that Joe must be a virgin. "That's odd." Susan spooned leaves into her little brown teapot. "He's not unattractive."

"*What*?" said Tim. He held the spoon like a lollipop, licking it.

"His face," she said. "He has a film-star face."

"He's *paralyzed*," I pointed out. The steam thickened into a spout, then began to whistle. "It wouldn't matter if he were a movie star. No one would get into bed with him."

"Can he even have sex?" Tim asked me. "Does the equipment work?"

"I think so." He'd said something once to another attendant about masturbating. Nothing untoward, just a mature, casual remark, which she maturely and casually passed on to me. The slight movement he had in his right fingers was apparently sufficient. "But, I mean, he can't roll around, obviously. And he's lacking a partner."

"No sex?" said Tim. "What would be the point in living?" The whistling grew shrill. "Aren't you going to turn that off?"

"It must be a full, rolling boil." Susan raised her voice over the shrieking kettle. "In San Francisco there are sex substitutes — no, surrogates. I read an article about them. They help people overcome sexual problems."

"Joe has the worst problem of all," Tim said.

Susan looked longingly at the ice cream Tim was eating. "God, I'm hungry." She poured the boiling water into her china teapot and clapped on its lid.

THE FOLLOWING WEEK Joe said he needed me to take him to Wednesday-night Mass because his evening attendant would be out of town. "What do I wear?" I asked. "The last time I went to church I was six, and we had doilies on our heads."

"Dress up if you care what people think," Joe said. "God doesn't care."

I couldn't decide if Joe's faith was amazing or conventional. I didn't see how he could believe in some supreme being after all his misery, yet I could see the appeal of an afterlife, complete with wings.

On Wednesday I wore my reliable pink number, but among the women in more elegant high heels and fitted waists I felt marked as an atheist. The church had sweeping wooden rafters going up to a point in the middle, like a circus tent. From the apex hung a cross, and on the cross hung Jesus, balancing with bleeding hands, looking sadly down his nose at us.

I wanted to go sit up front near the flowers, but Joe explained that he had to make his confession first. I rolled the gurney up to a little black booth with a curtain at the back of the room. I pulled Joe into the confessional and stood cramped between him and the wall. There was no light, except a sliver from where the curtain hung crookedly over Joe's long gurney. Joe looked at a little screen, which was right at the level of his mouth. I supposed most people knelt down to it. He said, "Bless me, Father, for I have sinned. It's been two weeks since my last confession."

"Should I leave?" I whispered.

Joe shook his head.

"Is there someone in there with you?" asked the priest.

"Yes, there is. This is Joe Beardsley. My friend helped me get in here."

The voice mumbled something I couldn't catch, and then Joe began. "Well, Father," he said casually, "I've been angry this week. I've been in a lot of pain, and I haven't accepted God's will. When I'm in pain, I swear a lot. I take the Lord's name in vain."

Jesus, I thought. *Jesus fucking Christ.*

"Also," Joe said, his voice suddenly more sincere, "I've had impure thoughts. I have wanted to be sexual with women, one in particular, and also with some men."

Men? He'd never hinted that to me. It seemed odd — I'd assumed heterosexuality for Joe in a way I wouldn't have for anyone else. And how was I supposed to react to the "one in particular"?

"I've watched pornography on television, Father. And I've masturbated — three times, I think. I've been trying to stop, but I can't." He sighed hugely and waited.

"Is there anything else?" The voice sounded strained and depressed.

"That's all, Father."

"Say one Hail Mary, and one Our Father. Let the peace of God enter your heart, and —" The voice stopped; the priest on the other side was trying not to cry.

"Yes?" Joe said attentively, and suddenly the whole situation reversed: Joe was ready to receive the priest's confession and to absolve him of guilt and grief.

"— and it is not a sin to be angry." The voice was calm again. "Go with God."

I STOOD, SAT, and knelt with the others at the appropriate times in the service, holding the prayer book in front of my face. At first I held it for Joe to read, too, but he shook his head; he knew all the responses.

After everyone else had gone to the altar to receive Communion, the priest who had given the sermon came toward us with a goblet, followed by another man with a golden tray. Their white robes swished on the red carpet. The congregation turned to peer until they saw Joe's gurney, and then all but the children swiveled their necks back.

The children stared, and I watched nervously as the first priest said, "The body of Christ," and placed a white wafer on Joe's tongue. I wondered if they'd need a straw, but the other man moved softly to Joe's head and with a silver spoon lifted a little wine to his mouth. "The cup of salvation," he said, and he and Joe smiled into each other's faces.

And then a small shudder went through Joe, and he closed his eyes like someone in love. The priests quietly returned to the altar to finish the service. I felt jealous — jealous that I couldn't have that magic.

When we were back outside again, heading for the van, I told Joe how the Communion had made me feel.

"You could always convert," he suggested. "Let's go get ice cream."

But I felt as if I had already been converted, because I had seen something in Joe that I'd never seen in anyone else: what I could only call a soul.

A WEEK AFTER the Mass, I agreed to help Joe get to the medical conference so he could give his talk to the doctors. I didn't want to take him, but I didn't want the guilt of refusing, either, so I enlisted Tim to help.

Getting Joe into the back seat of my car was relatively easy. We put his head on my rolled-up jacket, folded his legs, and secured him with a couple of seat belts. But then we still had the stupid gurney — half the length of the Honda — to squeeze in. It was supposed to fold up, Joe said, but he had no clue how. The contraption had little hooks and levers and handles, all of which we twiddled. Nothing worked. After ten minutes, Tim and I were flushed with exasperation, and the gurney was still the same size.

"Wait a second," Tim said. "We're doing this wrong." He walked two yards away and gazed at the contraption. "It's more Zen this way," he said. "Think about how it *should* go down."

"That's not Zen." I kicked a wheel. "Zen doesn't use *should*."

"Look." Tim came back and pointed at the cushions. "There's a seam in the middle."

The gurney bent in the middle so that the two halves of the mattress lined up with the legs. It folded down to the size of a large suitcase — a large suitcase full of books, judging by the weight. It was one frustrating inch too big to go in the hatchback, no matter how we pushed and at what angle. Finally, we tied it to the roof with some twine and trundled over the bridge, the gurney knocking and Joe flat in the back, asking what time it was every five minutes.

The San Francisco State University Hospital sprawls uphill, with about an acre of steep ramps between the visitors' lot and the hospital entrance. When we finally made it to the auditorium, several hundred doctors in suits were watching a video of an actor in an electric wheelchair cruising and pivoting around a tennis court, hitting killer returns and showing off his legs in white shorts.

Tim fetched the person in charge, a wisp of a woman with a lavender nose and sad eyes. She squatted to talk to Joe. "Thank you for coming out here today," she announced loudly.

Joe shut his eyes. "No problem."

"Sorry," she said, standing up. "I thought you might have a hearing aid."

"Nope," Joe said, not looking at her. "I hear just great."

She consulted her spiral pad, covered with florid handwriting. "You're on in about half an hour, but we're running a little bit behind. Is that OK?"

Joe had already been away from his lung for two hours; two more, and he'd be seriously uncomfortable. "Fine," he said, his breathing a bit labored. We waited for forty-five minutes, Tim and I standing the whole time. Only when she came back did we notice the long, narrow steps up to the podium. The woman pointed the way to the backstage entrance instead, which entailed bumping Joe back out the way we'd come in and going around the building to the loading dock. She trailed us the whole way, whining, "Oh, I'm so sorry, I should have thought about this." When we finally made it backstage, she said, "OK, thank you, I guess I'll take over now." She edged me away from the gurney with a martyred look and tried to push it with the brake on. The gurney lurched crookedly, and Joe frowned at me.

I put my hands back on the grips. "I got him this far, I'll take him onstage."

Blinking, she flapped onstage and began introducing Joe as she struggled to lower the microphone. I maneuvered him over the cables and around the chairs that littered the backstage. As soon as I pushed past the black curtains, the light blinded me. I stopped for a minute to let my eyes adjust. I guess the audience had been waiting some time for Joe's appearance because they started clapping. He wasn't even up to the mike yet, and the applause was bursting our ears. It felt as if they were cheering both of us, him for being in the gurney and me for helping him. I felt needed and glamorous; *he* was helping *me*.

From the wings, I watched Joe's face, assured and vibrant as he talked. His thick, wavy hair shone in the spotlight, which shadowed the gurney into obscurity. I watched all those people watching his film-star face. He made the whole audience laugh. I had been utterly wrong about the doctors. They loved his poem, and they asked questions about how he could use the Progress mover. I would have to apologize, I thought; we'd have to celebrate. A dinner on me, maybe the symphony. We could get a box; we'd glide the mover in front of the seats. With the mover, it could be an almost perfect date.

Tim nuzzled my hair with little grunts. "Let's fool around."

"Don't." I pushed him away. "There are other people back here."

"It's boring. And I'm starving. You want to go eat later?" He looked as he always looked when he was hungry: weak and confused.

"Maybe." I'd been thinking of staying at Joe's and talking after taking him home. I wished I could go out with Joe instead; I wished I could put Joe's head on Tim's body. "Maybe not."

I PLANNED THE date. I told Tim something about paying Joe back for the dinner he'd fed me, and I got busy ironing a dress, booking tickets for one evening of the Mozart fest, finding my almond massage oil. It was time, I thought, to give Joe a treat: an hour-long, full-body massage. That I could offer guiltlessly.

That Saturday night, Susan saw me putting on the new dress. "Black?" she said. "In summer?"

"I look great." I fluffed my hair up a little higher. "What are *you* doing tonight?"

"Touché," she said crossly, and flounced back to her room.

Joe had never eaten lobster before. They take a long time, I told him, but I willingly got my fingers buttery and accidentally spritzed his nose with lemon, feeding him the most succulent chunks. While I was cracking shells and teasing out pink wedges of meat with a little fork, Joe talked.

"It's the best thing that ever happened to me," he said of the mover, in which he'd been practically living for two days. "Yesterday I went to visit the Enabling Center on Telegraph. While I was there, three of us decided to go next door for coffee. It took me an hour to get to the center and another fifteen minutes to get next door, but it was great! All the streets between my house and the center have ramps, and —"

"Eat." I offered the tip of a shiny claw. "I'm so excited for you. Can you get the mover in a car?"

He nodded, chewing. "Yeah, and I can take more oxygen! I can go anywhere, almost. I still get tired, but I can be out all day now."

"The wine country," I said.

"The *beach*," he said. "I haven't been to the beach since I was six."

How terrible, I thought, remembering dried strands of sea-weed, the hiss and whisper of the waves, the salty, sunburned rides home. "You know," I said, "I haven't been to the beach yet this year."

"That's amazing," he said. "I'd go all the time if I could. But I guess it's hard for you, too, sometimes. Sometimes I think that if I weren't paralyzed, I'd be Superman, you know? I'd be a corporate executive, and get married, and stay married, and have two nice children, and take them to Europe and the beach a lot, and grow organic vegetables, and teach Sunday school, and get my master's and another BA in history, and learn how to cook like this, and exercise every day, and learn to sing, and maybe compose a little for the saxophone. And study more French, and Italian, too."

"In your copious free time," I said.

"But I guess maybe I wouldn't. Maybe I'd only learn how to fly a glider."

After dinner, we headed for the silver, shining marquee of the Palace Theater, joining the crowd in white dinner jackets and splashy sequins jostling the velvet ropes.

As we neared the entrance, more people crowded behind us, politely pushing. I let Joe navigate his mover with his mouthstick, instead of pushing him myself, so he could show off. But I kept my hand on his shoulder to establish myself as his date. As we got close to the door, I moved a little ahead of Joe, toward the ticket taker, who was looking at us. *Oh, my black dress is perfect*, I was thinking, reaching into my perfect purse for the tickets, when I heard a sharp, prolonged clatter of metal striking stone.

The mover had toppled off the curb and fallen sideways, throwing Joe into the street. Wheels screeched as limos and taxis swerved, honking. A dozen people surrounded him, shouting for help. I pushed through the elbows and hips as fast as I could. Joe was screaming for me.

I WENT WITH him in the ambulance, rushing back up the hills to the same hospital where he'd spoken. I stayed until after midnight, waiting while they set his arm. I wanted to know exactly how he'd fallen, and to apologize for not pushing him. But Joe was wired to so many IVs and monitors it was as if he were in a cage, and the combination of pain and drugs made him groggy and uncommunicative.

I returned the next morning and stayed past dinnertime, then returned the next day, and the next. I hauled his computer in and set it up, although he was too tired and dopey to care. At night I phoned his other attendants from home, demanding they at least call Joe. I reached his parents in Los Angeles, and his mother promised to come up. When Joe was not asleep, we played checkers; I made tired fun of the hospital food; I read aloud to him from *Rolling Stone* and *Mother Jones* and the ward's *Readers Digest*s. When my voice gave out, I turned on soap operas and game shows. "Quick, Joe! What's the capital of Colombia?" I took charge of understanding the doctors and physical therapists; I made charts of how he might expect to heal. By the end of the week, I was running on instant coffee; I'd lost weight and stopped washing my hair. When I could not be at the hospital, I slept like a stone or ran errands, buying books, tapes, sweets, and fruit, anything to cheer him.

Timmy stopped me. The day Joe's mother was to arrive, Tim took the keys to my car and made me go for a walk. "This is Saturday," he told me. "I haven't seen you for eight days. You look awful. Can we at least talk?"

Walking up the steep hill, I was panting. I felt lightheaded from lack of food. Tim's arm around my shoulders felt heavy. "Wait," he said. "We don't want you having a heart attack. Breathe."

I didn't want to stop; I wanted to hurry up and get back so I could rush to the hospital, grab the elevator to the fourth floor, take the second corridor on the left, and run into room 12.

But Tim brushed the leaves off the curb and tugged me to sit down. He pulled my head down against his chest and stroked my hair and held me until I relaxed.

Around us, people's gardens bloomed in the soft colors of late summer. A small, neat gray cat with blue eyes licked its front paw on a stone wall. In one of the seminaries, a choir was practicing. Squirrels ran up and down the phone lines; sparrows rustled the hedges. "It's all right," Tim said, rubbing my neck. "It's all right."

"What would you do?" I asked him. "I can't just abandon him because I'm tired."

"You don't have to abandon him," he said. "This is not black or white. I just think you're overdoing it, going there all day every day."

"He doesn't have any choice," I said. "He has to be there all the time."

"But you don't."

"I feel guilty if I don't!"

"And you feel crummy if you do! Do you really think it helps Joe for you to be as miserable as he is?" Tim broke a branch off a bush and angrily stripped the leaves from it.

I thought about the priest telling Joe it was not a sin to be angry. I remembered how the priest had seemed to ask Joe's forgiveness. And how Joe had given it, without even knowing.

JOE'S MOTHER STAYED even after he got out of the hospital, and she hired a nurse for him. I visited a few times, always with Tim. We drew rainbows on his cast. We talked of Tim's strange customers, of an odd coin I had found in a parking lot, of television movies. Joe said they had put better brakes and safety straps on the mover, and he could have it back as soon as his arm was better. "I've got to get back on that horse," he said.

Once, when Tim was out of the room, Joe asked if I would help him relearn to use the mover. "I think I'll remember the steering," he said. "It's using the mirrors that gives me trouble. If you could go with me and sort of tell me which way I'm headed, I think I could get the hang of it."

"Is that why you fell?" I asked. "You didn't get jostled?"

"I don't know." Joe closed his eyes. "I don't remember. There were a lot of people around. I was trying to stay close to you. I turned too hard to the right, and the back wheel went off the curb. Maybe I misjudged the distance. I think the mirrors really distort distances."

"Oh," I got out. "So it wasn't anybody's fault?"

Joe shook his head slowly. "Of course not."

"I'm sorry," I said.

"I just said it wasn't your fault."

"I'm sorry anyway."

WHEN THE TIME came for Joe to get back on the horse, I was starting a tough course in bioenergetics and working for Tim, too. He had gotten a hefty contract with a city office, and the job was so big that he hired me to do some of the grunt work. I did not miss being Joe's attendant, and I was generally glad that Tim was my boyfriend. I felt that it was not necessary to be in love; we had an effortless life together.

But Susan was gone, and I missed our Monday evenings. I wished our doorbell would ring and I could answer it to find a new person waiting there, someone I didn't know, someone who might turn out to be quite perfect.

Although I didn't work for Joe anymore, I thought about him and his mover. His new nurse helped him relearn it, and once I saw them together, progressing slowly across the campus. She had a round, thoughtful face and a strong, simple body. She looked like a wonderful woman, I thought, much kinder than I.

July 1993

ON SEEING
A SEX SURROGATE

Mark O'Brien

Mark O'Brien contracted polio at the age of six and spent most of his life in an iron lung. He died in 1999.

In 1983, I wrote an article about sex and disabled people. In interviewing sexually active men and women, I felt removed, as though I were an anthropologist interviewing headhunters, all the while endeavoring to maintain the value-neutral stance of a social scientist. Disabled and a virgin, I envied these people ferociously. It took me years to discover that what separated me from them was fear — fear of others, fear of making decisions, fear of my own sexuality, and a surpassing fear of my parents. Even though I no longer lived with them, I continued to live with a sense of their unrelenting presence, and of their disapproval of sexuality in general, mine in particular. In my imagination, they seemed to have an uncanny ability to know what I was thinking, and were eager to punish me for any malfeasance.

Whenever I had sexual feelings or thoughts, I felt accused and guilty. No one in my family had ever discussed sex around me. The attitude I absorbed was not so much that polite people never thought about sex, but that *no one* did. I didn't know anyone outside my family, so this code affected me strongly, convincing me that people should emulate the wholesome asexuality of Barbie and Ken, that we should behave as though we had no "down there's" down there.

As a man in my thirties, I still felt embarrassed by my sexuality. It seemed to be utterly without purpose in my life, except to mortify me when I became aroused during bed baths. I would not talk to my attendants about the orgasms I had then, or the profound shame I felt. I imagined they, too, hated me for becoming so excited.

I wanted to be loved. I wanted to be held, caressed, and valued. But my self-hatred and fear were too intense. I doubted I deserved to be loved. My frustrated sexual feelings seemed to be just another curse inflicted upon me by a cruel God.

I had fallen in love with several people, female and male, and waited for them to ask me out or seduce me. Most of the disabled people I knew in Berkeley were sexually active, including disabled people as deformed as I. But nothing ever happened for me. Nothing was working in the passive way that I wanted it to, the way it works in the movies.

In 1985, I began talking with Sondra, my therapist, about the possibility of seeing a sex surrogate. When Sondra had originally mentioned the idea, I had been too afraid to discuss it. She'd explained that first I'd need to see a sex therapist, who worked with a client's emotional problems concerning sex; then that therapist could refer me to a surrogate, who worked with a client's body. I rationalized that someone who was not an attendant, nurse, or doctor would be horrified at seeing my pale, thin body with its bent spine, bent neck, washboard rib cage, and hipbones protruding like outriggers. I also dismissed the idea of a surrogate because of the expense. A few years earlier, I had phoned a sex surrogate at the suggestion of another therapist. The surrogate had told me that she charged according to a sliding scale that *began* at seventy dollars an hour.

But now my situation had changed. I was earning extra money writing articles and book reviews. My rationalizations began to strike me as flimsy.

Still, it was not an easy decision. What would my parents think? What would *God* think? I suspected that my father and mother would know, even before God did, if I saw a surrogate. The prospect of offending three such omniscient beings made me nervous.

Sondra never pushed me one way or another; she told me the choice was mine. She gave me the phone number for the Center on Sexuality and Disability at the University of California at San Francisco. I fretted over whether I would call; whether I would call and immediately hang up; whether I would ever do *anything* important on my own. Very reluctantly, when no one was around, I called the number, but only after assuring myself that nothing terrible would happen. I never felt *convinced* nothing terrible would happen, but I was able to take it on faith — a frail, stumbling, wimpy faith. With my eyes closed, I recited the number to the operator; I was afraid she'd recognize it. She didn't.

"UCSF," a voice answered crisply.

Trying to control the shakiness of my voice, I asked for the Center on Sexuality and Disability. I was told the center had closed — and, momentarily, I felt immeasurably relieved. But I could be given a number to get in touch with the therapists who had once worked there; would I like that? Uh-oh, another decision. I said OK. But at that number I was told to call another number. There, I was referred to yet another number, then another, then another. I made these calls quickly, not allowing myself time to change my mind. I finally reached someone who promised to mail me a list of the center's former therapists who were in private practice.

About this time, a TV talk show featured two surrogates. I

watched with suspicion: Were surrogates the same as prostitutes? Although they might gussy it up with some psychology, weren't they doing the same work?

The surrogates did not look like my stereotype of a hooker: no heavy makeup, no spray-on jeans. The female surrogate was a registered nurse with a master's in social work. The male surrogate, looking comfortable in his business suit, worked with gay and bisexual men. The surrogates emphasized that they deal mostly with a client's poor self-image and lack of self-esteem, not just the act of sex itself. Surrogates are trained in the psychology and physiology of sex so they can help people resolve serious sexual difficulties. They aren't hired directly, but through a client's therapist. Well aware of the likelihood that a client could fall in love with them, they set a limit of six to eight sessions. They maintain a professional relationship by addressing a specific sexual dysfunction; they aren't interested in merely providing pleasure, but in bringing about needed changes. As I learned more about surrogates, I began to think that perhaps a surrogate could help even someone as screwed up and disabled as I.

When Sondra went on vacation, I phoned Susan, one of the sex therapists on the list I'd gotten from UCSF, and made an appointment to see her in San Francisco. I felt delighted that I could do something about my sexuality without consulting Sondra; perhaps that's why I did it. I was not sure whether calling the therapist in Sondra's absence was the right thing to do, or whether it was even necessary, but it felt good to me.

The biggest obstacle to seeing Susan turned out to be the elevator at the Powell Street subway stop, which went from the subterranean station to the street. Because of my curved spine, I cannot sit up straight in a standard wheelchair, so I use a reclining wheelchair that is about five and a half feet long. The elevator in the BART station was about five feet across, diagonally. Dixie, my attendant, raised the back of my wheelchair as high as she could and just barely managed to wrestle me and herself into the elevator. But when we reached street level, she could not get me out. This was ridiculous: if I could get in, the laws of physics should have permitted me to get out. But the laws of physics were in a foul mood that day. Dixie and I went down to the station level and discovered that I could get out down there. We complained to the station agent, who seemed unable to understand. We tried the elevator again. The door opened on a view of Powell Street. Dixie tried lifting and pushing the wheelchair out of that cigar-box elevator in every possible way.

"Well, do you want to go back to Berkeley?" she asked in frustration.

I thought what a waste it would be to go back now. I told her to raise the back of my wheelchair even higher. It put a tremendous strain on my thigh muscles, but now Dixie was able to wheel me out of the elevator with ease. Liberated, we strolled down Powell Street, utterly lost.

Eventually, we found Susan's office. Right away I realized I could trust her. She knew what to ask and how to ask it in a way that didn't frighten me. I described to her my feelings about sex, my fantasies, my self-hate, and my interest in seeing a surrogate. She told me the truth: because of my disability, it would never be easy for me to find a lover. She told me that her cerebral palsy, the only evidence of which was a limp, had repelled many potential partners. I found this hard to believe. She was so bright, so caring, so pretty in her dark and angular way. (I was already developing a crush on her.)

Susan said that she knew of a very good surrogate who lived in the East Bay, and that she would give the surrogate's name and phone number to Sondra when she returned from her vacation. If I decided to go ahead with it, Sondra would call the surrogate and ask her to phone me.

Because of our talk, doing that now seemed less scary. I had started to believe that my sexual desires were legitimate, that I could take charge of my sexuality and cease thinking of it as something alien.

When Sondra returned from vacation, she told me that she had a message from Susan on her answering machine. She asked why I had seen another therapist without informing her. Sondra seemed curious, not angry, as I'd feared she might be — actually, as I'd feared my parents would be. I said that I wasn't sure why

I'd gone to see Susan, but that I had felt odd discussing surrogates with Sondra, because she seemed to me so much like my idealized mother figure.

Meanwhile, I searched for advice from nearly everyone I knew. One friend told me in a letter to go ahead and "get laid." Father Mike — a young, bearded priest from the neighborhood Catholic church — told me Jesus was never big on rules, that he often broke the rules out of compassion. No one advised me against seeing a surrogate, but everyone told me I would have to make my own decision.

Frustrated by my inability to find The Answer, a blinding flash that would resolve all my doubts and melt my indecision, I brooded. Why do rehabilitation hospitals teach disabled people how to sew wallets and cook from a wheelchair but not how to deal with our own damaged self-image? Why don't these hospitals teach disabled people how to love and be loved through sex, or how to love our unusual bodies? I fantasized running a hospital that allowed patients the chance to see a surrogate, and that offered hope for a future richer than daytime TV, chess, and wheelchair basketball. But that was my dream of what I would do for others. What would I do for *me*?

What if I ever *did* meet someone who wanted to make love with me? Wouldn't I feel more secure if I'd already had some sexual experience? I knew it could change my perception of myself as a bumbling, indecisive clod, not just because I'd had sex with someone, but because I'd taken charge of my life and trusted myself enough to make decisions. One day, I finally told Sondra I was ready to see a surrogate.

ABOUT A WEEK later, my phone rang during my morning bed bath. It was a woman's voice I had never heard before.

"Hello, Mark? This is Cheryl."

I knew that it was the surrogate. She didn't have to tell me.

"I could see you March 17 at eleven o'clock," she said. "Would that be good for you?"

"Yeah, it would be. But I'm busy right now. Could you call me back this afternoon, when I'll be by myself?"

Now that I had decided to actually see a surrogate, I had another problem: where would I meet her? She couldn't come here, because I didn't have a bed, just an iron lung with a mattress barely wide enough for me. When Cheryl called back, she asked if I could come to her office, which was up a flight of stairs. I told her that would be difficult. Finally, we agreed to meet at the home of one of my friends.

I was terribly nervous when I asked Marie whether I could use her place. I had visited her often in her spacious living room, which contained a double bed. Marie, who uses a wheelchair, had made the cottage she and her lover share completely accessible. It was also within walking distance (or wheelchair-pushing distance). When I told her about Cheryl, she readily agreed.

As the day approached, I became increasingly apprehensive. What if Cheryl took one look at me — disabled, skinny, and deformed — and changed her mind? I imagined her sadly shaking her head and saying, "Oh, no, I'm sorry. I didn't know. . . ." She would be polite, but she would flee from me.

On the phone, Cheryl had explained that she would interview me for the first hour of the session; then, if I agreed, we would do "body-awareness exercises." I'd been too scared to ask what this meant, but had said I would give it a go.

When March 17 arrived, I felt unbearably nervous. I had to remind myself repeatedly that we were just going to talk about sex; then, in the second hour, we would do those body-awareness exercises, whatever they were, but *only* if I wanted to do them.

Vera, one of my morning attendants, dressed me, put me in my wheelchair, and pushed me to Marie's cottage. Vera tried to reassure me, but it didn't help. I felt as though I were going to my own execution.

We arrived at Marie's place at 10:45. The door was locked and no one was home. Vera sat on a bench in the yard, lit a cigarette, and chatted amiably as I sweated out the minutes. An eternity passed (seven or eight minutes) before I heard the buzzing sound of Marie's electric wheelchair.

Once we were inside, Vera put a sheet I had brought with me on the double bed. Then she lowered me onto it. The bed was close to the floor, unlike my iron lung. Since it's difficult for me to turn my head to the left, Vera pushed me over to the left side of the bed, so that Cheryl could lie next to me and I could still see her. Then Vera put the hose of my portable respirator near my mouth, in case I needed air. I thought it likely because I'd never been outside the iron lung for an hour without using the portable respirator. I was all set. I glanced at the noncommittal green numerals flashing on the nearby digital clock. 11:04. Cheryl was late.

Marie talked with Vera as I waited: 11:07; 11:11. Oh, God, would she ever come? Perhaps she had found out what an ugly, deformed creep I am and was breaking the appointment.

A knock on the door. Cheryl had arrived.

I turned my head as far to my left as I could. She greeted me, smiling, and walked to where I could see her better. *She doesn't hate me yet,* I thought. She pulled a chair up to the bedside, apologized for being late, and talked about how everything had gone wrong for her that morning. Marie went out the door with Vera, saying that she would return at one. Cheryl and I were alone.

"Your fee's on top of the dresser," I said, unable to think of anything else to say. She put the cash into her wallet and thanked me.

She wore a black pantsuit, and her dark brown hair was tied behind her head. She had clear skin and large brown eyes, and she seemed tall and strong — but then, I'm four-foot-seven and weigh sixty pounds. As we talked, I decided that she was definitely attractive. Was she checking out my looks? I was too scared to want to know.

Talking helped me to relax. She told me that she was forty-one, married to a psychiatrist, and had two teenage children. She was descended from French Canadians who had settled in Boston. "Boston?" I said. "That's where I was born." After talking about Boston for a while, I asked whether she was Catholic, like me. She told me she had left the Catholic Church during her adolescence, when her priest had condemned her sexual behavior.

I began to tell her about my life, my family, my fear of sexuality. I could see that she was accepting me and treating me with respect. I liked her, so when she asked me if I would feel comfortable letting her undress me, I said, "Sure." I was bluffing, attempting to hide my fear.

My heart pounded — not with lust, but with pure terror — as she kneeled on the bed and started to unbutton my red shirt. She had trouble undressing me; I felt awkward and wondered if she would change her mind and leave once she saw me naked. She didn't. After she took my clothes off, she got out of bed and undressed quickly. I looked at her full, pale breasts, but was too shy to gaze between her legs.

Whenever I had been naked before — always in front of nurses, doctors, and attendants — I'd pretended I wasn't naked. Now that I was in bed with another naked person, I didn't need to pretend: I was undressed, she was undressed, and it seemed normal. How startling! I had half expected God — or my parents — to keep this moment from happening.

She stroked my hair and told me how good it felt. This surprised me; I had never thought of my hair, or any other part of me, as feeling or looking good. Having at least one attractive feature helped me to feel more confident. She explained about the body-awareness exercises: first, she would run her hand over me, and I could kiss her wherever I wished. I told her I wished that I could caress her, too, but she assured me I could excite her with my mouth and tongue. She rubbed scented oil on her hands, then slowly moved her palms in circles over my chest and arms. She was complimenting me in a soft, steady voice, while I chattered nervously about everything that came to mind. I asked her if I could kiss one of her breasts. She sidled up to me so that I could kiss her left breast. *So soft.*

"Now, if you kiss one, you have to kiss the other," she said. "That's the rule."

Amused by her mock seriousness, I moved to her right breast. She told me to lick around the edge of the nipple. She said she liked that. I knew she was helping me to feel more relaxed, but that didn't make her encouragements seem less true.

I was getting aroused. Her hand moved in its slow circles, lower and lower, as she continued to talk in her reassuring way

and I continued my chattering. She lightly touched my cock —
as though she liked it, as though it was fine that I was aroused.
No one had ever touched me that way, or praised me for my
sexuality. Too soon, I came.

After that, we talked awhile. I told her about a woven Gua-
temalan bracelet a friend had given me for this occasion. She
asked me whether I had any cologne; I said I did, but that I never
wore it. That we could be talking about such mundane matters
right after an intense sexual experience seemed strange at first.
Another lesson learned: sex is a part of ordinary living, not an
activity reserved for gods, goddesses, and rock stars. I realized
that it could become a part of my life if I fought against my self-
hatred and pessimism.

I asked Cheryl whether she thought I deserved to be loved
sexually. She said she was sure of it. I nearly cried. She didn't hate
me. She didn't consider me repulsive.

She got out of bed, went into the bathroom, and dressed. By
then it was nearly one. Taking an appointment book out of her
purse, she told me that next time she wanted us to work on having intercourse. She asked me whether I had been afraid to see
her that day; I admitted that I'd felt spasms of deep terror. She
said it had been brave of me to go through with the session despite my fear.

The door opened. It was Marie and Dixie. They asked me
about the experience. I told them it had changed my life. I felt
victorious, cleansed, and relieved.

Dixie pushed me back to my apartment, through the quiet
neighborhood of small, old houses and big, old trees. It was a
warm day, which I hadn't noticed on the way over. I asked Dixie
about her first sexual experience. When she described it, I felt
admitted to some place from which I had always felt excluded:
the world of adults.

Back home, Dixie put me into the iron lung and set up my
computer so that I could write. Pounding the keys with my
mouth-stick, I wrote in my journal about my experience as quickly
as I could, then switched off the computer and tried to nap. But
I couldn't; I was too happy. For the first time, I felt glad to be a man.

When I saw Cheryl the second time, two weeks later, I felt more relaxed and confident. We chatted briefly, but there was no formal interview. After pulling down the window shades, she undressed me with more ease than before. I felt less afraid and embarrassed. As I watched her undress, I anticipated the sight of her breasts. There they were, full and rounded. Before she could even get into the bed, I had climaxed. I felt angry at myself for being unable to control the timing of my orgasms, but Cheryl said she would try to stimulate me to another orgasm. I didn't believe that she could arouse me again, but I trusted her enough now to let her try.

She lightly scratched my arms, which, to my surprise, I liked. I spent a lot of time kissing and licking her breasts. I asked her to rub the eternally itchy place behind my balls, which she said was called the perineum. The use of such a dignified Latin word to name a place that didn't even have a name, as far as I had known, struck me as funny. I screamed with delight as she rubbed me, surprised that my body could feel so much pleasure. Then I felt a warmth around my cock. I realized that Cheryl wasn't beside me anymore.

"Know what I was doing?" she asked a few seconds later.

"No."

"I was sucking you."

It wasn't long before I had another erection. Aroused and more confident, I said I wanted to try to have intercourse with her, so she quickly scrambled into place over me, her knees by my side. I breathed more rapidly, filled with anticipation, a feeling of *this is it*. She nearly stepped on my feet, which rattled me a little. Reassuring me, she held my cock and rubbed it against her, but when she tried to place it inside her, I panicked. For reasons I still don't understand, I felt that it couldn't fit. Perhaps I feared success. Perhaps intercourse would prove I was an adult, something I had never been willing to acknowledge. Perhaps it would suggest that I could have had intercourse long before, *if* I hadn't contracted polio, *if* I hadn't been so fearful, *if* . . . I did not want to contemplate this long chain of *if*s.

I insisted to Cheryl that I couldn't fit into her vagina. She said that was impossible. Then suddenly I came again — outside of her.

I felt humiliated. Cheryl asked me if I had enjoyed myself. I said, "Oh, yes, up to the anticlimax." She assured me that she had enjoyed it, which cheered me somewhat. And it was still pleasant for me, lying beside her, the two of us naked. I told her I wanted to recite a poem I'd memorized for this occasion, Shakespeare's eighteenth sonnet:

> Shall I compare thee to a Summer's day?
> Thou art more lovely and more temperate.
> Rough winds do shake the darling buds of May,
> And Summer's lease hath all too short a date. . . .

I stumbled through it, forgetting phrases, stopping and starting again, but I made it to the end:

> As long as men can breathe or eyes can see,
> So long lives this, and this gives life to thee.

Cheryl said that she was touched, that it was sweet of me to recite the poem. I felt glad that I was now a giver of pleasure, not merely a passive recipient.

An attendant came and took me home. I ate supper, exhausted and contented. But the next day I worried: Why had I panicked? Would I ever be able to have intercourse with Cheryl? With any woman?

MARIE TOLD ME that she couldn't let me use her house for the next appointment because she and her lover were going out of town. So I called Neil, a disabled playwright who lived in a large apartment building in my neighborhood. Although I hadn't known him long, he readily agreed. But he told me that his mattress was on the floor of his bedroom. This worried me because it would make it difficult, perhaps impossible, for an attendant to lift me back into my wheelchair.

On the day of the appointment, Dixie took me to Neil's

building. Neil has a rare disabling condition that impairs his speech, but allows him to stand and hop about on one foot. There he was, standing on one foot beside his wheelchair, which he had parked outside the building's entrance. Upon seeing us, he plunked himself into his wheelchair and led us to the elevators. Once inside the apartment, Dixie pushed me into the bedroom and eyed the mattress with skepticism, saying that she could easily put me on it but feared that she would hurt her back lifting me later. After a minute of mutual indecision, she picked me up from the wheelchair and set me down on the mattress. Once she'd made sure that I was comfortable, she and Neil left.

I lay there looking at Neil's clock and wondering whether Cheryl would ever arrive. Neil had told me he would wait for Cheryl outside the building to give her the keys. What if Neil had become bored waiting and left? Was Cheryl coming at all?

After waiting for forty minutes, I heard some noise in the outer room. It was Cheryl, who apologized for being late.

As Cheryl undressed me and herself, I noticed that I wasn't becoming aroused. I felt proud of my self-control and began to think of myself as a mature, sophisticated man, accustomed to being in a bedroom with a naked woman.

She got into the bed with me and began to stroke my thighs and cock. I climaxed instantly. I loathed myself for coming so soon, in the afterglow of my man-of-the-world fantasies. Undismayed, Cheryl began to stroke me, scratch me, and kiss me slowly. Reminding me of our previous session, she assured me that I could have a second orgasm. She said that she would rub the tip of my cock around her vagina. Then she would put it into her. I couldn't see what was going on down there, and I was too excited to sort out the tactile sensations. Suddenly, I had another orgasm.

"Was I inside of you?" I asked.

"Just for a second," she said.

"Did you come, too?"

She raised herself and lay beside me.

"No, Mark, I didn't. But we can try some other time if you want."

"Yes, I want."

After she got off the mattress, she took a large mirror out of her tote bag. It was about two feet long and framed in wood. Holding it so that I could see myself, Cheryl asked what I thought of the man in the mirror. I said that I was surprised I looked so normal, that I wasn't the horribly twisted and cadaverous figure I had always imagined myself to be. I hadn't seen my genitals since I was six years old. That's when polio had struck me, shriveling me below my diaphragm in such a way that my view of my lower body had been blocked by my chest. Since then, that part of me had seemed unreal. But seeing my genitals made it easier to accept the reality of my manhood.

Cheryl was still dressing when Dixie came into the apartment. Dixie dressed me and, lifting me with surprising ease, got me back into the wheelchair. Cheryl told me she would be out of town for a couple of weeks. She looked at her schedule book. "How would the twenty-ninth be for you?"

"It's OK with me," I said. "I'll just have to check with Neil or Marie to see if I can get a place."

"Well, just leave a message on my machine."

HAVING FAILED FOR a second time to have intercourse worried me. I became obsessed with this failure during the three weeks between appointments. What was wrong with me? Was I afraid that having intercourse represented aggression against women? Was it my lack of experience, or was it something deeper than that, something I would never figure out?

Before my next appointment, I was visited by Tracy, a former attendant who had worked for me in the early eighties while she studied at Berkeley. I had tried not to fall in love with her back then, but she was just too appealing. Young, bright, and pretty, she understood me thoroughly and was the wittiest person I'd ever known. Tracy was involved with another man; she maintained a warm friendship with me, but she made it clear that she was not interested in a romantic relationship. I felt awkward: a few years earlier, in a state of terrified, embarrassed passion, I had told her that I loved her.

I was waiting for Tracy in my wheelchair when she entered

my apartment. She leaned over so that I could kiss her cheek. Then she kissed mine.

"I love you," I said.

"I love you," she replied cheerfully.

We went to a cafe and talked about her boyfriend and my experiences with Cheryl. She said that she was proud of me for having the courage to see a surrogate. It felt terrific talking with her, and I tried to prolong the conversation by asking her everything I could think of about her graduate studies, her boyfriend, her parents, her brothers, her past, and her plans for the future. Eventually, though, we both ran out of words. She wanted to see other friends in Berkeley, so she brought me back to my apartment.

After Tracy left, I was saddened by the undeniable knowledge that she felt no sexual attraction for me. Who could blame her? I was seldom attracted to disabled women. Many young, healthy, good-looking men had been drawn to Tracy, who was in a position to pick and choose. My only hope seemed to be in trusting that working with Cheryl would help me in the event that I should meet someone else as splendid as Tracy.

WHEN I NEXT saw Cheryl, she said that this time she would minimize the foreplay and get on top of me as soon as I told her I was becoming aroused. She had the mirror with her again and held it up to me before she got into the bed. This time, I climaxed at seeing myself erect in the mirror. Cheryl got into the bed and adjusted herself so that I could give her cunnilingus. I had to stop it after a minute or so because I began to feel as though I were suffocating. But I had wanted to do something to give her pleasure, so I asked her whether I could put my tongue in her ear. She said no, she disliked that, but it was good that I'd asked.

"Some women like it. I just happen to hate it. Different women react differently to the same stimulus. That's why you should always ask."

When she started stroking my cock, I told her to get on top. Quick. I was feeling the onset of an erection. She got over me and with one hand guided me into her.

"Is it in?"

"Yes, it's in."

I couldn't believe it. Here I was having intercourse, and it didn't feel like the greatest thing in the world. Intercourse was certainly pleasant, but I had enjoyed the foreplay — the kissing, the rubbing, the licking — more. I came too soon, as usual. She kept holding me inside her. Then a look of pleasure brushed lightly over her face, as though an all-day itch were finally being scratched. Letting me go, she put her hands down on the bed by my shoulders and kissed my chest.

This act of affection moved me deeply. I hadn't expected it; it seemed like a gift from her heart. My chest is unmuscular, pale, and hairless, the precise opposite of what a sexy man's chest is supposed to be. It has always felt like a very vulnerable part of me. Now it was being kissed by a caring, understanding woman. I almost wept.

"Did you come?" I asked her.

"Yes."

I was exultant. She got out of the bed and went into the bathroom. Hearing her pee made me feel as though we were longtime lovers, familiar and comfortable with each other's bodily functions. When she came out of the bathroom and began dressing herself, I asked her if she thought I should buy a futon so that I could have sex in my apartment.

"I don't know if I should get a futon now," I said, "or wait . . . till something comes up."

"You may want to get one now because you never know when something will come up. And if you wait till then, by the time you get the futon, it might be all over."

I asked her whether she thought we should have another session. She said she would do whatever I thought best.

"Do you think there's anything to be gained from another session?" she asked.

"No," I said, relieved that I would not have to spend any more money. I had just enough to buy a futon. And besides, I'd had intercourse; what was there left to do? Later that year, I bought the futon, dark blue with an austere pattern of flowers and rushes.

I BEGAN THIS essay four years ago, then set it aside until last year. In rereading what I originally wrote, and my old journal entries from the time, I am struck by how optimistic I was, imagining that my experience with Cheryl had changed my life.

My life hasn't changed. I continue to be isolated, partly because of my polio, which forces me to spend five or six days a week in an iron lung, and partly because of my personality. I am low-key, withdrawn, and cerebral. My personality, it may be said, is a result of my disability, because of which I have spent most of my life apart from people my own age. Whatever the cause, my isolation continues, along with the consequent celibacy. Occasional visitors sit on the futon, but I've never lain on it.

I wonder whether seeing Cheryl was worth it, not in terms of the money but in light of the hopes raised and never fulfilled. I blame neither Cheryl nor myself for this feeling of letdown. Our culture values youth, health, and good looks, along with instant solutions. If I had received intensive psychotherapy from the time I got polio to the present, would I have needed to see a sex surrogate? Would I have resisted accepting the cultural standards of beauty and physical perfection? Would I have fallen into the more familiar pattern of flirting, dating, and making out that seems so common among people who become disabled during or after adolescence?

One thing I did learn was that intercourse is not an expression of male aggression, but a gentle, mutually playful experience. But has that knowledge come too late?

Where do I go from here? People have suggested several steps I could take: I could hire prostitutes, advertise in the personals, or sign up for a dating service. None of these appeals to me. Hiring a prostitute implies that I cannot be loved, body *and* soul, just body *or* soul. I would be treated as a body in need of some impersonal, professional service — which is what I've always gotten, though in a different form, from nurses and attendants. Sex for the sake of sex has little appeal to me because it seems like a ceremony whose meaning has been forgotten.

As for the personals and dating services, sure, I'd like to meet people, but what sort of ad could I write?

Severely disabled man, 41,
living in iron lung he can
escape but twice a week,
seeks . . .

Which brings up the question: what *do* I seek? I don't know. Someone who likes me and loves me and will promise to protect me from all the self-hating parts of myself? An all-purpose lover-mommy-attendant to care for all my physical and emotional needs? What one friend calls a "shapely savior" — a being so perfect that she can rescue me both from the horror that has been imposed upon me and the horror I've imposed upon myself? Why bother? I ask myself. I don't bother. Not anymore.

Which leaves me where I was before I saw Cheryl. I've met a few women nearly as wonderful as Tracy, but they haven't expressed any romantic interest in me. I feel no enthusiasm for the seemingly doomed project of pursuing women. My desire to love and be loved sexually is equaled by my isolation and my fear of breaking out of it. The fear is twofold: I fear getting nothing but rejections. But I also fear being accepted and loved. For if the latter happens, I will curse myself for all the time and life that I have wasted.

May 1990

A Priest Calls On Sunday Night

Edwin Romond

Tonight God is not enough,
so he calls too late
and my wife hisses,
"For Christ's sake,
tell him you're *busy!*"
But how can I
when I picture him
lost on his twin bed?
He says it's been one of those Sundays;
they thumbed through hymnals
as he preached
and he felt an eternity away.
What can save you
when you live all week
for just one moment,
then they sit
as if they're waiting for a bus?

So I stay on
even though she's pissed
and it's past midnight
and I'll hate myself in the morning.
He sounds on the edge
of panic, like a saint
who fears prayer
is just talking to himself.
I figure, what the hell;
when I hang up, I'll turn into her arms
but he'll have only the almighty
loneliness of speaking to Jesus,
his hands touching only his hands.

April 1991

The Heart
Of Compassion

An Interview With Ram Dass

Sy Safransky

I got a letter recently from a Sun *subscriber who said she enjoyed the magazine more than ever, because she no longer felt put off by its heavy spiritual emphasis. Had* The Sun *become less overtly spiritual, she wanted to know, or had she become more so?*

I asked myself a similar question after I interviewed Ram Dass [in 1987], Harvard-professor-turned-psychedelic-adventurer-turned-holy-man, whose name, during the sixties and seventies, became synonymous with Eastern mysticism and expanded consciousness and matters of the spirit. He seemed so much more relaxed and open-hearted and human than when I'd first met him fourteen years ago — so much less "spiritual." Had he changed, or had I?

Perhaps we've all changed, as we've learned there's not that much difference between being spiritual and being truly ourselves; and that some kinds of "spirituality" — because they're so appallingly self-conscious — do more harm than good; and that acknowledging our humanness may be the first and most difficult step on the path to truth.

Ram Dass has been walking that path for more than two decades, sharing with us every dip and bend in the road. He's a skilled storyteller with a flair for the dramatic and a keen sense of humor, as well as a surprising willingness to discuss his own fears and blunders and his sometimes overweening pride. Remarkably adept at translating Eastern ideas into language that's meaningful to Westerners, he's more than a mere popularizer; perhaps more than any other contemporary teacher, he's been a source of inspiration for other seekers, not a guru but a fellow traveler, a lover of truth with a deep commitment to serving others and a passion for God.

Born into a well-to-do New England family as Richard Alpert, he studied psychology, got a doctorate from Stanford, and became a professor at Harvard. That's where he met Timothy Leary, who introduced him to psychedelic drugs. They became outspoken proponents of better living through chemistry: Leary's injunction to "turn on, tune in, and drop out" became a sixties slogan; Alpert himself took LSD more than three hundred times, and shared the drug with ministers, prisoners, scientists, and — to the consternation of the Harvard administration — students. The two were fired in 1963 and spent the next few years traveling around the country, lecturing on the inner realms of consciousness that psychedelics had helped them explore.

Eventually, Alpert reached a dead end with the drugs. No matter how high he got, no matter how many doors were flung open, the experience always ended; the doors always shut behind him. Getting high wasn't enough; he wanted to be high. In 1967, he went to India to see if he could find a teacher. After several disheartening experiences, he was about to give up and go home. Then he met an extraordinary man called Neem Karoli Baba, or Maharaji. Despite the stories he'd heard about Maharaji's great powers and saintlike love, Alpert was skeptical. While the other devotees threw themselves at the old man's feet, he kept his distance. Maharaji teased him about the expensive car he was driving and asked Alpert if he'd give it to him. Then he motioned him to come closer and whispered to him that, the night before, Alpert had been thinking about his mother, who had died a year earlier. This was true, though Alpert hadn't mentioned it to anyone. "Spleen," Maharaji said. "She died of spleen." Alpert was stunned; his mother had, indeed, died of a ruptured spleen. He experienced a wrenching inside and broke down; here was someone who "knew." Before he realized what he was doing, he, too, had thrown himself at the guru's feet.

He stayed in India for six months, studying yoga and meditation, and returned to the U.S. with a new name — Ram Dass, meaning "servant of God" — which his guru had given him. When he stepped off the plane, he was bearded and barefoot and dressed in holy robes; his embarrassed father whisked him away before anyone could see him.

Soon afterward, his book Be Here Now (Crown) was published, and he became a New Age celebrity. At times, the robes were an embarrassment to him, too; he's talked of changing into jeans so he could slip out unnoticed for pizza. Stories like this, punctuated by his amazing chuckle — sly and intimate, as if he's laughing not just at himself but at the whole cosmos — have endeared him to audiences. His candor has meant more to many people than his purity, his revealing stories helping them make their peace with their own human weaknesses and failings.

In his books and lectures over the years, he keeps refining his message, but the core of his teaching — that the universe is a seamless whole, and that we share one consciousness — remains the same. Gone are his ponytail and beard and robes and beads and other paraphernalia. His audiences, too, have changed; he appeals now to a much broader constituency than the hippies and rebels who came to hear him ten or fifteen years ago.

He was in Chapel Hill to do a benefit lecture for SEVA, the international relief organization he cofounded. (Seva means "service" in Sanskrit.) Although originally created to end blindness in Nepal, SEVA has gone on to set up a national network of local groups whose members view service as a path to spiritual transformation. (SEVA, 1786 Fifth Street, Berkeley, CA 94710, www.seva.org.)

Ram Dass and I talked for more than three hours. Although he looked tired when we began — he said he'd been ill the day before

— his answers were thoughtful and his presence warm and inviting. His energy seemed to pick up as the interview went on.

One thing that hasn't changed is the way he talks: telling stories, and stories within stories; quoting holy books; digressing; laughing; gently mocking himself.

Safransky: As someone who spends a lot of time helping other people, how do you keep from feeling overwhelmed by all the suffering in the world?

Ram Dass: First of all, I'm not "someone who spends a lot of time helping other people." I just do what I do every day, and I don't feel like I'm busy helping people. That's how I don't get overwhelmed by it. If somebody calls and has AIDS, I see him, and we're together, and we do whatever we do together. And if I'm doing something for SEVA, and it has to do with blindness, or going into Nepal to see doctors, or whatever my part is, it's just what I do. I got over trying to dramatize my life; that's what burns you out or gets overwhelming — when you add the thought processes about what's going on, over and above what's going on. I mean, everybody gets up every morning. People say to me, "Travel must be very hard for you." But I just get up in the morning, go to the bathroom, brush my teeth, get dressed, and then I do what's on the plate for the day. It may be going to the airport, getting on a plane, writing letters, getting off the plane, meeting people. It's just what I do. If I thought I was busy helping people, I probably would be overwhelmed.

The deeper answer to your question has to do with balancing planes of awareness. It's the ability to stand back far enough to appreciate the lawfulness of events, including suffering. Plato says, for example, that to see the law and to honor the law as God, you must break your identification with your own suffering and pleasure. And once you've done that, you can just appreciate the law a little bit. Seeing the law and also feeling your human heart hurting about the suffering — those two balance each other. And that's the closest I get to what compassion's all about. Say I'm with somebody with AIDS, a young man who has to make these horrendous decisions, like whether to take this new medicine, for which he's got to stop taking his antipneumonia medicine — but if he gets pneumonia, he'll die; but if he doesn't take this new medicine, the AIDS virus may go through his brain barrier, and he'll go insane: deciding whether to die from pneumonia or go insane. He isn't prepared. I sit and hold him, and we cry, and we go through the decision, and there's another part of me that's saying, "What an interesting incarnation. What interesting work this guy's doing." I see a soul just going through this process, awakening through it. I'm meeting his soul. And very often he comes up for air, and we see the whole thing from this other place.

So there's a part of me that's perfectly allowing of suffering. And then there's the human heart that hurts like hell. And it's that balancing that's such a beautiful art form. The deepest line I work with, personally and in my lectures, is "Out of emptiness arises compassion." That's the one. Getting to the place where you do what you do, and you're not milking it for righteousness, and you're not trying to change the world.

Safransky: I wonder, though, when you're with somebody who's in a lot of physical anguish, does it ever get to be too much for you? I mean, are you always able to balance it with the deeper perspective?

Ram Dass: Sure, things get me. Last year, I was taking care of my stepmother, who was dying of cancer. I was carrying her to the toilet one night, and I suddenly realized that I should have had a catheter to put in her that night. The amount of pain I felt, because I was causing her suffering, was incredible. It was ripping me. Then I see that *that's* the work on myself: to see where my mind grabs with guilt.

When somebody gets me really angry, or throws me off balance, or something gives me a sense of revulsion, I use the reaction to go back inside and examine where the holding is. I consider how to open and soften around it, how to allow it, how to release it. In a way, you get to the point where, if it can get to you, good!

Now, you say, how much can you take? I used to think, if I had a whole day ahead of me seeing somebody with multiple sclerosis, a cancer patient, somebody who's just lost her child, and an AIDS patient, then shouldn't I have a little fun, too? But these encounters are such living truths, they pull me into such immediacy because these people are demanding so much truth and presence —— what more do I want of life? I mean, this is the stuff of life! This is the richness of the moment. And to me, the fun is in these moments. I often go to a party and the superficiality of it just bores me. I'd really rather be alone or be with people who are purely seeking, whether it's seeking surcease of their pain or something else. I come into a city to speak, and people who are dying want to see me, and then there's the chance to go see this mountain or that river. And there's not even a comparison. The depth of a human spirit is, for me, like a doorway to the infinite. I work with Gandhi's line "Think of the poorest person you've ever seen and ask if your next act is of any use," and it embraces so much. So when somebody says, "Are you happy?" I say, "Yeah." "Are you sad?" "Yeah." "Is this enough?" "Yeah." It's not either/or. You don't deny the pain for the pleasure.

Safransky: There's the temptation, I would think, to distance yourself from difficult emotions, to imagine you're relating from a higher consciousness when you're really just denying pain. How do you keep a check on yourself to know which is which?

Ram Dass: You ask yourself whether you lose touch with a feeling through the process. See, if you dissociate, you lose touch with a feeling; you're not feeling the pain of it anymore. When you add the other level, however, the pain is still there. It didn't go away; you just added another plane to it. It's a little different. It doesn't have a pushing-away quality. I mean, I really loved my stepmother, and when she died, it hurt like hell. And I didn't want to go away from the hurting like hell. It was very rare for me to be this attached to somebody. And at the same moment, I was cultivating that other part, which was that it was all right for her to die. This was a process that she was going through, and it was quite beautiful. I could feel both of these all the time, and I didn't want to push one away, and I couldn't really get lost in either of them anymore.

What I say about grief to people is "Don't be done with your grief too soon. Go back into it." I'm always pushing people back into it. "Are you really done with this?" I say to myself. I keep going back to look and to scrape and to see. And sometimes it takes a long time for a thing to come around again. I'll say something that was a little too glib, and I'll think I'm done with it. Then, later on, I can feel a soft spot inside me, like a sickness or a little pus underneath the skin, and I've got to go back. I wish I were wise enough to know the moment I dissemble, but the ego's so slithery. Sometimes what you feel when you start to dissemble is a thickening of the vibratory field, a kind of a denseness, as if you've protected yourself, but life isn't as alive anymore. That's one of the clues I use.

Safransky: Sometimes it's guilt that calls us back to some pain we're not finished with. How do you view guilt? Is it ever useful?

Ram Dass: Well, there can certainly be a feeling that one has done wrong or has done something that's caused suffering. But guilt, which plays on your unworthiness, isn't very functional. It tends to narrow your perception of the universe, reduce your effectiveness of action, close you down from your beauty, from life and truth. Guilt and fear are really corrosive in terms of the human spirit.

Safransky: So you never feel guilty?

Ram Dass: What would create guilt in me is if I used another person to satisfy my desires in a way that would not bring them closer to the spirit — in other words, if I exploited somebody in a way that left them more separate, more isolated, more paranoid. If you lust for somebody, you see them as an object; and if you look at somebody as an object, you are reinforcing their separateness from God, from spirit. It's interesting: Would you then have guilt about lust or guilt about the actions that stem from lust? Or would you get to the point where you could acknowledge the feelings in yourself, but couldn't act on them because you wouldn't want to be that separate from another being? You live with all of this inside yourself, where there's the desire as well as the awareness of what you get from living out that desire. There's a poignancy in it, because it's part of your human condition. Yet, as you keep working on your consciousness, you're less and less likely to create new karma. You just don't want to do it, because you've got to live in the world you create.

Safransky: Is lust still an issue for you, then?

Ram Dass: Well, I'm fifty-five. It's getting less so.

Safransky: You mean getting older is the only solace?

Ram Dass: It may well be. [Laughs.] It hasn't gone away completely, but I don't lose my consciousness into it. There's a thread that stays aware now. There's a part of me that says, "Oh, there it goes again," and, "How empty it is," and, "Ah, God, you're going to run that off again?"

Safransky: That's not the same as guilt?

Ram Dass: No, it's just observation. It's not even judging. See, the judging is the guilt, the reaction. You learn to cultivate the witnessing mind, which is just noticing it — noticing you're human, and that these are the phenomena of humanity. Not that it's an error or bad. It's not wrong that I'm human, not wrong that I have all these feelings. That's the difference. I am much more allowing of my humanity, allowing of my lusts and my laziness and my greed and my hungers and my overeating and my angers.

Safransky: How much of this has to do with being Ram Dass, and how much of this has to do with getting older, since as people get older they become more accepting?

Ram Dass: It is clear that not all older people get these perspectives. And the culture itself does not have this perspective, so you can't say it is intrinsic in all aging. But it is certainly available with aging if you just quiet down a little and listen. In the Eastern cultures, where there are extended families and certain clear roles for aging, it's built into the system. When you start to lose your hearing and your eyesight, you can meditate more. That's the stage of life you're at. Well, having lived in that way, which was part of becoming Ram Dass, tempers the way I experience my own life. So it is partly Ram Dass and it is partly aging. I'm really learning a lot of patience. I'm not so interested in how soon I'll get enlightened, nor even in the future that much. And the past doesn't have any hold on me. I find this moment as interesting as anything that has ever happened in my life.

Safransky: Do you have a sense of unfinished business?

Ram Dass: None. I feel like I'm right up to date. If I were to die this moment, it would be perfectly fine. I don't have a sense of omission. I don't have a mythic sense of myself. I don't milk the astral story line at all. You can't believe how simple it gets. I used to read about the simplicity of Zen monks, where they just did dishes and walked, and I couldn't understand what was going on. And now it's just getting so clear to me how simple it all can be. And I think it's going to get much simpler, much quieter.

Safransky: Did you have to go through a complicated stage before you could get to that simple stage?

Ram Dass: It's like going from knowledge to wisdom. Knowledge is very complicated, and wisdom is very simple. My guru, when he wanted to chide me, would call me clever. When he wanted to reward me, he'd say, "You're simple." It's the simplicity of just being so in tune with things that nothing seems special anymore, nothing stands out. It's all just part of a flowing process. It's such a refreshing release not to be cultivating my specialness. I mean, whether I'm doing the laundry or dancing or paying bills — it's what's on the plate today. People say, "Well, isn't that dull? Don't you want to be excited? Are you excited by *anything* that's coming up?" I'm not! Because if this isn't enough, what the hell is? I can't think of anything coming up in my life at this moment that excites me. Or you could say everything in my life excites me, but it doesn't *excite* me. It's enough.

Now, sometimes things will still get me. Like yesterday, when I was sick, the thought of lying down in that bed and going to sleep excited me. Fatigue gets me; sickness gets me. And there are times when my heart doesn't open as fully as I'd like. I might be up in front of a lot of people, and my heart closes up with fear. If I see myself and the audience as a group of souls reaching for light together, trying to figure out what it's all about, meeting like a club to share notes, then my heart's wide open. But if I think, *Here is a group of people in Chicago who have paid money to hear Ram Dass*, it's a different vibratory place inside myself. What I've learned now is to slow down enough to let my heart catch up. Because initially, when you come out on stage, you start with a routine or something to get the game going. And then you're off on a certain tack, and it's very hard to come back in your heart. You can't get the rhythm shifted. Once, in New York, I came out on stage, and I wasn't quite ready. I started with a routine, a story of some sort, and after two or three minutes somebody in the balcony yelled, "Ram Dass, my heart is hurting!" When somebody does that, you have a choice. You can, with one flick of a facial expression, turn the whole audience against that person, as a heckler. But I heard what he was saying because my heart hurt, too. I thought, *Well, if your heart hurts, all of our hearts are hurting. Let's start breathing together and see if we can get here together.* And the lecture just took off.

Safransky: Do you meditate?

Ram Dass: No, service is my yoga. I'll take off now and then for a meditation retreat, but lecturing itself is a meditation. Once I sit down in that chair cross-legged, I really go into another space of quietness in which I'm just watching the forms. It's very interesting, like a two-hour meditation every other day. What I'm trying to learn to do is experience and articulate what karma yoga is all about, which is using the stuff of life as the vehicle of awakening. For years, what I did was live my life, then go back into my room to prepare myself to live it again. Now I'm seeing whether the stuff of life can do it for me. And when the toxicities build up and it's too much and I can't handle it anymore, I try to examine it. I do almost go under at times. And those are the times when I'm hanging on by my fingernails and saying, "OK, am I just going to jump back into a meditative place to get my center back, or am I going to keep converting this stuff and keep coming back through it?" And I'm really working at it. That's the most exciting adventure of my life, in a way. And I expect that by next April, when I've finished this tour, either I'll be a basket case ready for a meditation retreat somewhere, or I'll be full of energy and light, playful and empty — depending on how much I can transmute it, how much it is true that I only want to be free.

Safransky: Why do you carry a picture of Jesus? What does Jesus represent to you?

Ram Dass: He represents something I'm beginning to know very well. It's a feeling of loving somebody incredibly, knowing that you can't take their suffering away, wishing you could, and knowing that it's all right, all at once. It's the bittersweet poignancy of the moment.

Safransky: The cross represents that same poignancy.

Ram Dass: Exactly. Jesus' crucifixion and resurrection is such a brilliant, spiritual statement for the relief of suffering of other beings. He's just saying so clearly, "I am not this body. I took a human birth. You aren't this body either. If you hear this, you're free. Look, I'll go get crucified and I'll be back in a few days and show you." In that sense, he did make a statement that could take everybody's suffering away. Let those who can hear, hear. I love it in its symbolic, exquisite sharpness, its clearness. And how it was misused! Jesus would turn over on his cross! Yet I can sense in him the knowledge that this misuse would happen, too.

Safransky: What distinction do you make between Jesus and Christ?

Ram Dass: I think of Jesus as a story line, and I think of Christ as that consciousness, or that love, that is the same as the Buddha mind or pure awareness or Brahma. I think of it as the One. Jesus is a semihistorical story line that has a certain teaching in the story itself. To say that Jesus is the only way is a perversion. To say that Christ is the only way is a certainty. We are all the Christ and it's the only way. I just tell people, "Focus on Christ, not on Christianity."

Safransky: What does the picture of your guru mean to you?

Ram Dass: Sometimes it means nothing to me. And sometimes I feel him loving me so deeply that he allows me to love myself. Sometimes I experience the grace that he allows me to do his work. And sometimes I experience him as laughing at me. Sometimes he's laughing with me. Sometimes he's chiding me. Sometimes he's just snuggling up to me. Sometimes he's very remote and cold and distant. Sometimes he's looking at me and saying, "Aren't you ever going to finish with all that shit?" Sometimes he's just an old friend. Sometimes he's like a doorway into this infinitely vast space. He's always my closest friend, even when I hate him.

When it was discovered that Phyllis, my stepmother, had a growth, we were waiting to find out if it was malignant. The doctor called, and we were both going to listen on the phone because I was going to be with her through all of this. And Maharaji's picture was there. So I said to Maharaji, "I could never ask you for anything, because you know what's best. But if it's all the same to you, could you make it benign?" And the doctor came on and said, "It's malignant, and it's the worst kind. If we don't get it out in a few weeks, you'll be dead." And I felt my heart close, just like that, and I thought, *You son of a bitch*. And the next moment I was flooded with such love and such light. I can't tell you what happened! It was like *whoosh!* And then I realized this was an expression of his love, too. That's another way we are in love, through the unbending law of the universe unfolding. And I was closer to him after that.

Once you've touched something as real as Maharaji is for me, it's like a protective shield against your own impurities and everybody else's. You can get sucked in for a moment, but you keep coming back to the touchstone. I would like just to die into him. But I realize that I can't die into him — I've got to die into me. I can't become a little Maharaji. I've got to become a perfect Richard Alpert or Ram Dass, and I will be the same as Maharaji. But that means honoring the unique form, not imitating another form. It's like feeling your way into your humanity without denying your divinity.

This interesting thing happened in Burma. I went to Rangoon to sit. It was the most intense meditative space I'd ever been in. There were eight hundred Burmese and five Westerners meditating from three in the morning to eleven at night — no surcease. I mean, it was full-time. You didn't walk anywhere; you didn't read anything. Well, I had snuck into the place with my picture of Maharaji and one hundred poems of Kabir. I'd planned to be there for three months, so I figured one poem a day. And I'd brought two one-pound bags of M&Ms, one with peanuts and one without peanuts. So each day I would allow myself a poem of Kabir, a look at Maharaji's picture, and two plain M&Ms and two peanut M&Ms. That was a little edge in my personal life.

About the second week, I woke up one morning and I started to hear, "Sri Ram Jai Ram Jai Jai Ram, Sri Ram Jai Ram Jai Jai

Ram," which is the mantra I'd learned in India in '67 and chanted for years. I couldn't stop hearing it. At first, I thought somebody was singing it, but then I realized it was my mind, because the traffic was singing it, the cows were singing it. I thought, *This is not what I came here for. I'm supposed to be following my breath.* I'd try to follow my breath, but between the in-breath and the out-breath, I'd hear it! I thought, *Oh, my God, I'm going insane!* I couldn't hold my ears because it was inside. I didn't know what to do, and I didn't want to tell the teacher. I was embarrassed by it. But after about a week I went to the teacher and said, "I'm noticing this chanting going on all around me and in myself." And he said, "Oh, you've been in India?" And I said yes. And he said, "You've studied in some of those disciplines?" I said yes. He laughed. "Well," he said, "take refuge in the Buddha. It will help."

So I went back to my cell, and — can you imagine? — I'm going to take refuge in the Buddha to get rid of Ram. I'm not being distracted by lust, I'm not being distracted by power or greed; I'm being distracted by another spiritual method. I mean, that's bad! Can you imagine using Buddha to get rid of Ram? There's something obscene about this. And here I am, *Ram* Dass! So I thought, *I am seeking truth. And I believe Maharaji's truth and Ram's truth and any method I use purely will bring me to truth and will bring me closer to Maharaji. So I'll put away the picture and I'll put away the poems. I won't put away the M&Ms, because they're not a part of that. But I'll put away these other things, and I'll just do the method.*

It was the first time I came to grips with the Buddhist teaching about Right Effort, where you take full responsibility for something. See, I'd been operating under "Not my will but thine" and "I can't do anything; you do it for me." And I'd always been afraid of personal power because of my fear of misusing it, because of my history of misusing it, because of my lust and my greed. When I was given power, I used it to hurt people or to get something. That's a really deep neurosis inside me.

So I accepted personal responsibility for my spiritual practice. I set Maharaji aside as a father figure who'd do it for me. The minute I did that, the chanting stopped. And I started to experience this tremendous force coming down into me. The fear of letting it come into me — because of my identification with it as *my* force — had made it get weird. And I was so busy saying, "It's not me, it's Maharaji, it's not me," because I was afraid that, if I said it was me, it would corrupt me, or I would corrupt it. But the minute I said, "OK, let it be, let it come through me; it's OK, and it is *my* power and *the* power, and *my* power is an instrument of *the* power," there was something that clicked in. It was the first time I had done an extensive meditation retreat and come out more moist than when I went in. Usually, I'd come out dry as a bone and wanting love and warmth. But I came out soft. I came out feeling closer to Maharaji than I'd been before. And I think I've been much more comfortable with my own power since that time. I don't think I'm nearly as afraid that I'm going to misuse it. I'm now willing to use my discrimination in ways I never would before.

Safransky: Just imagine what will happen when you give up the M&Ms.

Ram Dass: [Laughs.] I've been through pizza and root beer — I can make it!

Safransky: Concerning the use of discrimination: In the past, you've been reluctant to comment critically on other spiritual teachers. In recent years, there have been a number of scandals that have left people feeling they had been led astray, whether by Rajneesh or Muktananda or whomever. Do you feel any responsibility to speak out for the benefit of others?

Ram Dass: My way of dealing with that has been not to spend my time becoming a connoisseur of clay feet, but rather to take whatever teachings I could get, and to realize that if I came to a teacher with a pure heart, I'd get a pure teaching. If they were doing something impure, it would be *their* karma. I didn't have to protect myself from them: the purity of my intent would protect me. If I wanted power, then I'd get sucked into the power games. But if I wanted God, I'd just take from them that which would help me get to God. I think people get what they get because of what they want.

I have no idea who these guys are. I really don't. I mean, I look at Rajneesh, and it is inconceivable to me that anybody has an ego that needs all those Rolls-Royces. I can only see it as an exquisite chiding of the culture: by taking the symbol of opulence in our culture and making a joke of it, he has created a beautiful Tantric game. At the same moment, the paranoia that attended his scene and all of that ugliness with the people in Oregon turn my heart off. And I wouldn't have anything to do with it. I trust my own intuitive heart about that, and I trust everybody else's intuitive heart to know what they're getting into. So I am inclined to say to people: "Trust your heart, because I don't know any more than you know." It might be the greatest Tantric teaching. I mean, Trungpa Rinpoche is a rascal. He's a hell of a teacher, an extremely profound guy. But he's also quite a wild man. Parts of him turn me off, but I'll sit at his feet and take tremendous teachings from him. It may be that some parts turn me off because there's too much attachment in me yet to hear those parts, or it may be that it's his corruption. I have no idea! So I've learned to try not to judge these guys. And I don't feel I have to protect others. I think the protective mechanism is in individuals. I think they get what they get because it's what they want. I don't think there's any Indian saint who hasn't had some people cast aspersions on him for something or other. So it's hard to know. I mean, Maharaji is thought of as a rascal, as a dirty old man. And maybe he was, I don't know.

Safransky: What is your view of the AIDS epidemic? How do you think it's changing the awareness of the whole culture?

Ram Dass: I know what it *isn't*: it isn't punishment. In a way, AIDS is reconsecrating sexuality, making sex and death connected — making sex precious again, rather than something trivial. It's like living in the moment with death over your left shoulder. It's that quality that has come into sex now. "Does this act mean enough to you that you would die for it?" I think it has weighted the balance toward more commitment in a monogamous relationship so that people will go deeper into relationships, rather than jumping from bed to bed. I think it's gotten people to weigh the meaning of the sexual act more, and possibly to be able to tolerate a lot more horniness without expressing it as immediately as they had in the past. I don't see any of that as bad. Or good. It's just what is. It's not better or worse, just what is.

What I noticed in the gay community was that AIDS brought to the surface a tremendous amount of compassion that had absolutely not been there before. I was teaching a workshop for a group of buddies who were going to be AIDS helpers — all young, mainly gay guys. And I looked around and said, "You know, five years ago I couldn't have been in this room without being distinctly uncomfortable, because everybody would have been focused on everybody else's crotch, and whom they were going to go to bed with, and who was sexually desirable. And suddenly, here we all are, meeting out of caring and compassion for other human beings. We can see other qualities in each other." We can hate AIDS as a terrible blight, and it certainly has brought the gay-rights movement to a screeching halt, but at the same moment, it is having this spinoff.

In my own life, I watched that place in me that goes with my particular sexual patterns, to link sex and death, to be attracted to the sexuality that would increase the risk. I could see that place in me. But it isn't strong anymore. Ever since I've known that, my sex has been what's called "safe sex." I saw that tendency in me to take tremendous risks in sexuality, risks that could get me killed. I mean, dark nights in strange countries looking for sex in the middle of the night. That kind of obsessive behavior, where the more exciting and the more threatening, the more sexually exciting it was. I felt that particular link in myself very much over the years. And I could see that AIDS brought that to the light. The desire isn't strong enough anymore to lead me to be promiscuous. I've not been tested for AIDS, which interests me. I feel that if I were going to be promiscuous, I would have to be tested for AIDS. I don't think I could take the karma of spreading something to somebody else. But I think that I would almost rather just not do either of those things.

Among homosexuals, the most likely way to spread AIDS is anal intercourse, and that has never been anything that appealed to me anyway. The lowest probability is deep kissing. And mutual masturbation is not a major way of transmitting. So that's enough for me. That's fine. See, as you know, I'm bisexual. I have relationships with a woman and with a man, and I have had these relationships with the same people for years. I think we're aware that we may be at minor risk, but we know what we're doing and we're willing to live with that possibility.

There is such a deep desire in this culture to deny death, yet, from a spiritual point of view, AIDS may be one of the greatest gifts to humankind. And that's a hard one to even mention aloud. The same with the Holocaust, or the Bomb. We almost can't stand looking at God's work because it's so scary.

In San Francisco, there's a whole list of AIDS patients whom I see. It's a strange experience to be with a young guy, somebody I am attracted to, and he's got the worst symptoms of AIDS, and I'm loving him so much and holding him. And it's right on the edge of everything all at once. It's the holding on to death, and it's the lusting and the desiring. God, I can't tell you how much of everything it is at that moment. And anything less than that would be giving him less than my truth. And you've got to give your truth to another human being. And I'll say to him, "I'd love to make love with you now, but I really don't think I should die for that." It's interesting. Many of these people with AIDS want to test the limits with everybody: "How much do you love me? Die with me."

Safransky: I'd like you to address the general question of relationship. Wendell Berry, in a poem called "Marriage," wrote, "We hurt and are hurt and have each other for healing. It is healing, it is never whole." Is it possible to feel completely whole with another person?

Ram Dass: I think it's possible, but I don't think it's probable. It may be possible for a moment, but to keep it continuously that way takes such conscious effort. The love that's beyond romanticism has to permeate the relationship so that all the deceptions and posturings and judgments bow before it. And that love is cultivated through the yearning of those two beings to use their relationship as their major spiritual work. It's a full-time job.

Safransky: Do you yearn for that kind of relationship yourself?

Ram Dass: I really love my friends deeply, my lovers and my friends. My relations with all of these people have incredible beauty, and yet the idea of living with any single one of them doesn't appeal to me at all, because every relationship brings out certain qualities in me but not others. And I like the kind of changing panorama that keeps bringing up different parts of my being. I notice that some of them just go into the shadows when I'm with any one individual. I mean, it's not a question of whether I can live with one person or not. It's a question of why would I do it? I don't yearn for it, no. I did, or I thought I should. I thought I must be missing something and that I must be terribly afraid. But then I was with people, and I went in so deeply with them, and I saw that we could keep exploring, but that wasn't enough for me. I can't tell . . . I can't even hear in my voice whether that's dissembling or not.

Safransky: Five years ago, you were describing a relationship you were in with a man, and you had just come through a difficult winter that had sent you back into therapy. There was a lot of jealousy and difficult feelings associated with the relationship. And you said at the time that a desire for a kind of intimacy had been very great in you, and you were finally acknowledging that.

Ram Dass: What happened was that I gave up my romantic images of that relationship, and then we started to become friends. As our friendship has deepened, we have become lovers again, but in a whole other way — not out of any grasping, but out of celebration of just this delight in being together. And yet we place no temporal or spatial demands on each other. I mean, I could see him twice a year, and it would be fine; talk to him maybe once a month on the phone. It would be all right. There's no "I miss you." There's just a tremendous joy in being together, and an appreciation that we really are together. So, in a way, all that was part of that process.

Safransky: So you don't miss anybody and don't miss missing anybody?

Ram Dass: That's exactly right!

Safransky: It sounds pretty good.

Ram Dass: Yeah, I know, but . . .

Safransky: Where are the M&Ms?

Ram Dass: [Laughs.] I shouldn't have gotten off so easily. Maybe I'll end up a bitter old man, alone in an old folks' home, rocking back and forth, everybody having forgotten me. My Jewish training isn't letting me off the hook just yet. "Sure [in a Yiddish accent], he didn't need anybody! Sure!"

Safransky: There are other aspects of relationship I'd like you to talk about — for example, the relationship between the individual and society. "Changing society" and "working on oneself" are often viewed as quite different and incompatible. Yet I know you have spoken about how they aren't really separate. How do you reconcile them within yourself?

Ram Dass: I've been directing a lot of attention to how they are not separate. Recently, I was walking on the Great Peace March, and one evening I said to the marchers, "You're making this brilliant statement, just by doing the thing you're doing. But use it to work on yourselves. If you want peace by the end of this march, something about this march should have made you more peaceful, so that the means and the ends are peace. Because at the end of the march, peace isn't going to appear. You're going to go out into life, and the march should have prepared you to carry forth the torch of this march, so that the rest of your life is the Great Peace March. To the extent you're filling your mind with anger because there's not enough media coverage, you are perpetuating the problem you're trying to get rid of. So it really behooves you to work on yourselves, to become an instrument of the thing you represent."

Dan Ellsberg and I have had this very intense dialogue. Dan was chiding me because I kept saying how important it is that activists sit and learn how to be quiet. And he said it's not only important that you learn to sit, it's important *where* you sit. I think it is exquisite when you can integrate both of those.

I have a sense of the shared consciousness in the world. For example, I'm helping to support a friend who's doing a three-year meditation retreat in southern France. Now, is he helping the world or not? My knowing that he's there, committing himself to that — I'm a part of him doing that, and he's a part of me doing what I'm doing. In a way, we're standing on each other's shoulders. It's too narrow to think everybody should be doing any one thing.

You can't wait until you're enlightened to serve people. You know that your service, because it's coming from a less-than-enlightened place, is relieving the suffering only a certain amount, and in some way digging the hole deeper. In view of that, you use your service as a way of working on yourself in order to become pure, in order that your service might become pure. So there is a circle. It can come through social action, which is what I'm trying to figure out how to do. Social action is a vehicle for working on oneself. It's like when I say to an audience, "I am really not here for your well-being. I am here because this is what I do and I'm trying to learn how to do it purely. What you get out of it is your karmic predicament; it's not mine. If I get caught in yours, that's not going to help you or me." It's a hard thing to say or to hear. People say, "Thank you for being here." Why thank me? I didn't do it for them.

Safransky: You say you're not here for them, but it's also true you *are* here for them.

Ram Dass: I know. In fact, I used to say, "I don't care about you," and it was cruel and it wasn't true. In the sense that the Bhagavad Gita enjoins us not to be attached to the fruits of our actions, what I'm really saying is "Look, I'm doing this because I'm doing this. If I'm attached to what you get out of it, it is a lesser doing."

Safransky: Does this have anything to do with the fear you spoke of earlier? Isn't there an element of that in pushing away other people's appreciation of you?

Ram Dass: I hear your point. I'm a little sloppy because I'm still counteracting the feeling I get when people say, "What courage you have!" In my lectures, I talk about Mother Teresa in the streets of Calcutta, picking up a leper lying in his own vomit. And you look at that and say, "What courage she has!" But if you listen to her, she says, "I'm taking care of Christ in his distressing disguises." She's in love with Christ, so she's taking care of her beloved. She's being with her beloved at that moment. Now, can you imagine lying in bed with your beloved, and somebody comes up and says, "What a courageous thing you're doing"? It's as bizarre as that. But people look at one level and not at another level. And there is a level at which I look out at the audience, and they are so beautiful, and their reaching is so pure, and my reaching is so pure, and how can we do anything but help each other as much as we can? The highest compliment most people pay me is "Thank you for being so human." Isn't that an extraordinary compliment? I mean, if I put something on my tombstone, it would be "He was human." Isn't that bizarre? After all these years of trying to be holy?

Safransky: That's really what everybody hungers for, isn't it? People are so afraid, so ashamed. We put a lot of energy into hiding.

Ram Dass: What I'm trying to do publicly is not hide my humanity. And that gives other people license. You know, it's interesting: every time I, as a public figure, get up and say, "Look, I'm bisexual," or, "I have homosexual relations," something happens in which people suddenly feel it's all right for them to be whatever they are.

I went into a hall the other night and people said, "Oh, Ram Dass is coming." Loving, smiling, happy people. Now, we put out a brochure telling the people who are organizing the lecture all the things they should prepare. One of the things I really like is a miniboom, which is a microphone on an arm. I can put it behind me or have it low, and it doesn't get in my way. I sit cross-legged, so if there's a straight-up-and-down mike, it doesn't get in close enough and I've got to spend the whole evening leaning forward. So I walked into this hall and was full of love and happiness, and I looked at the stage and there was a straight-up-and-down mike. I said, "Where's the miniboom?" The manager of the building said, "Well, that's all we have." I said, "Well, it was specified that there be a miniboom." I was getting really tight, because I saw that all evening I would be reaching for the mike, and it changes the consciousness. "Well, if that's all we have," I said, "we'll have to use it." But I was really making everybody pay. And I was watching myself being caught up in it. As I got up to lecture, and I was speaking about love and the heart, I said to the audience, "I've got to tell you, as I walked in here, I turned into the worst shrewish bastard in the world. I just didn't realize the microphone was my guru in drag, coming to say, 'I'll get you this time.' "

Safransky: It's been twenty years since you met Maharaji. The sixties and the seventies came and went. When you look back on that time, what seems significant and enduring?

Ram Dass: I think what happened in the sixties in terms of shifting consciousness is much more profoundly entrenched in the culture than we yet realize. Something very deep did happen to the culture. The reaction against it in the late seventies and eighties — the yuppies and all that — was partly in response to the impurity of the sixties, because, besides its virtue, there was impurity in it, too, as there is in any human venture. But the deepest change I saw was the loosening of a kind of monolithic, absolutist way of looking at the universe, into a kind of a relative-reality framework. Then that got institutionalized through rock-and-roll music, Dylan, the Beatles. It fed into the consciousness of the generation and permeated the culture. I know it's happened because, when I look at my audiences in most cities, I would say that 70 percent of them have never taken psychedelic drugs, never been to the East, never read philosophy — and yet there they are! And I'm saying the same thing I was saying twenty years ago. Who are they? Where did they come from? Why would they want to hear Ram Dass?

When you consider the Bomb; the media, which allow us to transcend space; computers, which allow us to get out of real time; modern transportation, which makes the world shrink; psychedelics; the change in the power balance in the international political scene — all of these things are great for the awakening consciousness. I mean, if I were playing God, this might be the way I'd do it. Throwing in AIDS along the way, too. It's really far out to consider these possibilities, that all this isn't some monstrous error, but an exquisite process of forced awakening.

Safransky: Do you have a sense of what might be emerging from this? Do you look into the future?

Ram Dass: I'm inclined to think of evolution individually, not culturally. I would be naive if I didn't acknowledge that we're more sophisticated about a lot of things, but greed, hatred, and lust all seem to still be around in profusion. And working with that stuff seems to be the essence of what a human birth is. So maybe the forms are changing, but I'm not sure the essence of it is changing at all.

I'm asked night after night, "Is this the New Age, or is it Armageddon?" And I say, "I used to think I should have an opinion about this, but as I examined it, I saw that if it's going to be Armageddon and I am going to die, the best thing to do to prepare for it is to quiet my mind, open my heart, and deal with the suffering in front of me. And if it's going to be the New Age, the best thing to do is to quiet my mind, open my heart, and deal with the suffering in front of me. It turns out it doesn't matter. So I don't care." That really is the way I feel about it.

Safransky: What thoughts do you have about drugs and the concern today about drug use?

Ram Dass: I think cocaine is a lousy social drug of choice. It's like alcohol. Our culture seems to have an amazing skill for picking lousy politicians and lousy drugs.

Safransky: Do you do drugs anymore?

Ram Dass: I've taken MDMA three times this year. I took acid last year. I've smoked marijuana two times in the last six months. That's about it. I'm not busy *not* taking them; I don't seem to have much yearning to take them.

Safransky: Did those experiences feel useful? Did you have any regrets?

Ram Dass: I'd say both. They were a little useful, and I also regretted doing it. I don't like what they do to my body, first of all. They're very hard on my body, and as you get older you really notice it more. And afterward they always seem like . . . it's like sex that came out of lust: "What the hell did I do that for?"

Safransky: Have you ever taken a drug, believing it was the best way to achieve a certain state of awareness, and then concluded, before the experience was over, that taking the drug wasn't necessary at all?

Ram Dass: Yes, that's exactly the way I've felt. But I won't make any general conclusion about it. Because now and then, when I take a drug, I get some direct experience that's worth something to me.

Safransky: Do you feel vulnerable acknowledging that you do drugs occasionally?

Ram Dass: Well, my audience usually includes a good smattering of people who are in drug-rehabilitation programs and Al-Anon and AA, and they don't want to hear this. I don't really like to push against people and antagonize them. I don't talk drug talk in my lectures. But in the questions and answers, it always comes up, and I always answer the question as directly as I can. I don't say more and I don't say less, but I answer the question. I don't feel it's my cause, any more than gayness is my cause.

Safransky: Are there any questions that are significantly embarrassing or uncomfortable?

Ram Dass: I think it has to do with who's in the audience. I mean, there are certain kinds of groups where a drug question is hard for me to deal with, because I want to be truthful, and yet most of the audience doesn't want to hear it. Or sometimes my sexuality would be uncomfortable for the audience to deal with. I mean, I'm a biblical affront to them, and yet they like me, and it's really pushing them. I don't want to push. I don't have an ax to grind. Other than that, I kind of like the ones that push my buttons. I like it when I'm slightly off balance. That's nice, because I feel it's the most living moment. I also like it when I'm completely in control and I'm just prancing, high as a kite. I love that, too.

Safransky: Are you always honest with everybody?

Ram Dass: No, I think I'm too enamored of a good story. I'm too much of a storyteller. I notice the way I cut edges to make the story fit a more beautiful image. I am truthful about the big things, but not about the little things. I don't know how to deal, for example, with feeling lousy. I don't know how to deal with coming before an audience that has come out of the hills and driven hundreds of miles and paid money to hear me, and really feeling that I don't want to talk. If I say to them, "I don't want to talk to you," it's not right, and yet that's my truth, and if I don't say my truth, it's all a lie.

Safransky: But when you're asked a direct question, do you ever find yourself not being truthful then?

Ram Dass: I'm sleazy. I won't tell a direct lie, but I will infer something that could be a lie, make it appear a certain way by how I use words, take the edge off, take the pain away from myself. I feel myself do that. I'm working to get straighter and straighter. But I've got to feel safer and safer. And sometimes I just don't feel safe enough to be that truthful. I lie only to protect myself. That would be my only reason to lie.

Safransky: Are there particular areas where you feel less safe — drugs or sex or money?

Ram Dass: They're changing so fast, it's almost as if we're talking about a historical thing. I almost don't care about these anymore. But I would say, yes, those three things.

Safransky: How is money controversial?

Ram Dass: A guy came up to me the other night. He looked like a street person. He was an older man, about fifty-five; he had a certain kind of twinkle, a certain look of wisdom. But he was all smelly and seedy, and I didn't know how he got into the lecture hall. I couldn't imagine he paid to get in, but there he was. And he came up front and said, "Very interesting. You know, you should give up your money and your credit cards." I said, "Thank you. I hear you." And he smiled and shuffled off. I mean, it wasn't heavy-handed. He just laid that on me, and I heard it very clearly.

I'm almost at the point where I'm just ignoring money and using what I need and assuming there'll be enough and not worrying about it. I think that's the way I'm free of it. But I don't really know, because it's so easy for me to come by. I really don't know what I'd do if I were thrown out in the street without my wallet and without being Ram Dass. I assume I'd look for a job and get a room in a rooming house, and what would happen would happen. And if there weren't a job, I'd be one of those people in the soup kitchen. Or, if not that, I'd be sitting in a park, and I'd be cold and hungry, and then I'd be dead. I think I could go through all those stages and stay pretty conscious through all that. But I think it would be phony to do it intentionally.

I live in a rented house with my father. It's a very wealthy neighborhood, a Republican-alcoholic neighborhood. It's on the ocean, very posh. The rent is exorbitant: twenty-two hundred dollars a month. My father is so old, he goes from the toilet to the bed. But he's familiar with this place.

It looked like they were going to sell it out from under us, which I thought was great. I didn't have the guts to move him; but if they sold it, we would have to move. Then a friend suggested that we all get some land together with a couple of houses in western Massachusetts. So we found a house with beautiful acreage and a beautiful river. It was $260,000, and I was going to have to put $100,000 down. I — who usually have $300 in the bank — was suddenly figuring out where I was going to get $100,000. Well, I started to scrape and pull, and suddenly it was appearing. I had control of my father's estate, so I was borrowing money from him, and also holding on to money instead of letting it go through me so fast. The more I was doing it, the less pleasant it felt. Then a structural engineer said this house was no good. And my friend found another house that was even better: $400,000. Well, I figured if we could spend $260,000, why not $400,000? It's all numbers anyway.

Then I noticed that the thought that I was going to be carrying a mortgage of so much a month was influencing my decisions about my lecture schedule. Usually, my whole decision is based on: Is it in my best spiritual interest? Is it the most dharmic thing to do? Will it be timely? Suddenly, it all had to do with my mortgage. And I was accepting things that I didn't want to accept, just because they were paying a lot of money and were going to cover that mortgage. And I got sick to my stomach. The house was assessed to be a steal for $400,000. It had a pond, a hundred acres. We could have big meetings there. It would be lovely. I mean, it was everything.

In the meantime, the people who owned the house we were

living in pulled it off the market and said they'd give us a new lease. So I pulled out of the other thing. I thought, *I just can't do it*. And everybody was very perturbed with me. Everybody was all ready for the move. We'd gone through the whole exercise, down to the lawyers and the financing. I know now I'll never own a house. And I have a lot of compassion for people who have families and who have to own houses and who have to pay mortgages. That was good for me to do.

I've wondered sometimes whether my lifestyle has insulated me against pain. When somebody says their child has been raped and murdered, I can sit down and write a beautiful letter, but I wonder . . . *my* child hasn't been raped and murdered. I wonder how compassionate I really am. But what can I do? I can't manufacture it. It's just an opening of the heart, deeper and deeper. Often people can say to me, "It's easy for you to say." And I think it is. It *is* easy for me to say. But then, how can you equate sufferings?

Safransky: Viktor Frankl, who survived the concentration camps, said human suffering is like a gas: just as gas pumped into an empty chamber will fill the chamber completely, so will suffering fill the soul, no matter whether it's about something great or small.

Ram Dass: Somebody has a blemish on his face, and somebody else has multiple sclerosis, and both of them are suffering. It's really hard to judge. I just address the suffering itself. "Suffering is suffering," you could say. If you've had great pain in your life about anything, you know what suffering is, and you're speaking from that place.

Safransky: Then there's the kind of suffering that arises simply from being caught in the separateness of an incarnation.

Ram Dass: That's true. But as you realize that and open your heart to it, you can then allow everybody's suffering to be your suffering. As long as you're not caught in your own separateness, you don't have to push it away; you don't have to do all those things that create more separateness and more suffering.

February 1987

CHRISTMAS IN SEATTLE

Fred Hill

I had been laid off by Boeing, along with thousands of other employees. With that many people on the street, jobs were few and far between — and, that winter, Seattle would experience its worst snowfall in three decades.

Just before Christmas, after I'd been out of work for six months, I landed a seasonal job at the post office delivering mail to branch offices on the midnight shift. I liked that shift because there was no traffic and I worked alone, except for a crew that helped me load the truck, tossing in bags and boxes with gleeful abandon.

Every night I walked the five miles to and from Central Distribution, holding my breath until my first paycheck arrived. Then I bought a 1957 Plymouth for ninety dollars. The tires were bare, but the flathead six always started.

I moved to a flophouse in downtown Seattle to be closer to work: Nine dollars a week, payable in advance, cash only. A lumpy bed, sagging in the middle, old sheets provided but washed only once a week. A dresser with a mirror, a wooden chair, threadbare carpet, a steam radiator popping under the window.

IN THE PREDAWN hours of Christmas Eve, I fled my cramped room, walking from Fourth Avenue toward Pike Street Market. Light from a jittery red neon sign on an all-night diner bounced off the glass canyon walls and the wet street. The sole patron of the diner turned to study me through the window with bleary eyes. At the docks I watched fishermen ready their boats. Merchants were just opening their doors, turning on lights. The cast-iron figures in Pioneer Square peered at me as I waited for the time to pass.

Other men like me were on the street and in doorways. We glanced at each other suspiciously, careful not to get caught at it.

TURKEY. DRESSING. GRAVY. CRANBERRY SAUCE. So proclaimed the hand-lettered sign outside the Salvation Army kitchen. I stood in the street a long time, looking at the sign, trying to see through the window. I thought about the crackers and water in my room. Pride and weariness battled in my mind. How had it come to this? Just months ago I had been a well-paid, respected professional.

Practicality won. *I am where I am*, I thought, and I opened the door and slipped inside. I quickly scanned the room through hooded eyes. One or two people looked my way. I headed for the service counter at the far end of the room, my eyes straight ahead.

But the men behind the counter did not judge. They served me with a welcome and a smile. The smell of turkey and gravy was intoxicating. I breathed deeply.

There were three rows of tables lined tight on both sides with metal chairs. Almost every chair was filled, yet the room was eerily silent. The slight clatter of silverware, a low mumble, and an occasional clear word or two were all I heard. Everyone studiously ignored me. At first I thought it was aloofness, but I came to realize it was a rigid respect for boundaries, for privacy.

The men sitting there looked just like me. None of us wanted to be there, but we were glad to be there all the same. We were both embarrassed and defiant.

I sat next to a man who was about my age and dressed very much like me. We nodded a simple greeting.

The other men and I didn't wolf down the food, though there was plenty to be had. We savored each bite, rolling it around our mouths, chewing thoroughly, swallowing slowly. We drank little of the water, so as not to wash out the taste. We ate as much as we could with the fork, then sopped up the rest with the buttered roll. Few of us went for seconds. It was as though accepting too much would be an insult, a kind of thievery.

The man next to me rose, nodded again, and carried his plate and utensils to the busing table. Then he slipped quietly out the door. Not long after, so did I, a toothpick waving ostentatiously between my lips.

I stood on the street with a full belly. Other men had left just before I did, but I didn't see any of them. I wondered where they'd all gone, and how they'd disappeared so quickly. We all seemed to be good at slipping away.

I SLEPT LATE on Christmas morning — a long, sound, dreamless sleep. When I finally awoke, thoughts of family and decorations and gifts intruded, but I reflexively packed them away, deep into a corner of my mind. Instead, I concentrated on getting a can open with my Swiss Army knife. It was a little like opening presents. I sliced the meat out of the can and spread it ceremoniously on the crackers I'd lined up in a straight row on the dresser. Then I spread French's mustard over the meat. The bright yellow gave a tang to the bland dish and added a festive color.

I slept again. The day passed.

THE CARD WAS just a form letter from the state, but I read it again and again, memorizing the date and time of the interview. I calculated and recalculated the speed and distance and the lead time required to get there. On the appointed day, I brought in my dress clothes from the car and cleaned them up as best I could. I polished my shoes with toilet paper. I asked the desk clerk for some Scotch tape and used it to pick the lint off my jacket. I sprinkled water on the wrinkles behind the knees of my pants and smoothed them out over the back of my chair. I showered in the bathroom down the hall and trimmed my hair with a combination comb and razor from the five-and-dime. I triple-checked my tires and the oil and gas in my car and left an hour early. I had already spun out once before on the snowy streets, so I was especially cautious on the forty miles of freeway.

The person who interviewed me was noncommittal about my chances, and I returned to my room late that afternoon feeling depressed. The gloom of the hall and the smell of stale air and cigarette smoke pushed away any small speck of hope. I reluctantly pulled off my good clothes, folding them carefully into my suitcase. I sat on the edge of the bed a long time.

I spent the next several days waiting for a phone call. I treated my room like a job, forcing myself to be there between eight and five, Monday through Friday. When five o'clock came, my hope faded, and I went out to walk the streets.

THEN ONE DAY, the phone rang, echoing down the long hallway. I froze, listening. It stopped in the middle of the third ring, and a few seconds later I heard heavy footsteps coming my way. I sat up and swung my legs over the side of the bed, the springs and frame squeaking loudly. I stared at the scarred door. Feet scraped to a stop just outside it. Then there were three loud knocks, and a deep male voice demanded, "Are you Fred Hill?"

"Yeah."

"Phone call for you."

I leapt up and grabbed my shirt, slid back the locks and yanked open the door. The messenger was already disappearing around the corner as I stepped into the hall and carefully locked up. Breathing rapidly, I trotted down the corridor in my bare feet, mincing around shards of glass. As I turned the corner, in the dim light I could just see the pay phone on the wall, the receiver swinging by its steel cord. I picked it up. "Hello?"

"Mr. Hill?" a crackling voice inquired.

"Yes?"

"Mr. Hill, we'd like to offer you the statistician job."

Somehow we must have held a coherent conversation, because I remembered to accept the job, and I was especially careful to write down the start date when I got back to my room.

Then I lay on the bed, shaking uncontrollably. My mind jittered like the neon sign outside, flitting among scenes from my six-month exile. I was suddenly overwhelmed by the urge to leave this place. I sat up on the bed and for the last time took in the shabbiness and dirt, the sounds and the smells and the damnable blinking light.

In five minutes I was hurrying down the hall, down the stairs. The men in the lobby stared at me, but I didn't look back. I jumped into my car and sped down the highway, tail tucked under tight, as though some prankster might reach out and yank me back. I urged the old flathead onward, to the south. I, who had welcomed the brotherhood of our common fate, now wanted only to get away.

December 1993

Sonderkommando

Ivor S. Irwin

Everybody considers dying important
but as yet death is no festival.
— Friedrich Nietzsche

I hadn't thought of my father in at least two years. Being a bad Jew, I don't light yahrzeit candles. The truth is harsher than that: I am so frightened by the intensity and mediocrity of my hate for him that I have blocked him out. It's truly sad.

When I was small he would beat my mother mercilessly and methodically with a sewing-machine strap. Her face was like a club fighter's: broken nose, cauliflower ear, scar tissue around her eyes. Poor, *nebechal* Mama. There is a black-and-white photo at the bottom of the trunk I use as a coffee table in the living room. In it she is young and beautiful, blue eyes glimmering so bright a shade of pearl gray that my father might have taken the tip of a knife and scratched away the pupils. She sits on the hood of a British Army Land Rover in an Austrian refugee camp, her bare legs dangling, her knees soft and round. The reason I recall the photo so well is that she looks happy in it. I never saw her look happy for a moment, except maybe when she recalled stanzas of morbid German poetry. I have always wondered what it must have been like for her to lock herself in the bathroom, look in the mirror, and see beauty beaten black and blue. She loved him; I have no idea why. She never received a kiss or a smile or a word of love from him. There was nothing but hatred in his eyes.

They are both dead now. I can't think of one without the other; therefore, they are both equally condemned. There are disparate moments, like slivers of light under a door in a darkened room, when I recall her satiny good-night kiss; but ultimately, he is always there, somewhere on the periphery, slapping the machine strap against his palm, eyes shiny as obsidian, black as his *Sonderkommando* heart.

THAT DAY BEGAN so beautifully. The rain spit against the window as Carla and I made slow, sweaty love; and then, when we were finished, the sun came out, as brilliant as a newly hammered nail. I smoked a cigarette as she slept — an obstinate, almost hostile sleep, her arms flung up in abandon around her head, her legs bent, her face pushed into the pillow. I kissed her ear goodbye. "I love you," she said, her voice muffled.

I walked to work through Lincoln Park, ecstatic with the sight of the mottled sheen of wet leaves and the rich smell of rain-soaked asphalt, beaten into a pleasant concussion by the sight of thousands of office girls on their way to work, promenading up and down the avenue wearing their summer pastels and stiletto heels. Chicago is wonderful in September. You want summer never to end because here there is no autumn — winter comes down your neck like Madame Guillotine.

I own a hot-dog joint on Eleventh and Wabash. My clientele

are the students at Columbia College, policemen from the Eleventh Street Station, and the applicants and workers from the unemployment office next door. Not a great business, but it pays the rent. You'd be surprised by the number of hot dogs I sell in the morning, especially to cops. People love dreck for breakfast.

At 2:15, after the lunch crowd had gone, I was sitting reading *Winning through Intimidation* when a big, hulking fellow walked in the door, a Polish or Ukrainian cat with high, flat cheekbones and deep-set blue eyes. He said, "I want a triple cheeseburger with grilled onions and an order of fries."

Up close, he had red veins in his eyes. The look of a man who stole from his mother, a man with a gift for petty burglary. The gift of being able to rummage efficiently through closets, search the correct drawer, find the cash and jewelry underneath the sweaters and socks. He frightened me.

He looked at the book I was reading, lit a cigarette, and sat down. The cigarette, a filterless Camel, made him cough.

"How can you read that shit?" he said.

"Read what shit?" I knew what he meant, though. I always buy these books — *Winning through Intimidation, The Greatest Salesman in the World, Looking Out for Number One, Restoring the American Dream* — and I don't know why. Ambition does not flow strongly in my veins. My failure as a businessman is on some days an embarrassment, on others the source of a clear, nihilistic pride. There is within me a nagging gap between the mind that desires and the world that disappoints, like a speck of sugar permanently trapped within a rotted tooth.

"That shit," he said, "rots yer brain like the fucking TV. They got you so jealous with greed for stealing what the next guy's got that you forget why you went into business in the first place."

Appearances deceive, I know that. "You in business?" I asked.

"Nope."

The guy had that proto-Polish look of brush-cut blond hair turned white by the summer sun, fat factory hands like a bunch of bananas, and, as I said, a scary face. "You ever been in business?"

"Yep."

He ate like an animal, noisy and fast, as if he knew for certain, right then and there, that this was his last meal. On the wall over the griddle, I've got a Seka picture calendar. Seka is a beautiful porno actress who comes in sometimes for a Polish and fries. I think she takes serious acting lessons over at the Eleventh Street Theater. It's kind of amazing; she makes a good living, I'm sure, but there she is doing all that Stanislavsky shit. Maybe she wants to be legit, but I doubt it; her face is hard, sharp all over. I think she does it as a hobby, like I read these books. The guy stared at the calendar as he ate. When he finished, he wiped his fingers with a napkin like he was masturbating.

"I fucked a chick who looked a lot like her," he said.

"Yeah?"

"It was," he said, lighting up again, coughing terribly, "the crowning moment of my whole existence."

"She comes in here," I said. There was a look in his eyes as sad and distant as ten miles of telephone poles on a New Mexico highway. I looked away and up to Seka. She was kneeling on an overstuffed red couch, gripping a white windowsill. Her nails, shining and red, were the exact same hue as the couch; you can always tell good nail polish by the way it shines, my wife, Carla, says. I pointed at the squiggly writing just above her mons veneris. "That's her autograph, personally addressed to me: 'To Marty from Seka — keep up the grease.' "

"It's not the clothes that make the woman," he said. "And it isn't the difference between beautiful and pretty, pretty and attractive, attractive and plain, plain and homely, homely and ugly. . . ." He had smoked the cigarette down to its end and made grand gestures with both hands, like a desperate soap-opera actress. "It's the way they save themselves up. They wait."

He lost me there and I was glad when a customer came in. A fat black man, his skin a caramel shade, ordered a pizza puff and a hot tamale: a dreck aficionado. He was wearing a cheap imitation Walkman and I could hear the warped treble of the speaker like the moaning of some wounded creature.

"I love all women," I said to the first man. "I love my wife."

He was not impressed. "It's the ones who get looked at, the ones whose skin is as white as the belly of a fish, who are the most desirable. The girl who looked like her," and he pointed up, "had skin like that and left me to go to Florida. I hope she's happy on the beach."

I felt sorry for Seka. No amount of money is solace enough for the kind of verbal abuse she must be subjected to. In those acting workshops they go for the jugular — I remember from my years in film school.

"Did you get laid off?" It's not a question I normally ask; you look for trouble with such inquiries when your business is situated right next door to the unemployment office, but his unhappiness was distracting.

"I got laid off nine months ago, brother," he said. "And now the motherfuckers are gonna cut me off. I only got through the *F*s on my job-search sheet." He was taking an agitated draw on his cigarette when he crossed his arms over his chest and began massaging his biceps — soft, then hard — crushing the still-lit cigarette between his knuckles. "This. . . ," he said. "This . . ." He stretched his arms out, made fists, and began desperately pulling in and out, as if he were rowing a boat. The look in his eyes was queer and desperate, a miasma of fear. *The last man in the regatta*, I remember thinking. "This is it, man," he said calmly, "the big-bang theory." With that he emitted a low moan and fell backward off his stool with a resounding crash. The crown of his head smacked against the floor, making a noise like a fresh scoop of ice cream hitting pavement. Then he rolled over onto his side and lay still.

I vaulted over the counter and began screaming for help as I turned him onto his back. "Call an ambulance! Call an ambulance!" For the first time in my life I gave mouth-to-mouth resuscitation, all those classes at the YMCA finally proving useful. Of course, what the instructor never warns you about is the taste of your own drecky food. "Suck the windpipe clear," you are told, but all you get when you suck in on a dummy's throat is the vague taste of rubber.

There I was, sucking away, spitting improperly masticated food onto the floor, ranting hysterically; from behind me I could hear the steady, throbbing treble from the imitation Walkman. I began taking turns sucking and blowing into him, thumping his chest with the meat of my fists. Harder and harder I hit him, alternately gasping for air and then slithering forward, breathing into him and counting to six. Suddenly, his head began rolling and he puked into my mouth and over my chest as I arched back in horror. As I spit puke onto the floor, I heard the big black man from behind me, like the voice of God.

"What the fuck are you doing, man?"

"What do you think I'm doing?" I screamed without turning. I'm afraid to face black people if I'm going to say anything nasty to them. "What the fuck do you think I'm doing? I'm trying to give him mouth-to-mouth resuscitation." I began thumping at his chest again.

"YMCA, am I right?" said the voice behind me.

"Right. Would you please call an ambulance?"

"You hit like a sissy, man."

"Will you call a goddamned ambulance? Please?"

"There's no phone here, you fool," he said. "If I run next door to a pay phone, will you take the quarter off the price of my tamale?"

"There's a phone in the back! You can have the tamale for free!" I screamed.

"It'll take 'em a half-hour to get here, man. Let me do it. You're doing it wrong."

"Are you trained? How do I know I —" I felt a big, meaty paw pull at my collar and lift me. I went with it. I did not struggle. I was wearing a brand-new shot-silk Givenchy shirt under my apron; it was bad enough that I'd gotten puke all over it.

"You don't know what the fuck you're doing," he said. His mouth was filled with chipped and broken teeth. He thrust me aside and got down on his knees. "Just go call an ambulance," he snapped. "I gots to be crazy. I gots to be fuckin' crazy."

I vaulted over the counter and ran into the back room.

"If I get AIDS," he shouted, "I'm gonna come back here and shoot your ass!"

I called emergency. The woman who answered had a bubbly telephone personality, like something from a tacky car commercial. I was hypnotized by the sight of the back of the big black man's head bobbing up and down, his bald spot shiny as a freshly buffed brogan. My gullet was as still as a pebble as I tried to give her the address. I had to light a cigarette and breathe deeply before I could talk at all. Then I closed the door and called Carla. "What should I do?" I asked after I'd explained everything. "I want to come home for a hug. I don't like today."

"You can't come home yet, silly," she laughed. "You've got the night crowd yet." Love like a razor.

"Carla," I said, "I hate death. I can't stand suffering. It makes me ill."

"Marty," she said, "you're thinking about your mother and father, aren't you?"

"Yes." Actually, I wasn't, but it was easier not to argue.

"The Holocaust is boring, honey. I lost it with that last Louis Malle film. It's as old as platform shoes. They trivialize it." Carla isn't Jewish. "You oppress yourself, honey." I nodded. I opened the door a crack and saw two sets of shoes — one vertical, one horizontal — both pairs severely in need of heeling. "Tonight," she said, "I'll give you a bath and we'll eat Chinese."

When I put the phone down I felt worse than I had before. Whenever I felt bad, Carla always brought up the Holocaust: the psychology of comparison/contrast, I think — I'm not sure. Looking up at the grease-spattered walls, I realized in one jolting moment, the way it always is in cheap novels, that without her my life was as cold and empty as Greenland. I walked around the counter and found my two customers. I knelt next to them. My Slavic customer was breathing desperately, his eyes red, but I was overjoyed to find him alive and conscious.

"You shouldn't have saved me, brother," he sighed. "Really, I ain't got a fucking penny to my name. They'll just put me in Cook County with the dregs. I'm gonna die any-fucking-way."

He then began a sad recitation:

Oh God, cut short my agony.
Hasten the muffled drum.
You know I have no talent for
The art of martyrdom.

"What was that?" I asked.

"Hiney," he said. "You're a Jew, an' you never heard of Hiney?"

"Hymie?"

"God, but you're stupid," he sighed. "I don't want to live in this world."

"Ssh!" I said. "Don't talk like that. It's a sin to talk like that. In my religion, what I just did for you is called a mitzvah. It does me good, too."

They both sneered at me, he and the black man. "You fuckin' Jews, man. You make me laugh."

"I've hired you," I said. "You are now part of the staff at Marty's Mouthful. You have Blue Cross. They will at least put you in Swedish Covenant, where the nurses are hot and they don't make you beg for drugs."

There was a strong smell. The hot tamale was burning. Before I could get to it, the man had me by the wrist. He may have been sick and dying, but he had a grip like a vise. "My father," he said, "was in a concentration camp."

"Oh?"

"He fell off the guard tower."

They started to laugh, he and the schvartze, like devils. And there was that awful burning smell.

My FATHER WAS a *sonderkommando* at Auschwitz. *Sonderkommando* means something like "special soldier" — which sounds good, but the Nazis were really clever at making jobs sound good; in the end, my dad was a prisoner who, because of his size and strength, was chosen to man the doors of the gas chambers and crematoriums, and so allowed to live — what we Yids called a schlep. Still, he was lucky to have a job in a death camp, even a schlepper's job; but in his case, the value of the work ethic upon the soul was quite overrated.

Every day of my youth, Monday through Friday, my mother worked at a sewing machine, killing herself to keep up with the newly arrived and illegal Chinese and Mexican girls. And every day, from the time I could talk until I began first grade, my father explained his one and only job very clearly and methodically. It was the only time we were ever in the vicinity of being close.

We would sit in the kitchen of our brand-new bungalow in Lincolnwood, watching either each other or the garden. It was my mother's dream kitchen, yellows and oranges, porcelain, steel, and formica, like living inside a bowl of fruit. That garden, I remember, was Kelly green, smooth and square with irrigation holes sloping down from the corners and center so that it resembled an enormous pool table. He would sit and stare through the slats of the Venetian blinds, elbows pushed into the table, shoulders sloped, knotted muscle showing through his shirt as he held himself up, cocked like a hammer with a hair trigger. His body never moved, his head twitching economically up and down or side to side. If the sun dazzled at too sharp an angle, he would swear in Polish (the secret language he could deny me and my mother), flip down the clip-on sunglasses that were fixed to his spectacles, and recommence his reverie. He never shut the blinds.

Next to his elbow would be a bottle of Wyborowa and a shot glass. The silver label with the royal blue pinstripes and red piping reminded me of the uniforms of Napoleon's chasseurs — pride of my tin-soldier collection. He was always at his best at around one o'clock, when the bottle was half gone: two shots past Polish melancholy and two hours short of the rage he would build up to coincide with my mother's homecoming.

"Martin," he would say, "you are beautiful like *ein* girl." An attempt at a kiss would follow, a flat dryness and then the burning of my cheek with his whiskers. "You should *shtup* lots of shiksas, especially." He would laugh and point at me. "German and Polish girls, in the *tuchis*. So that they'll remember you for a long time after you leave their beds." Sometimes he would laugh until he burst into tears; the only thing that held his misery in abeyance was the misery of others.

"Don't you worry about what you do for a living, Martin," he would add with the next drink. "Let the ass-lickers call you a schlep. You do what you like." As far as I know, in all the time he lived in America he never had a job. He worked with diligence, however, at drinking himself to death.

Then there'd be the two o'clock scenario.

"In Oswiecim, you work or you die." He spoke of those war years in a curt present tense, often running his fingers slowly through the slats of the Venetian blinds, making particles of dust float around him. "There are artists, pianists, violinists, forgers, pickpockets, writers, actors, mathematicians, doctors — oh, *lots* of doctors, and they all die, die, die. All shit themselves when they die, too. All burn well." The idea that he had outlived the artisans gave him a sense of joy. And suddenly he would jab at the air with his fists, slap the table, rock the bottle and save it like a goalkeeper, then massage my Adam's apple, close as an eyelash to turning his acrimony and guilt upon me.

Then our game would begin. "How long to gas 'em, Martin?"

"Fifteen minutes before you open the doors." By the time I was four, he had trained me to parrot questions and answers to fit his whim.

"We'd pull 'em out with ropes and grappling hooks. *Then* what?"

"Pull their teeth." I'd never get the answer quite right.

"Pull their *gold* teeth," he'd correct. "Pull their gold teeth with *these*." He'd reach behind to his back pocket and come up with a pair of steel pliers. "Pull their teeth faster than you can say Hans Frank."

"Hans Frank!" I'd say as he grabbed me in a soft headlock. "Hans Frank! Hans Frank! Hans Frank!" And each time I yelled the name I would make my voice shriller and shriller, knowing my fear was an aphrodisiac to him. He would grab my jaw in his left hand and reach with the pliers in his right. He would take a grip on one of my tiny incisors, cackle, and make crunching noises — as if he would really go through with it. I would scream and bang the backs of my ankles on the crosspiece of the chair until the bones throbbed. Finally, he would kiss me, replace the pliers in his back pocket, and, as an afterthought, give me a popsicle or something sweet from the refrigerator. Now and again, if he'd imbibed too much too early, he'd begin to cry.

"I love you more than I love your mother," he would say, as if the knowledge that filial love meant more than passion might tilt my love in his direction.

To this day I cannot stand the smell of vodka. And visiting the dentist guarantees a month's lost sleep, before and after the fact.

THAT NIGHT CARLA and I made love again. Afterward, when she lay there smiling, loose as a dead cat, I told her that I'd hired the Pole. She was livid. I deliberately didn't shower and didn't brush my teeth before we went out to eat. I sat in the Hunan restaurant sniffing my fingers, running the tip of my tongue against my mustache, looking at my distorted self in the gold waiter's bell.

"You're so nice, Mr. Nice Guy," she said, looking at her fingernails, looking for the waitress, then grabbing the bell to break my concentration. "Nice like a sucker."

The walls of the booth were padded like a lunatic's cell: harsh, shiny vinyl walls that made words bounce back at you.

"I'll pay him minimum wage," I said.

Carla just sneered, her thick black Sicilian eyebrows twitching in time with her agitated fingers. A pitcher of daiquiris arrived along with the soup and spring rolls, and Carla calmed down. She forgave as quickly as she angered. "You've got to stop being so nice," she said between mouthfuls of egg-drop soup. "We'll have nothing when we're old."

"The meek shall inherit the earth," I said, for want of something more erudite.

"The meek shall inherit a hole in the ground."

"It's a mitzvah."

"Don't give me that mitzvah shit," she laughed. "You just can't say no. You let the world walk all over you."

These rebukes were a magnificent ritual to me. They set my heart off inside my T-shirt like a pillowcase full of puppies. "I want to do it to you again."

"I've not finished telling you off, Marty."

The moo shu pork was wonderful. We stared at each other across the table and played footsie. "Can we? Can we?" I whined.

"Your dick'll be the death of us both," she said.

Too true, actually. When we met in film school, I fell for her instantly; the thick black hair on her arms, I think. I told her she looked like Claudia Cardinale. Carla didn't have a clue who she was. Visconti's *The Leopard*, I said, ogling her, Visconti's *Rocco and His Brothers*. She shook her head and told me that all Italian men were assholes; it seems her ex-husband, Pino, was the missing link between Cro-Magnon and Neanderthal man. Somehow we lost Luchino Visconti, left screenwriting early, and went back to my apartment. We would ball six, seven, eight times a day, running off in the middle of lighting exercises and editing periods to rut on the fire stairs. One day I didn't tighten the nut properly on a gobo head. The gobo was holding up a five-hundred-watt, twenty-five-pound, baby-Fresnel key light over the soundstage, eighty feet up. It fell and missed the teacher's head by three inches while I was on another floor, *fressing* Carla as she sat on the six-plate Steenbeck. The administration demanded remuneration and threatened to throw us out. I was as broke as any working-class film student ever is (that's why there are none), and Carla can't handle threats — so we quit.

Thus, my penis can be a weapon of fate. Twelve years later, my brain is filled with shards of information about filmmaking, all useless because of this monster between my legs. I am an unrepentant prisoner of my own lust, trapped like one of my David Berg hot dogs in a bun too long on each side.

"I'll keep him," I said, "for two weeks after he gets out of the hospital. So's he can get back on his feet."

"You're a saint."

"I told you. I'll pay him minimum wage."

"You're a Jewish saint."

"I'm a horny saint."

"You don't even know his name," she snapped.

"It's something Polish." I had the urge to pull it out and slap it on the table.

"Polish people hate Mediterranean people."

"You believe in clichés," I said, not meaning a word. "Poles are all bigots, Jews whine about the Holocaust, and Italian men are all macho."

"It's true."

I noticed that I was bouncing back and forth off the padding in the booth. "My father," I said, "told me to *shtup* Polish girls in the ass."

"You'd get shit on it, Marty. And they'd call you names afterward when they found out you were Jewish." Carla laughed and lit a cigarette. When she was amused, the small bit of fat on her arms moved with her.

"I'm burning," I said.

"It's not your fault that you were born with him as a father." She blew smoke at me, her mouth perfect and slightly fleshy — the kind that WASP girls have to paint on.

"I'm burning," I said again. "When we talk about sex it becomes unbearable." The hairs on my arms were stuck to the vinyl booth; it was painful for me to pull away.

The fortune cookies were stale. "Plan on having many children," mine said. "You will be reunited with a rich relative," said hers. I tore mine into tiny pieces and put it in the ashtray with all the dead butts.

We paid the bill and Carla left a 10 percent tip. I insisted on throwing down another two dollars without a word; cheapness is an embarrassment to me. Carla turned around at the cashier's desk and went back to get the cigarettes she said she'd forgotten.

I let Carla drive. I was exhausted and groggy from the daiquiris. I put on my seat belt, as usual, and Carla didn't, as usual.

The car hit us as we were going north through the intersection of Lawrence and Clark: an English tourist driving a Hertz rental down the wrong side of the street. There was a flash of yellow and black and the sound of breaking glass, like a little boy biting down on boiled candy.

I opened my eyes to a crowd gathered around the car; I saw them as a series of fuzzy circles, like hair-covered softballs with red zigzag stitching moving in and out across their diameters. *It's going to be a curveball*, I remember thinking, raising my arms, feeling for the bat, and realizing I was in restraints. I managed to undo my seat belt but couldn't move.

"Keep still," someone said. "You've been in an accident."

So I began a slow inventory of my body with the tip of my finger, slowly touching my face and neck and lap and arms. And feeling nothing warm or wet.

"Are you OK, man?" some Puerto Rican kid asked, his head moving in and out, up and down, like some junk-baller's change-up.

And then it struck me. "Where's Carla?" I began to scream. "Where's Carla?"

Carla was dead. Her head had gone through the windshield.

"I'm sorry, mate," a voice shouted. "I thought *you* were on the wrong side of the street." A tired voice, like one of those second-rate English actors who work over here on sitcoms as the butler or the effete snob of a neighbor. "I'm really sorry."

By the time they came to pry us out, the inside of the car smelled like a butcher's shop. A fireman delicately unfurled Carla's fingers from the steering wheel and the two dollar bills, folded into a tiny square, were embedded in her palm. The last thing she'd said to me, just ten feet before the intersection, was "But I don't have any rich relatives." I didn't wash for a week after I buried her.

The Englishman was reprimanded and fined. In Chicago you can pay anything off. He was decent enough to pay for the funeral, though. I'm sure he wasn't really a bad person: when I was crying after the funeral he came up, put his arm around me, and offered to pay for a free vacation anywhere; a "holiday" is what he called it. "If you holiday in England," he said, "I'll take you to a football match. You've never seen anything like it." He slipped something into my suit pocket and walked away. It was a roll of bills, five thousand dollars in crisp new notes, wrapped in a rubber band. I didn't notice it until a week later, when I was taking the suit to the cleaners.

It turned out the Englishman's name was Glickstein. "I didn't even know there were Jews in England," my sister-in-law Carmella said. "Goddamned amazing. You people look after each other."

Carmella was Carla's younger sister. She had an older sister named Carol. Not a very imaginative family. Carla saw them only at Christmas and Thanksgiving; they were an embarrassment to her. I never dealt with any of them after the funeral.

My MOTHER TOLD me once that I was such a handsome baby that when she pushed my carriage around Sauganash Park, people would stop to admire me and give her a dollar to buy me a new rattle. That was her story. My father told another. He said that my mother was so pretty, men gave her money simply as an excuse to make conversation with her. I chose to believe my mother.

My father gave me that nugget of information on my twelfth birthday, celebrated while he was lying in a private room at Michael Reese, waiting to die of cirrhosis. His every waking moment was spent in a *Götterdämmerung* of complete, relentless, indefatigable obnoxiousness to everybody but me. He spit at doctors as if they had maliciously engineered his exit. He would grab Filipina nurses, mauling their breasts and buttocks, terrorizing them. "Did you *shtup* General MacArthur?" he'd ask. "Or," and he'd pull the corners of his eyes into sloe shapes, "did you *shtup* Yamamoto and Tojo?"

He grew thinner and bug-eyed. He would grip me by the wrist and beg me to bring him a bottle of Wyborowa. My mother, iron-willed, would refuse. "You Vienna whore!" he'd scream. "You *fakaktah courvah!*" She showed nothing on her battered face as she sat next to him, her icy blue eyes blank. "Kaltenbrunner's *courvah!*" he screamed. The only betrayal was a whitening of the knuckles on her entwined fingers.

"Ssh!" she'd say. "The only one I loved was you. You!"

It took him three months to die. During the last month I began smuggling in pint bottles of cheap Hannah & Hogg vodka. I was only a kid, but it was easy to bribe the man at the liquor store around the corner. I prayed every night for my father to die, but he held on to life like a mollusk to a rock.

On the day he finally died, my mother sat next to him, using a tiny round mirror to apply her makeup. He grabbed the mirror; he wanted to see himself. By then his eyes had receded so deeply into his skull that he resembled a starved rodent.

"Once again," he whispered as he stared at himself, "once again, I am *der sonderkommando*. How long to gas 'em, Martin?"

"Fifteen minutes before you open the doors," I murmured — Pavlov's dog become an elephant.

My mother grew wide-eyed and gave an involuntary squeal that bypassed her closed mouth, and went up and out her nose before she could stop it.

"Then what?"

She bit her lower lip and turned to face the bare white walls of the room. "It's OK, Mom," I said. "It's just a game we played when I was a kid."

"Then what?" he broke in, merciless.

"Pull them out," I mumbled.

She screamed.

"Pull them out with what, Martin?" he went on.

She jumped onto the bed and began pummeling him about the face. "Pull them out with grappling hooks!" I shouted, trying to drown out my mother's weeping. He was laughing uproariously when he died, his hands spread on his lap as he exposed his face to her blows. There wasn't even a gasp, just a last *ha*, and he was gone.

There was no one to say a farewell prayer for him. Only Eddie Rabinowitz, the owner of the raincoat factory my mother worked at, showed up. Rabinowitz went out and hired a minyan: a bunch of shabby old men with hair exploding from their ears who davened with a splendidly convincing grief.

I thanked Eddie Rabinowitz and did the mensch-of-the-house number I felt was my duty, stiff as an SS officer on a parade ground in the shirt, shoes, suit, and tie that Rabinowitz had purchased for me.

"You must visit us more often," I said, as the starched collar cut into my throat.

"Oh, I will, Martin," he said, too good-humoredly to suit me. He gave my mother a quick wink. There was something depraved about Eddie Rabinowitz: his hair dyed jet-black, thin waves forming stripes across his pate; that, and the clear varnish on his fingernails and the whorishly grand ruby that shone from the gold ring on his pinkie finger.

"I don't like him, Mummy," I said after Rabinowitz left.

"I don't like him a lot myself," she said with a shrug, "but he's been good to us all." She put her arms around me and let me mash my face against her breast.

"Daddy said that you do it with him," I mumbled against the soft threads of her cotton dress, feeling the contour of her nipple against my lower lip.

It took a long second for her to digest what I'd said, thrust me away, and then rear back and slap me backhanded across the face. Her flat, weary blue eyes betrayed nothing, neither anger nor sorrow, just an aquamarine indifference that hurt all the more.

"You do do it with him, don't you?"

She raised her hand again but let it slump dramatically and glance off her hipbone. "Who do you think paid for this house? Who do you think paid for your bike and your books and your tennis racket?"

"You let him stick it in you." It's too easy to say such things when one is twelve years old. There was something too easy to mock in that beautiful, battered visage.

"Your father was never a husband to me," she said. "Never. You know nothing." She reached out, and I backed away as far as I could. The drawers of the fitted kitchen furniture, cold and steely, dug into my spine as I watched her come slowly undone. "The only skill I have is sewing, Martin. I learned to sew in the camp. I was brought up in a house with servants."

"So you do it with Rabinowitz?"

My mother sat down and began playing with the string of pearls around her neck as if they were worry beads. "Being poor is a terrible thing," she said in her haughty German. She began to weep. "I didn't come from a Polish village like your *shikker* father; I grew up on Franz-Josefsstrasse in Wien."

I walked away and left her to weep alone with her beads. Upstairs, I turned on the television and watched blue-eyed, fresh-faced American boys burn down a Vietnamese village on the news.

The next day in the schoolyard, I beat up Milton Rabinowitz. He was three years older than I but round and soft like his father. I broke his nose, broke his glasses, beat him with my fists until my knuckles ached; then I picked up a metal garbage-can lid and hit him repeatedly until my arms were too tired to lift. "A Jewish boy," moaned Mr. Glass, the principal, as I sat in his office staring down at the bald spot in the rug. "A Jewish boy. How could you hit someone who wears glasses?"

A month later, my mother and I were suddenly evicted from a home I had always assumed we owned and forced to move into a cockroach-infested one-bedroom apartment at Devon and California. It wasn't the apartment my mother hated so much as the army of Russian Jewish immigrants all around us; their loud conversation, the hissed *z*s and *s*s like fast-running water over pebbles, oppressing us through the thin walls. She hated them: women in babushkas who wanted to barter for everything in the stores, or else stole, stuffing the pockets sewn into the lining of their heavy, ankle-length winter coats with canned seafood while their men, dressed in cheap, ill-fitting polyesters, stood around with their cigarettes pinched between their thumbs and pinkies, all malodorous, unwilling to wash themselves or their clothing. Mother would get so angry, her cheeks red, the scar tissue around her eyes purple-pink. "They bring the stench of Kiev and Odessa with them, Martin," she'd say. "That's why the Germans murdered us." She was as rancid with hatred as my father.

She died at her sewing machine, doing piecework at home. It was one of those hundred-degree Chicago days when the heat rises from the concrete and is reflected back by the skyscrapers and the wind off the lake. I got beet red that day, sitting by myself in the bleachers at Wrigley Field as Ernie Banks and Billy Williams both hit line-drive home runs, despite the wind blowing in.

I found her slumped back in her chair, eyes wide open, the Singer humming a sad, metallic mantra in front of her. The machine rasped and jerked for a couple of long, hard beats after I switched it off. I sat down, smoked one of her cigarettes, kissed her face all over, and then called an ambulance.

Two weeks after Carla's death, I got up early. I had to be at my store to meet a series of reps. There was a relentless war of attrition going on between David Berg, Kosher Zion, Kosher Vienna, Romanian Kosher, Best Kosher, and Palestine Kosher. I had to say no to all but one. There was that and the knowledge of my promise to Carla to terminate Slobodan, the Pole, that day.

In Lincoln Park I saw children everywhere: bicycling, throwing frisbees, and skateboarding, while the very young ones entertained themselves in the playground, carefully supervised by mothers and nannies who sat with sloped shoulders, smoking cigarettes, chattering at each other and the children, displaying a shared torpor of the mind that made me turn away and watch their charges. Four children caught my eye as they wailed and wobbled back and forth on their swings, their bright-colored dresses and shirts flashing and disappearing, flashing and disappearing, a constant rhythm in the sunlight.

I pulled a zip-lock bag filled with Carla's pubic hair from my pocket, opened it, and sniffed myself a quick fix. I kept a bag at work and a couple at home in my sock drawer. It was a habit I had begun years before when Carla had gone to Pensacola to visit her sister. It calmed me. The mortician had stared at me when I made my odd request for a cutting, but I hadn't cared; having experienced all the degrees of rage and despair, I'd achieved a bird's-eye view of the ludicrous and could sneer at those more innocent than myself. I walked the rest of the way to work, happy with the memory of Carla still in my nostrils, whistling and singing Marvin Gaye's "Too Busy Thinkin' 'bout My Baby."

Outside Marty's Mouthful, curled up into a fetal ball in the corner doorway, his pockets bulging with the excessive weight of a bottle and a book, smelling of sour whiskey, was Slobodan. I was about to say something insulting, something like "You drunken scum of the earth," when the Berg rep pulled up in his bright yellow Subaru — the kind that looks, from a distance, like a BMW. He climbed out of his car, alligator attaché case in hand, and went straight into his schtick as he shook my hand. He was a slick-looking Indian with liquid brown eyes that exuded a kind of relentless mirth. "Chatterjee's the name, Kosher meat's the game," he began. "I sell to anybody, to me it's all the same."

I realized I had a rapping Hindu on my hands. "Do you mind if I open the door first?" I said sarcastically. "Can I get myself a cup of coffee?"

"Oh, yes, yes, yes, certainly," he said, "but hadn't you better remove this tramp fellow from your door first?"

"This 'tramp fellow,' " I began, putting the key into the first of three deadbolts, "happens to be my one and only employee." I gave Slobodan a couple of soft kicks in the ribs. He moaned and cursed me as I opened the door and switched off the alarm. "This 'tramp fellow' is an absolute expert on your product. I trust his taste buds totally."

As Slobodan kissed the porcelain in the back, I ordered three dozen Polish and a gross of red-hots. I wanted simply to be rid of the Indian. I signed on the dotted line but was forced to endure another rap on behalf of cocktail wieners.

"I don't sell cocktail wieners, Mr. Chatterjee," I said. "I don't have the clientele, and I don't do weddings or bar mitzvahs."

The Indian ignored me; closing his eyes, he began to hum, slapping out a beat on his thigh before going into another rap:

Brisket and pastrami are the meats for me.
That's why my name is Chatterjee.
I sell good cold cuts everywhere,
From Cabrini Green up to O'Hare.

"You're pushing your luck, my man," I said. "There's nothing more silly or confusing than a Hindu who tries to sound black."

"Salami?"

"Begone, or I shall cancel my order forthwith!" I bellowed, pointing toward the door.

When Mr. Chatterjee left, Slobodan was already preparing a hot dog and fries, bent precariously over the side of the griddle with all his weight thrust onto his elbows.

"I like the Romanian Kosher best, Marty," he said.

"I know, but I just wanted to get rid of him."

"Can't say no to nobody, Marty." He took the bottle out of his pocket and drank a healthy gulp of whiskey. "Can you?"

I came around and shook the deep-fry basket a tad. "Did you spend the night here?"

"Yeah," he laughed. "I don't know how I got here. I must be like a Polish lemming or somethin'."

"I have to let you go," I said.

"Why?"

"I can't afford to keep you. Even you can see I don't make enough money to —"

"You goddamned liar. You lie like breathing," he hissed. It was a good excuse to take another slug of Jim Beam. "I'm sorry." His voice came down a notch, suddenly coy to the point of simpering. "I really am sorry."

"I'm sorry, too. I really can't afford to keep you." I served up his fries and put mustard and onions on his hot dog. "I can't stand to watch you destroy yourself either. I've seen enough —"

"You really do care about me, Marty, don't you?" He sighed. He began to eat his hot dog, alternating each bite with a wash of Jim Beam. The effort of switching his balance from arm to arm began to tell on him; his face ended up an inch or so above the griddle.

"You're like Hiney," he said. "Poor little Jew boy."

"Don't start with the slurs, Slobodan," I said.

He struggled to take the book from his pocket with his hands already full. The effort made him topple to the floor. I reached for him, fearing one more heart attack, one more death.

" 'M OK," he sneered, brushing away my helping hand. His left cheek was red from the griddle. He opened the book and found a page folded over at the corner. Despite the fall, he had managed to keep the bottle upright with all the loose dexterity of a bomb-disposal expert holding on to a detonator. "You don't know Hiney?"

"Nope," I said, "I don't think so. Did they film any of his books?"

"Jesus, Marty," he said, "you should be ashamed." He took another sip, flipped to a page toward the end, and began:

In the grave we're warmed by fame.
Foolish words! A silly claim!
There's a better warmth to seek
From a milkmaid, though she reek
Of manure from toes to tips,
If she kiss with loving lips.

I interrupted him with a laugh. "Heine," I said. "Not 'Hiney.' Heinrich Heine. He doesn't translate very well. I mean, human beings don't have 'tips.' You know what I mean?"

"You know, Marty, I thought you was a goddamned philistine for a while there; you had me worried." He took a long hard belt of booze and winced.

"My mother was from Vienna. She was very cultured."

"I'm cultured," said Slobodan. "I really am, but my dad and her —" he pointed to the Seka calendar — "knocked it out of me." He offered me the bottle but I refused it. "*Take some!*" He was either ordering or begging. I took a very small sip.

I clambered onto the counter, carefully pulled the tape from the corners of the calendar, and dropped it to Slobodan. "It's all yours," I said, climbing down. I helped him to his feet. He kissed me on the cheek and began to cry. I felt a cool, ticklish vacuum inside me, as if my organs and bones might evaporate.

"I hate her," he said, kissing the calendar.

"You can't like Heine and hate women, Slobodan. It's not possible. The milkmaid is warm . . ."

". . . even if she smells of manure." He ate what was left of his hot dog and fries. "Jesus, Marty, the guy died of goddamned syph. Poisoned by love."

"My mother put me to bed every night with a never-ending series of morbid German poems," I said. "Goethe, Heine, Kleist. Death, death, death. I remember them the way others remember nursery rhymes." And I quickly considered the moment they must have set eyes on each other — desire behind the barbed wire. I took the zip-lock baggie from my pocket, opened it, and inhaled. "There are worse ways to be poisoned than by sex," I said.

He looked at me strangely and wiped the corners of his mouth with his sleeve. "Time to go," he said.

"Don't go," I said. "You can stay." Slobodan shook his head from side to side. "What is it?" I asked. "Do you think I'm a sicko or something?"

"No, it's time to move on." He tossed me the volume of Heine as if it were a frisbee. "It ain't that it's sick, Marty. It's the fact that you flaunt your love that way. It embarrasses people like me 'cause, you know, we wouldn't know." He stepped toward the door. "You sure you don't want the calendar?"

"No, I'm sure she'll come in again sometime."

He pulled the calendar from his back pocket and unfurled it. "I can't believe that she lets them take photos of her with a pecker in her mouth."

"An exhibitionist, I guess." I shrugged. Then I gave him a hundred-dollar bill, a crisp new one. He kissed Ben Franklin, folded the note down the middle, then made a thing of rubbing the bill between Seka's legs and licking his lips.

"You won't say anything bad about anybody, will you, Marty?" I shook my head. "What about her parents? Don't you think it makes them feel just awful?"

"Like you said, Slobodan: it embarrasses people when, you know, they wouldn't know."

"Goddamned Jews, always gotta have the last word." He carelessly stuffed Seka and the money into his pocket. Without another word he lifted his hand, waved, pulled his pants up past his hips, and opened the door. I never saw him again.

I got rid of the Heine anthology on my way home. I put it in a Salvation Army box. Share the misery, that's what we should all do.

May 1991

Even Hitler

David C. Childers

Even Hitler,
moonlight on the crumbling dreams between us,
even bloodstained misanthropic
souls with hatred on their last acts,
even users, child
killers, torture
artists, devil
worshipers, even
Hitler, as we kissed
in our embittered longing, trapped in our
regret,
our tiny, personal, tragic
frustration blooming like twelve
billion
nocturnal orchids, even
as the Milky Way slides down, as
heaven stands up to pass above
the shadowless pattern of our
embrace, even Hitler,
by that great, kind
God who set us on this
planet, in this crazy, lethal darkness,
burning, burning, even
Judas, even the unforgivable, and
even (I thank you, love you)
we,
you said, and
Hitler
are forgiven.

October 1990

JUDAISM'S MYSTICAL HEART

AN INTERVIEW WITH DOVID DIN

Howard Jay Rubin

O*ver the centuries, something has been lost. It must have been, because when my family's religion was passed to me, neatly packaged, the box was empty, or so it seemed. Oh, I sang the proper songs, heard the history and the heartache; at thirteen I recited the proper Scriptures to be declared a Jewish man (quickly, so the catered affair could begin), and then retired from Jewish life. As with most suburban American Jews of my generation, what was meant to be the beginning seemed to me the end. No need for active rebellion; when all you see are the laws of bagels and lox, you simply learn to look elsewhere for inspiration. And look I did. I stared at the world until it seemed to sparkle, embraced spiritual traditions East and West, and began to sense in each the same spark of holiness — except one: I couldn't quite forgive the empty box.*

. *Still there are nagging questions: Where is the lifeblood of a tradition that has sustained so many for so long? Where is the frightening joy that set Jews dancing even in the cattle cars bound for Auschwitz? Where is the spark in the peculiar and idiosyncratic ways of traditional Judaism? Three thousand years without a spark? Unlikely. And all the time, stooped over his prayer book in one inward corner of myself, there's a white-bearded Jew, head covered in reverence, eyes ablaze with messianic fervor — waiting. He has boxes, too — on his head and arm, ritual tefillin filled with scrolls of holy Scripture, symbolically binding him to God's service. I peer at him apprehensively and lean forward to hear his whispered words. "Just remember," he says, his body shimmering and fading, "the bush still burns. . . . Have you looked?"*

This was an odd trip to be taking with my father. I was born in Brooklyn; he had lived most of his life there, but had hardly looked back since moving the family to New Jersey more than twenty years ago. We got a bit lost on the way to Rabbi Dovid Din's house. Dad remembered the street names, but couldn't quite place them.

"I have one question," he said, as we wove through streets lined with double-parked cars and overfull dumpsters, past Italian groceries, kosher meat markets, and beer joints. "Why do they have to dress that way?" He meant the Hasidic Jews and their distinctive broad-brimmed hats, long black coats, black shoes, and long locks of hair over their ears. But behind the question was much more. There's little love lost between the isolated communities of Hasidic Jews and their more assimilated brethren. The traditional ways and ultra-orthodox stance of the Hasids make them embarrassing reminders to many, like vestigial organs whose very purpose has been forgotten. Though Hasidism was founded in the early eighteenth century as a revitalizing current in Judaism — stressing joy, heartfelt prayer, and mystical fervor — it now seems more like the rigid keeper of the flame. I had never met Rabbi Din and had no answer to my father's question. But I would ask it — and many more questions of my own.

The interview had been recommended to me by a friend in Berkeley, whose glowing words of praise were hard to ignore. He had told me that Dovid Din was a kabbalist — one who studies the inner teachings of Judaism — and had recently opened a public center for Judaic and mystical studies in Manhattan called Sha'arei Orah ("Gates of Light"). My friend said Rabbi Din was a man whose very presence on the street made heads turn.

Two blocks from Din's house, the neighborhood changed. It was much cleaner, obviously well kept. Children wearing skullcaps played on the sidewalks, bantering in Yiddish. My father dropped me off, and I climbed the narrow steps to Rabbi Din's second-floor apartment, where his wife, Bracha, met me with a sweet smile. "Dovid's resting," she said. "He's been out working since early this morning." She poured me tea while I waited. The rabbi's youngest child joined me at the kitchen table and, repeating after his mom, in Hebrew, the blessing for bread, gobbled a piece of cake. (There were four children in the family, ages three to eleven; in the small apartment, they seemed omnipresent.)

Soon, Rabbi Din appeared and escorted me into his book-lined study. What struck me first about him was a certain gracefulness. Tall and slender, he seemed to glide rather than walk. When he talked, his hands danced, tracing gentle elaborations in the air. His words were carefully chosen; he seemed constantly aware of the slipperiness of words and their inability to adequately express deeper meanings. His perspective was sharply honed, his manner light, modest, and exactingly courteous.

Dovid Din was born in 1941 in northern England, to Jewish but nonreligious parents. Over the years he carved his own path in Jewish tradition, and eventually studied in yeshivas (religious schools) both in Israel and in America. In 1977, he brought his family to this Brooklyn community. Beyond that scant outline, Rabbi Din preferred to be none too specific about his past. While he drew great inspiration from the Hasidic master Reb Nachman of Bratzlav, he claimed no allegiance to any specific Hasidic branch.

Although Rabbi Din was at the forefront of a Jewish mystical renewal, there was nothing updated or New Agey in his approach. He stressed the importance of not compromising the laws and norms of the tradition. "It's not as people think, that there is a normative Judaism and then a mystical side," he said in our interview [in 1984]. As we spoke, the picture became clearer: in Rabbi Din's understanding, the laws of Judaism are precise mystical teachings designed to bring humanity into alignment with God. "You are not apart from your environment, from the cosmos," he said in one of his classes, "and to align yourself with it is 98 percent of the work toward enlightenment." While acknowledging that "we all want to lick the spoon from the honey jar of mystical experience," Rabbi Din attached the greatest importance to individual actions in the world. "All the mechanisms of daily life are the hinges of holiness," he said. This understanding creates "a truly religious awareness that doesn't force one out of the context of reality." It is this grounding of illumination in the details of daily life that gives Jewish mysticism its distinct flavor.

I recorded part of the interview on the front porch, with the elevated trains roaring overhead and children scurrying noisily around us. ("Raising kids is my life," Rabbi Din said. "Teaching is just a side job.") As sundown approached, I still had more questions; Rabbi Din invited me to come with him to evening prayers — perhaps we could talk more later. It was bitter cold as we walked briskly through the neighborhood toward the temple. Rabbi Din cautioned me not to feel intimidated if the other worshipers ignored me; they'd have nothing to say to an outsider, especially one who didn't speak Yiddish. I marveled at the ease with which Rabbi Din straddled both worlds. Obviously, this was his home, this community in the heart of New York, living almost entirely by its own rules, relatively untouched by the tumult all around it. Brought up outside the community — with both an early secular and a later religious and kabbalistic education — Rabbi Din was in the unique position of knowing both languages, and of thus being able to translate the inner workings of the tradition for those outside. While this in itself was a trifle unorthodox, it seemed apparent that the people there loved him and respected what he was trying to do. His sincerity was unquestionable.

At the temple I clumsily washed my hands in ritual fashion and read the prayers through in English, watching the swaying, impassioned praying of the others. Afterward, I wanted to ask a few, more personal questions, but there wasn't time. "Besides," Rabbi Din said with a smile, "I'm sure I would have found a way to fill them with hot air also." To me, however, abstract as his ideas were, they were anything but hot air. The glimpse he'd given me of a living Jewish *spirit and wisdom had moved me profoundly. No, the encounter didn't answer all of my questions, but it encouraged me to take a look.*

Rubin: I'm told that the two intersecting triangles in the Star of David mean "As above, so below."

Din: That's a phrase from the Zohar [The Book of Splendor].

Rubin: Could you speak about the idea that every action and event on earth has enormous implications in Judaism: there are laws governing the way you eat, the way you wear clothes. . . .

Din: Judaism conceives of the universe as a vastly integrated system. That's a philosophical point that the scientific tradition also endorses, but it's still a platitude there. The scientific and psychological tradition has worked very hard to fragment the image of man, with very disastrous results. People have sensed that fragmentation and have come away with problems of identity and alienation, not knowing who they are and where they are going.

Now, the system of the Torah [traditional law] places the greatest possible emphasis on life in a very real context. It insists that all the mechanisms of daily life are the hinges of holiness. So there is a vast system for the sanctification of food, called kashrut, concerned with what is kosher and what is not. The Torah is also very concerned with the sanctification of time — through the Sabbath and the holidays, through certain periods of prayer — which is true, to some extent, of all religions, but Judaism has it worked out to a dance. And thirdly, it is concerned with the holiness of various kinds of actions and functions and objects of daily life.

The purpose behind all of this is that a person must deal with the world as it is, and out of that distill a true religiosity. Across the board, among the common people, this system will produce an extremely high level of ethical sensitivity. For this I can only give you a word in Yiddish — *Ehrlich-keit* — a certain basic, high-level integrity, a kind of purity. Not what you or I, from our deeper studies in other areas, would think of as enlightenment, but something that approaches it, and that has about it one extremely conspicuous authentic characteristic: it is not self-conscious. It's not a postured "I have to sit and look spiritual. I have to move like a lily." That's also a discipline, but the problem is that when such things are too self-conscious, there is too much attention on achieving something, rather than having it flow out of the core of one's real being.

Judaism is very concerned with the natural rhythms of things and is therefore prepared to deal with all the realities of life, like crying children, and the pulse of family life. It insists on family life and is very wary of the ascetic or celibate life. Normally, life does not flow like that. Orthodox Judaism is very careful not to tamper with the naturalness of family life, not to inhibit the birth of children, because it believes that the family is a sacrament of divinity. The Western religious traditions, and the Far Eastern also, have implied that a person who seeks religious depth must isolate himself in order to commune with God in the pure realm of spirit. Judaism would say no; maybe at some point, but the first step is through the processes of nature, which are, in effect, the processes of the Torah. The Torah is an extrapolation of the laws of the universe in a kind of artistic mosaic. The vicissitudes of life in a large family — the crying, the screaming kids — are really a means toward achieving the balance necessary to perceive the divine.

The Torah's spiritual discipline is like yoga. The person comes to a yoga master and says, "Swami, I want to learn yoga." And the Swami — whose name might well be Schwartz — looks him straight in the eye and says, "Go home and do this or that exercise. Then come back." The boy says, "But I don't want to do this exercise." So the Swami says, "Fine, don't do the exercise, but this result you won't get, for you didn't do what it takes. You won't get water unless you turn on the faucet."

The Torah as a system compels one to do the exercises and guarantees the result. *Guarantees* it. And the result is a truly religious awareness that doesn't force one out of the context of reality.

Rubin: What is kabbalah — or Jewish mysticism — and what is its relation to traditional Judaism?

Din: Kabbalah is the esoteric side, while Torah, as we normally think of it, is the exoteric. But kabbalah is not other than

Torah; it is part of the Torah. The Zohar says that everything possesses an esoteric nature and an exoteric nature — like the soul to the body. Judaism understands that it is inherent in the nature of man to turn inward; that man, by his nature, tries to understand the universe and everything that happens in it by reflection on the self — at best, a somewhat unbalanced method. It is the goal of kabbalah, as an applied theoretical spirituality, to turn the being outward, Godward.

Now, you have to remember: When we try to understand Judaism, we are reading its truths in terms of the Occidental mind. But Judaism, when all is said and done, is an Oriental tradition — not from the Far East, but Oriental, nonetheless. The underpinnings of Western culture are from the Greek, the Latin, and finally from Christianity, but Judaism is very different in its understanding, not simply of what reality means, but of how it works. The Western mind has largely arrived at a dualistic competitiveness — the idea that everything is subject to judgment and that judgment means something is better or worse than something else. And Judaism doesn't really know from that. Now, even the best of practicing Jews, and many more who are not practicing, think of their Judaism in terms of Western culture. That creates many problems. The difference in cultures is vast. It has to be understood that the Hebraic mind, like the mind of the Far East, works as a totally different camera taking totally different pictures and producing totally different impressions. The Semitic mind sees the integration of paths of knowledge that the Western mind sees as compartmentalized. The Hebraic mind understands truth as essentially the face of God, who is truly one presence in all knowledge and meaning — though we may have come to see knowledge in a fragmented form. We should not lose sight of the fact that each truth is only a spark in a bigger fire.

Rubin: In kabbalist terms, all the actions and things in our lives are said to contain holy sparks, so by the way we approach our life —

Din: — we release the sparks of holiness. Meaning that holiness is inevitably hidden under the appearance of nonholiness.

There are an abundance of texts in kabbalah, but they are really discussing something that can be known only in a direct, experiential mode. I don't mean that it can all be reduced to contemplation or practices or postures, but that it is knowable only in an intuitive manner. Texts and other things can stoke the engine, or turn up the gas, but they can't light the fire. The fire can be lit only by the intuitive spark. All the rest is to bring the rational mind into gear and to disperse the process throughout the corridors of the being.

Rubin: What kinds of practices does kabbalah teach to bring the mind into this gear?

Din: It uses practices that are deeply meditative in character. It has a vast array of meditative practices that are the equal of practices in the Far East. They are of all different sorts, different hues and colors, and there are a vast array of different steps in the process.

Rubin: If somebody wants to find a way into Jewish mysticism, is there a direction in which you can point them?

Din: True entry has to come gradually and with genuine commitment to the path of the Torah. It is absurd to think of trying to immerse oneself in the profundity of Jewish mysticism without accepting and integrating the normal tools — practices, disciplines, and postures of spirit and character — that the Torah system offers. There are many guideposts where the journey can begin, but a true teacher is necessary to complete the process — one who can contain and reflect the material and who is solidly and soberly linked to the tradition he represents.

If you go to India, you'll see that the great mass of Hindus are involved in their tradition in a rather petty way. This is not a problem particular to Hinduism or Judaism; it's a problem of human nature. The number of illuminated ones in any tradition is a minute minority. Now, the Hinduism or Buddhism we find in America is, you might say, packaged for export. People here were searching deeply and discovered that the East had great spiritual riches, and a great effort was made to make Eastern religions accessible to them.

Now, in Judaism, there's no question of making it a little different so it will be more palatable for people. Consequently, the main structure of Judaism presents something of a problem of entry. There is, however, a rising sensitivity within the main structure to people's need to get to the meat of the thing. The authentic structure is not being undercut to produce some sort of device for the moment, but that structure is being largely overhauled to bring up its riches and match them to people's needs. This must go hand in hand, though, with the fact that you cannot divorce kabbalah from normative Jewish practice. While I understand and respect the intentions of those who would try to do this, I have to disagree with the attempt. To depart from the norms of the tradition is to do violence to its fabric. It is a bit of an arrogance. I don't say that of the people, but of the intrinsic posture of a culture that empowers people to sit in relatively uninformed judgment on any great, venerable spiritual system.

Rubin: You're saying that it's not appropriate to approach kabbalah without also embracing the laws and holiness of the Torah.

Din: Right, it will not work. As I said, it's not as people think — that there is a normative Judaism and then a mystical side to Judaism. The Torah is, by nature, a mystical system. It's not as if there were separate compartments. Perhaps it is just that most do not penetrate to that level. The desire to cut away part of it, like Shabbos [the Sabbath], is simply the influence of contemporary America, which is afraid of commitment and discipline. People mean well by it, but it's an arrogance. You can't expect the harp to produce its music when you pull its strings out.

Rubin: Let's look a little at the inner meaning of Shabbos.

Din: First keep in mind what I said before: that Judaism seeks to evince reality's spiritual texture. So, as one of its peculiar practices — parallel to the peculiar genius of every tradition — there is once a week an immersion in holiness. That's really what Shabbos is: an immersion. It creates a space in which things are no longer happening. Since nothing is happening, there is no demand on us to define ourselves in terms of *what we do*, so we must then define ourselves in terms of *who we are*. We cannot hide behind having to go to work or do this or that. I can't busy myself with all kinds of activity; I just have to *be*. If you've ever spent Shabbos in a religious atmosphere, then you've seen that Shabbos doesn't have at all the restrictive character that people would suppose it does when they hear that you can't do this and you can't do that. You really don't notice what you can't do. You run up against it occasionally and realize, *Oh, I can't drive around in my car; I can't smoke a cigarette today; I can't turn on an electric light*, but you don't want to do all of that. You might at the beginning, but that recedes into the background, as do the pegs that hold the tapestry on the loom. The colors of the tapestry are much more important than the pegs holding the threads.

Judaism creates this space. It understands space and time as the sacraments of God in the universe. Now, what happens when a person enters Shabbos? First, everything stops; I stop feeling that I can hide from who I am or who God is becoming in me. And next, I am released. I am released from the necessity of doing anything, and therefore released from anxiety, because I know the character of the next twenty-five hours. It is simply whatever is. There's no compulsion to do this or that; everything just is. This is the essential posture of a meditative practice. Judaism always wishes to understand that a meditative practice can never leave behind the physical component. You can never say to the body, "Shut up and sit in the corner while the spirit goes someplace else." It must always take the whole being with it. Always. The body must also be trained in the path of spirituality to contribute whatever it has to the process of meditation or contemplation. And that's magnified many times over in Shabbos, which is repeated week after week. Now, people need not carry it that far. Most don't. But still it produces a level of awareness. For the common people, that is a basic transformation of consciousness — at a level they are able to digest.

Rubin: In kabbalistic thought, the phrase "There is no place empty of Him" keeps coming up. How does one get to see this?

Din: Something happens in the process of being propelled by the energy the lifestyle of Torah creates. One gradually comes to see that there is a richer texture to ordinary events than previously it seemed. And once that is seen, the next step is wondering what gives them this greater depth. What is it but the presence of something supernal within? Then it is a question of growing with and nurturing that basic insight in moments of stress, using it even in times where it would suit us to avoid it. That is hard work, but the supportive structure of the lifestyle does that work for you. It doesn't release you from responsibility, but it provides the oars for the boat by insisting on the sanctity of everything — sanctity not because the thing possesses anything outside of itself, but because it is what it is. Since it doesn't possess anything outside of itself, it can't frustrate you; it can't tantalize you with a vision of things being other than they are and then let you down when you see that it's not so. It insists that what is, is, but is deeper than it looks. You must invest the effort to seek that insight, and when you do, it in turn will support you. The framework is the lifestyle — the rhythm of prayer and observance and holiday, all of which are saying, "This is not just one more ordinary moment in time after another."

But there is more. Gradually, as the spiritual process unfolds, a totally new dimension of being and awareness emerges. The truth is that this is a new metaphysical perspective: the prior state of being was one in which the self formed the lens of perception. The self was perceived to be the subtle focus of reality and meaning. Now the being has been transformed by turning outward toward the cosmos and the divine presence. It cannot be adequately emphasized that this is not a romantic vision, but a description of a precise metaphysical process of transition.

Rubin: Were you raised in this tradition?

Din: No.

Rubin: How did you get involved?

Din: I got into it by myself twenty years ago. Of course, I was born to Jewish parents, but my upbringing was probably not unlike yours. I had a very broad education. I grew up in England and was always interested in religion and spirituality. I became more and more interested in my own religion's roots, studied a great deal, and put it together for myself over a long period of time. On that path, I met Rabbi Shlomo Carlebach at a formative moment, and Rabbi Zalman Schachter also. I don't mean to say that my formative period is over, but my position differs from that of some of my companions and teachers in that they have taken liberties with the ritual and Orthodox belief structures. My own path is clearly aligned with tradition. This is because I feel the Torah is oracular in nature. The path that I have chosen and that I teach my students seeks to restore the traditional patterns of belief and practice to their full and pristine integrity.

The problem with most traditional structures is that they have maintained the forms but have fallen short of sustaining the deep, inner spiritual content. Nevertheless, the seemingly secondary vessels of a societal nature that characterize traditional Jewish lifestyles shouldn't be dismissed; they have successfully translated the profound values of the spiritual inner core to the group as a whole. And they have passed the tests of cultural endurance, sustaining profundity of vision and context over centuries of remarkable adversity and difficulty.

Rubin: Were there any particularly important turning points or moments of awakening for you?

Din: I can honestly say that all moments are important. I don't mean that as a flip rejoinder. I have many weaknesses of many kinds, but I have one strength: a very strong spiritual focus that is central, and that seems always to percolate.

Rubin: It seems that in many newly developed spiritual traditions and communities in this country, the majority of members were born Jews. It makes me wonder why so many Jews are seeking, and why they are looking for the spark outside of Judaism.

Din: There is a very deep teaching in kabbalah that I'll try to simplify, one that's really at the heart of the kabbalistic idea of how the world and God are put together. The Ari, a very great kabbalistic master of the sixteenth century, taught that, in the proverbial beginning, there was something like a tremendous explosion of Godness that at once created the universe and filled it; it sort of created it one moment before it filled it. Now, looked at one way, this was a disintegration — as the language of the text says, "a descent of sparks of holiness into the outer darkness." Looked at another way, however, it was the expansion of holiness to enlighten the darkness. The difference between these two ways of looking at this is that, one way, it's as if you are in the mind of the darkness; what you see is that the light is being fragmented. But if you think it out calmly and are patient with what's going on, then you see that it's not necessarily that the light is being fragmented; it is that the light is spreading, in order to redeem and reintegrate the darkness.

Now, again we have to keep in mind the vast difference in cultures. The Ari was not really erecting some kind of quasi-scientific explanation of things, but this is perhaps the closest we can come to describing it. This was his grasp of a vastly intuitive mythos — a mythos that is a truth in cosmology, in sociology, in philosophy, in psychology. In this way, the Ari understood the presence of the Jews in the world as the presence of meaning in the universe: In one sense, they have been scattered and dispersed and exiled; that's the plain history of the thing. But in the other sense, it's only from the perspective of history that it is a tragedy; from a more transcendent perspective, it is a redemption. What appears to be spreading apart is in fact gaining momentum in order to return.

To bring that down to the practical aspect, I think that in many ways Jews are a people of light — not by what they know, but what they are. (That's not to say that no one else has any light; that's not the point.) The light at once seeks the darkness in order to illuminate it by its spark and bring about its redemption. And that is, in fact, what is going on. Jews are a very spiritually conscious people.

Now, what remains is the question: why didn't Judaism in America provide this understanding? That's a problem in history that we could discuss. At the time of the Second World War, Judaism in Europe, just at the point when it might have produced an Aquarian Age of Judaism, was obliterated. So who was left on the scene? There were practically no Jews except in America. And the ones who had come to America were not the cream of the crop. They were those who had come a generation or two earlier, trying to get away from Europe. They were people who needed to make a better living, or found it too hard in Europe, or whatever. And the Judaism they brought was basically a "circumcise them, marry them, bury them" brand. The generation that came after wanted the deep things, but it did not possess the wherewithall to search out its own storehouses. That, I think, is the practical answer.

I think in the next generations we will see many Jews returning from where they have been — which is not to imply that where they have been is so bad — enriched in the peculiar way that darkness can enrich light. I don't mean to say that over there is darkness and over here is light, but darkness is the metaphor that I'm using. Today there is a tremendous resurgence of practicing Judaism. A tremendous number of young people are returning to the practice of Judaism, though nowhere near the number who might.

Rubin: It's hard not to hear that metaphor — of scattered light coming back enriched by the darkness — as judgmental. When I spoke to Shlomo Carlebach, I also had the feeling he was saying that Judaism is the best way, the right way, the chosen way. Is there really any validity in calling one people a chosen people?

Din: Of course there is, only it has to be said without arrogance and with a great deal of love. Everything in nature is chosen. This blade of grass is chosen to grow in this place, the other is not. This person is in this place. This happens on this day. Choice is the mechanics of the universe. Our culture is terrified at the notion of judgmental selectivity, but what we're not hearing is that, just as the blade of grass is chosen for this, the other is chosen for something else. The choice of this is not the rejection of that; it's the choice of that for some other purpose. Everything is chosen to fulfill its purpose.

Now, if what you need for your nutrition is grass that grew in this place, with this kind of soil and this kind of moisture, then this blade of grass is the best for your purpose. Are you, then, prejudiced against the other blade of grass? No. It is a judgment, but what's wrong with that? Perhaps it doesn't feel right to you, but that feeling won't stand up under scrutiny. When you choose to wear a blue shirt, you are in essence saying to the world that you believe that blue shirts are the best. If you don't, then why are you putting it on? But you do not mean that it has to be the best for someone else. Maybe he needs a yellow shirt. Well,

Judaism says the same: Judaism is for the Jews. It's not for anybody else. We would not throw people out, but we're not coercing anybody to come in. Judaism says that all men have a path to God. It will affirm and protect the right of others to go to God in their way. It doesn't discount the other venerable traditions of mankind; it respects them.

Nonetheless, Judaism affirms that the Torah is the clearest focus on God. Implicitly, all faiths are saying that theirs is the clearest focus. What they are not saying, and what Judaism is also not saying, is that no one else possesses any validity. Let me give you a little insight into this. Judaism is adamant, almost fanatical, in its insistence on monotheism. To suggest that there is more than one God or that God is tangible in any way is not even discussable. Reform, Conservative, and Orthodox all agree that God is one, transcendent, and nontangible. That is, in effect, saying that the image of God must remain accessible to all. As soon as I say that God is wearing a green shirt and a blue jacket and combs his hair this way, then everyone who likes that will get along well with that image of God, and everyone who doesn't won't. Judaism is saying that the focus of the Torah is clearest. What is that focus? That absolutely nothing may interpose between God and man — no images, no idols, no coloration, nothing. The lens must be absolutely clear. To carry the analogy to its conclusion, the implication is that the lenses of other religions, though they let in an abundance of light, are somewhat cloudy. The Torah is, however, adamant in its condemnation of idolatry, because it sees the idol as a projection of the self, of man — sometimes a very dark or questionable aspect of his nature — and thus a perversion of the religious process.

Rubin: In the Eastern traditions, there is a striving for union with God. I have the impression that, in Jewish thought, there is not this union in the same sense.

Din: Not in the same sense, I think. But then again, I would stress that unification is a very dominant concern in kabbalah. It's very clearly expressed in the language. The system of the Torah presents a path and its practices as an end in itself. It requires that a person come to it and say, "I submit to it. I do it because I realize that it has a power to do something to me, but I have to give myself over to it." What happens in that practice is that anything that would otherwise be turned inward is turned outward. All the energy goes into it. I have to do the exercises like this and not like that. Why? I don't know, exactly. "Why?" is the Western mind's question. But I have to do it if I want this result.

Now, going a step further, all the practices in Judaism are designed to take the person out of the arena of self-concern and into another arena. If not carefully understood, this could seem a platitude. It doesn't mean simply that I shouldn't be concerned with the self. The Torah understands that the implicit, unconscious illusion under which man operates is that he is the measure of the universe. It is an illusion on the existential, philosophical, and metaphysical levels. I feel that I am the universe, and therefore everything goes inward. I walk around for my whole life thinking that's how things are — "me" cars to drive, "me" food to eat, "me" things to do. The system of Torah comes and forces everything outward, which means that the unity that Judaism is looking for is the point at which, without losing this individual identity, a corridor of correspondence is opened between it and the transcendent identity. But it isn't that I become annihilated and flow into the great river; it's that I am maintained in a scale-model relationship to the transcendent. I stay here, but I grow outward. I become that without ceasing to be this, and that becomes this without ceasing to be that. This sets up a perpetual interflow; both are the same, and yet they are not.

There's one point that requires a little digression. If you examine the whole thrust of the Occidental culture that we live in, you see that for the past two or three hundred years there has been a great emphasis on the individual. That came out of the rationalist Enlightenment of two hundred years ago and all subsequent intellectual and philosophical history in the West. Today, the only commandment is "Thou shalt not say no." Everything is good, everything is cool, everything is wonderful, everything is fine — just don't hurt the other guy, and even that's OK sometimes. Why is everything good? Because I want to do it, and my primary task is the fulfillment of my individuality. That means we gear everything to the fulfillment of the individual self, and therefore we interpret and align all experience and all the inner workings of reality toward that unspoken end.

Following that, the danger would be that I would conceive of my spiritual practice as a means of achieving fulfillment. This sounds wonderful and just as it should be, but the truth is that it's a tremendously false position. I would go around arranging my spirituality so that I get a charge from it all the time, so that I'm being "fulfilled." Then the whole superstructure is in danger of being discarded. We say, "I don't have to do all this. It isn't turning me on. All this is from the Dark Ages." Now, the superstructure is, in effect, ensuring that the development is balanced and whole and that all parts of the being — the intellect, the emotions, even the physical being — are synchronized in their spiritual development, and that the lust for the sweet taste doesn't get out of hand and distort the whole. Intuition is the heart of the whole, but that heart needs the rest of the circulatory system to work. The deep, essential truth of spirituality is present in the other, less glamorous levels of reality — the cognitive and functional levels.

Rubin: To pin it down, if it's not to fulfill myself, then what is it for?

Din: It's to align me with God. That mode that I induce for self-fulfillment, which could be made to sound very good, matches the kabbalah's essential understanding of evil, which is illusion. Illusion is the root of evil.

Rubin: But the kabbalistic idea that the world is illusion isn't like the Hindu concept of Maya.

Din: Well, the world is illusion, but it is, so to speak, a necessary illusion: it's true, but it is ultimately not true. It is a functional necessity.

Let me illustrate this a different way. There is no one whom we can conceive of as evil. Let's say — God forbid — that someone kills someone else. In the moment that they pull the trigger, they have to be saying that this is good. In their twisted mind, they are saying it is good that this person should be dead, and it's so good that they will do it. Now, that is a very strong illusion.

Other illusions that are not so powerful, but are also quite vicious, are the little illusions with which we grease the corners of our lives and manipulate people to our own ends. What is the root of such illusions? What makes me think that shooting this guy is a good thing? Because it is what I want. The self sees itself extended to become the universe, to become reality. Rather than having to force myself out of me, to accept reality as it is, to align with the universe, I'd rather live the illusion that I *am* the universe, and in my universe you should be dead.

Notice that the crossing point there is between what is good and what gives me pleasure; that's the hitch in the train. When we want our mind-frame to be the universal mind-frame, we prevent ourselves from being at rest in who we are, from coming into alignment. To avoid that, we people our illusion with all kinds of phony evidence to convince us that we are everything. It's easy to extend yourself and say that the world should run according to what you say, but to return to the truth of one's being, in which one is at rest and therefore aligned, is more difficult.

Rubin: When you speak of Torah stories in your classes, it's from a seemingly deeper level at which it's easier to see them as precise mystical teachings. How is the Torah meant to be understood?

Din: The Torah is perhaps not meant to be understood as much as it is meant to be experienced. The Torah is fundamentally a means of effecting this process of turning man's vision outward, toward God. The preponderance of Judaism is actual, experiential, though it is also well endowed with the conceptual. We in the West essentially regard the conceptual before the experiential: we think, and then we act. We conceive of what we want and then act upon it. Judaism sees that as having a high probability of lending itself to illusion, because there is a great distance between conceptualization and actualization, a long process between the two that does not always go smoothly.

Rubin: It seems that inner Judaism is both a path of the heart and also quite intellectual. Can you talk about the relation between mind and heart?

Din: The kabbalah understands that what we know as the heart is the synthesis of awareness and sensitivity — the point at which I know something. I may know intellectually that God is one, but I still need to move into that truth and be completely aware of it. The heart is the doorway, and also the end — the way in, and then what you ultimately come back to.

Now, the path of the heart very often involves what we think of as emotion. Kabbalah is very careful to maintain the proper balance between the mind and the heart; otherwise, the heart will pull the mind away into an emotional vertigo of one sort or another, or the mind will freeze the heart into a theoretical abstraction. Interestingly enough, the Hasidic movement is set up to deal with that problem exactly — to bring the richest of ideas down to a safe, tangible, emotional level. Hasidism preserved these values, producing an amazingly rich tradition of music, stories, and life.

Rubin: How do you keep your heart open and in balance with your mind?

Din: My family keeps my heart open. In Christianity there is a whole system of sacraments, acts that are construed as bringing access to the divinity. I think that, in a very special way, the family does that in Judaism. The sacramental dimension of the family is very precise in Judaism: family life is a recognized and exalted spiritual practice, and protected as such in the details of Judaic life. Tampering with the primacy of the family and any form of artificial control are strictly forbidden. The union is sacred, and broken only in circumstances of the greatest duress. The family sometimes drives you crazy, but it holds you, embraces you, makes you laugh. There's nothing like being crawled over by all your children while they're tickling you. They are the framework of the life that Torah creates, which is centered in the family and which the family healthfully, wholesomely reinforces.

We talk about the abstraction of the Torah and its system, but the Torah is, in fact, a way of life. That is part of the uniqueness of Judaism. In the West we have the notion that religion is a system of beliefs that is superimposed on a given reality. Judaism does not understand religion as a system of beliefs, but rather as a practice of being. Judaism is a way of life, not a religion at all in the theological sense. Its emphasis is on creating a lifestyle, and, at certain points, it does that by sharply dividing from the world. So it has produced an insular culture, especially among the Hasids. It is, by choice, an isolated culture that has very little to do with the world outside. And in return for that, what does it have? It has a world where there is no crime; there is virtually no psychological disorder or breakdown; there is no pushing of the old into rest homes — the old are maintained in the family and treated with respect in the community. It produces a very wholesome existence. The rituals and nuances of this cultural enclave — Shabbos, celebrations for the seasons of the year, observances and weddings and bar mitzvahs, circumcisions — stoke the heart and keep it very healthy, very responsive, very integrated. And all of it is done in terms of the community. One must always function in terms of others.

Now, the cost of that is the exclusion of the world. As you see right here, I talk with my children in Yiddish. In our home there are no newspapers, no magazines, no television, no radio, no influence of the secular world — because we do not need to teach our children about such a disturbed and troubled civilization. They can see what the outside world is all about. They walk on the streets; they see what is going on. But that does not mean it should be in the fabric of the home.

Rubin: What is Hasidism all about?

Din: Hasidism is a popular, pietistic movement that arose in Judaism about two hundred years ago. It attempts to encapsulate the truths of kabbalah at an experiential level accessible to everyone. Accordingly, it is possible to be a Hasid who knows nothing of kabbalah, but nonetheless lives it. This movement has certain characteristics that have endeared it to a very large segment of the Jewish people and, mainly through the work of Martin Buber, to the world at large, to some extent. It is characterized by a philosophy of joy and an attempt to make the transcendental immediate — not by hocus-pocus or recourse to the metaphysical, but in the texture of life and the fabric of everyday things.

Now, two hundred years have passed since the movement began. A historian might look at Hasidism and say, "Well, like all movements, it had its heyday when it began, but today it has ossified." Well, that's undoubtedly true, to some extent, but it misses the point. Everything that Hasidism started out to do, it is accomplishing today. The molds have shifted somewhat, because the sands of time and history have shifted, but it has not lost its vigor. It may have lost a vigor that makes it romantically attractive to the outside world. That is the coinage with which the outside intellectual world approaches such a thing.

Rubin: Still, there seems a large split between the Hasidic community and the assimilated Jewish community. Where I grew up, Hasids were definitely looked on with suspicion.

Din: That's a nice way to put it.

Rubin: Why do you think this is the case?

Din: Hasidism today represents the ultraconservative point of view — and we're saying this about a movement that was innovative in its time. While it is true that Hasidism has become rigid, it has done so because it has chosen to resist certain changes in the society at large. We could largely characterize those changes as promiscuous — not promiscuous in the sexual sense (although that's surely included), but promiscuous in the intellectual, the emotional, the cultural, and the spiritual senses.

For example, it is fashionable in the spiritual realm to flit around from here to there and there to here, all in the name of the liberty of the human spirit. Hasidism, whose spiritual underpinnings are very authentic, regards that attitude as suspicious. It therefore appears to have adopted an ultraconservative stance, and that accounts for its antipathy to the society at large. It has retained habits of life and dress and ways of being that appear to be archaic and limiting. They are, rather, conscious choices to preserve a certain way of life that Hasids feel is spiritually healthier. This is no different from taking a plant that won't grow in the snow and putting it in the hothouse. Admittedly, the hothouse is artificial, but the alternative is the death of the plant.

Rubin: You mentioned a philosophy of joy. David Zeller said something recently about always remembering the *oy* in *joy*. In Judaism there seems to be an embracing of sorrow, even within joy. So you don't move above or beyond the sorrow —

Din: — you move into it. There is a holiday called Tishah b'ab, which is the anniversary of the destruction of the Temple. It is an all-day public fast and comes at the climax of a period of three weeks that, at the most external level, is intended to redramatize the historical period that led up to the destruction. From the beginning of the three weeks, we don't have weddings, we don't make music or cut the hair, we don't make a blessing over anything in joy, we don't buy any new clothes. When the Hebrew month of Ab begins — in the third week — then not only those restrictions are enforced, but we also don't take showers or baths (if you can stand it), and we don't sing, drink wine, or eat meat. Finally, Tishah b'ab afternoon we don't even study Torah — because the texts say that the study of Torah rejoices the heart — and we eat what is called the meal of consolation, the traditional meal that is served to mourners. We sit on low stools on the floor, ashes are put on the head, and we sit as if we were in mourning that whole day. It's very dramatic; you cannot get away from the heaviness of it. In the synagogue they intone the Book of Lamentations the whole day, and the next morning also.

Then, in the afternoon, at the end of this long series of public lamentations, suddenly the cantor gets up, goes to the reading desk, and intones a melody. Then we straighten up the benches, straighten up the house, in preparation for the coming of the Messiah. The text says that the Messiah will be born on Tisha b'ab: the day that is the very heart of sorrow is the day on which redemption will come. The psychology of the observance apparently is that you cannot get to the joy unless you enter the sorrow. It is the only way out, and the coming out is splendid. That afternoon the silverware is put back, the tablecloths are put back on, the pictures are put back on the walls, and at night we light candles and have the Feast of the Messiah.

That's one answer. Now I'll tell you a story. On Passover, I prayed in a synagogue full of European Hasids who came here after the war — very simple, unsophisticated people, very "unspiritual" people, but golden people. At the holiest part of the service on Passover morning, a point where everyone stands perfectly still and you're not allowed to do anything, there's a dialogue between the congregation and the cantor saying, "Holy,

holy, holy is the eternal of hosts. His glory fills heaven and earth."
Now, that little statement is sort of the consummate statement.
That is the basic reason for all religion — not because God will
do good things for me, but because God is holy, altogether removed, altogether unapproachable. They all sing this to a melody,
a very beautiful, lilting melody. I noticed that when they were
singing, all the men in the front row were crying. At the end of
the service, I asked my friend next to me why they had cried. He
told me that this melody had been composed by the rabbi in a
concentration camp on Passover afternoon when his two sons
were taken to the gas chamber. To what words did he compose
it? He composed it to a prayer that was sung in the temple when
the Passover sacrifice was brought. He understood in the moment that his sons were being killed in the gas chamber that it
was like a sacrifice, an occasion of joy.

I have nothing to say about that. I cannot fathom how someone could be able to do that. I can fathom how someone might
want to, or think it an exalted level to reach, but it takes something that is extremely far-out to do it. This man obviously knew
that only in the midst of tremendous sorrow could true joy come —
free from the sorrow because one has a direct connection to God.

Rubin: I'm told that there's a Jewish teaching that one should
live as if the Messiah were always about to walk in the door. Are
you personally awaiting the Messiah?

Din: Every day. There's a story of a great teacher who always
had his Shabbos clothes by the door, so that when the Messiah
would come he could quickly throw on his Sabbath finery and
go out to greet the Messiah.

There is another anecdote that preserves an important aspect of this. It is the law that when the Messiah comes, he may
not disturb the small children learning Torah. Even though the
Messiah comes to rebuild the Temple and gather the exiles, he
may not take the small children away from learning Torah. Also,
it is said that the Messiah will not come on Shabbos. Why not
on Shabbos? On Shabbos, of all days, the Messiah should come.
But he will not come then, lest he throw us into confusion as to
whether the Shabbos law applies to the coming of the Messiah.

Now, Judaism means to say by this that the messianic dimension, dear as it is to the hearts of the people who have suffered so much in history, will nonetheless grow organically out of the existing structures of spirituality. Life won't be reversed in midstream. So the observance of the normative law is not something that goes against the coming of the Messiah and all that will mean. Rather, the coming will flow into it.

Rubin: It seems that one of the major blockades to peace in the world is that every religion feels it has the one way to God and will not accept anyone else's. What steps would it take to bring people from different paths together?

Din: I think that the first step is to recognize the validity of each tradition for its people. It does not do a service to spirituality to suggest to everyone that they should all blend into a homogeneous spiritual milkshake. It's not realistic. It is not in the nature of man to do that. People and cultures differ. So to tell everyone to put away their differences is not realistic and not true. It must be that the equation works the opposite way: when a person is truly grounded in the best that his tradition has to offer, he will from that point be accessible to others, having passed the first gate of being true to himself.

Rubin: Where do you look for a way of averting nuclear war?

Din: I'll tell you an interesting thing I was discussing with someone this morning. There is a law in the Torah that if you wage war against a city, you must leave an exit for people to get out. You cannot close the town off and just kill everybody. And in nuclear holocaust there is no exit. That, in a way, summarizes the attitude of Judaism — that a destruction so lethal and so total cannot be valid. War might be. There are times when war may be justified, however much we may wish that conditions did not warrant it, perhaps even an offensive war, provided it has a defensive mentality. But it cannot just be destruction.

As to avoiding war, I think that the machinations of politics will never avoid it, because politics means the abuse of power, putting the power of the people in the hands of the tyrannical few. Under whatever guise, it always boils down to the same neurotic distortion resulting in despotism. Avoiding war can only happen, I think, if the people of the world prevent it. That, too, does not seem likely. It seems desirable, but it doesn't seem as though that will work, because tyrants will always get into power and will always manipulate. So maybe we will never solve this problem. It may be in the nature of man's fallen state that he has to endure the scourge of war. I don't like to say that, but it may be so. We could have a world of goodwill, but need only one deranged tyrant someplace with his finger on the button — only one — and the probability is that there are more than one.

Now, Judaism seems to answer this question by saying that the ultimate resolution will be in the hands of the Messiah, which means that supernatural intervention is the only answer.

Rubin: Let's look at modern-day Israel. Its image in the media has gone from paschal lamb to neighborhood bully. What are your feelings?

Din: That is a complicated question. It would seem clear to me that contemporary Zionism is wrong — not wrong because it happens to be pushing the Palestinians off the land they have lived on for centuries; I don't mean to say that's all right either, but that is not the whole reason. Zionism is wrong because it is a rather clever attempt to trade on the thousands of years of aspirations of the Jewish people. It asks Jews baptized in a sea of blood and yearning and suffering to jump into a small pool of cheap Third World nationalism.

Zionism, by nature, is the translation into political philosophy of the ideals of secularism and materialism. That's almost what gave it its impetus. The cultural milieu of assertiveness and acquisitiveness and the capitalistic emergence in Western Europe reached the Jews, at which point they said, "What are we sitting here getting hit on the head for? We should get our act together, get ourselves a piece of land, and do the same thing that the English, French, and Americans are doing." The next question was where. The answer seemed to them obvious: our historic land. If they took that option, instead of getting a few thousand square miles on the coast of Africa or South America — which were some of the options proposed by original Zionism — they could capitalize on the spiritual thrust of two thousand years of exile. To the little Jews in the ghettos of Europe, Africa wouldn't have the appeal that the Holy Land would. There was a very subtle process by which the secular proponents of Zionism took over the spiritual armament of traditional Judaism to fan their own fire. A lot of problems arise from that.

Now, while that is obviously something of an indictment, at the same time we have to realize that God moves in mysterious ways. Apparently, a number of Jews who would not otherwise have been affected by their religious heritage have been affected by their relationship to Israel. So it might be that, in a way that we cannot fathom or chart directly, thousands or millions of Jews have now been brought back into relationship with their heritage who otherwise would have run themselves off the rolls.

To regather the Jewish people to their land (at the time of the Messiah) is the messianic vision, and we see that perhaps God's ways are not ways that we can fathom so easily. Purposes that appear to be at odds, in the greater wisdom, dovetail into each other. It appears that secular Zionism is the antithesis of spiritual Judaism, but it may be that for his own purposes alone God put the two together to reap a harvest of spirituality out of the seedbed of materialism.

Rubin: In working with kabbalah, there's a great deal of looking behind the words of the Torah to find the inner meaning. Could you speak about the relation of inner truth, which is unspeakable, to the words we can speak?

Din: You should have asked me that first, because then we would have begun and stopped. The Ari, the great teacher of kabbalah, wrote almost nothing; he wrote three poems, set to beautiful melodies that we sing on Shabbos. But he wrote nothing on any of the texts. And yet there are eight very large texts that are called "The Writings of the Ari," which his disciples wrote. When he was once asked why he didn't write anything, he said that he couldn't — he would open his mouth and it would be just like [shrugs] *Whewww!* That says it. All the rest is to prime the engine.

June 1984

A Clouded Visit With Rolling Thunder

Pat LittleDog

So sometime or another, when i was living in east texas trying to figure out how to make a living doing nothing, i decided i would be a promoter. except of course i wouldn't be a promoter of rotten stuff, nothing porno or bad for the environment, only healthy, good-for-you kinds of promotion, the educational, let's-learn-together kinds of things, everybody learning and expanding their consciousness while i did the same but took a little money off the top for my extra trouble of bringing in the guru/rolfing instructor/ecstatic-religious counselor, who would get the rest of the money after my percentage was taken off.

i was mostly interested in native americans. i wanted to hear for myself what they had to say about balanced relationships. and the first people i wanted to promote were sun and wabun bear from washington state, because they had written a book that i had enjoyed called *the medicine wheel*, and it pleased me very much when i thought about the possibility of meeting them. and it pleased me even *more* when i figured i could meet them and not have to pay for it, because at the time i didn't have any money, just as at many other times in my particular life.

so then the possibility of my even *making* a little money began to thrill around in my head. i began writing letters and designing copy and making a poster and checking out places where the bears could appear and scheming and thinking and putting pen to paper and guesstimating. i figured there could be a *huge* texas tour — el paso to dallas to houston to austin. and of course somehow edgewood, the town where i was living, would be along the route. after some heavy negotiating, i lined up a new age center in el paso that would host the bears. then a ranch woman outside of austin who was a devotee of muktananda and who wanted to learn how to build a sweat lodge. then i started looking around for a place in dallas.

i looked around and called, but nothing seemed to fall into place. so then i remembered some people i'd met at the rainbow family gathering in west virginia — that free gathering the love family commune puts on for thousands of people once a year — and i decided to go up there and find the commune and ask if they would like to set aside a weekend for the bears.

so i drove my truck up one day and found the commune. there were a half dozen or so concrete buildings in various stages of construction, and i walked up and down the mounds and looked at the plans and saw how all the families were putting time into their own houses and how each house was unique; even the theories of how to build a house underground were diverse — no spirit present of super unidevelopment.

and there were some rainbow family people camping there — hippie gypsies in a wooden camper-truck. and just about sunset, when the sun was dumping itself behind the hobbit-style turrets topping one of the house-hills, i found myself talking with a commune man named george and one of the rainbow family men whose name was majo. george was telling me that this year might be a little premature for the commune to have something like the sun and wabun bear workshop, but that he himself was interested in them and knew some other people would be, and that maybe next year, when more of the houses would be complete, something like that could happen.

majo had joined us in the middle of this conversation, and he was quiet while george was talking. i knew his name because he had been the first person who greeted me when i drove up — a thin man with tinkling little bells on ribbons and chains draped around various parts of his body. and even while george was still talking, majo distracted me. i kept looking at him because he and the sunset were more or less mingling together — majo with long and wispy gold hair and wafting beard, no shirt, purple pantaloons, red fingernails at least three inches long clicking on his crossed arms while george and i were talking.

then he said, sun bear . . . yes . . . sun bear, i've heard of him — don't you have to pay for his workshops?

and i said, well, yes, but it's not very much.

but as soon as i heard majo's question, with the rainbow family point of view stamped on it, i knew i had been caught — caught by one of the love family sleuths in the middle of wheeling and dealing in matters of the heart.

and george and i kept talking, but george had been caught, too; he had as much susceptibility to majo as i did — once you've danced in the woods with the love family, you've danced in the woods, there's no erasing that. and majo didn't say much more; he knew he didn't need to. in fact, he said, oh, but there are some things i would pay money for, and i have heard a lot of good things about sun bear and wabun. . . .

but those words didn't matter. in five more minutes i was back in my pickup truck headed for edgewood.

when i opened up the *dallas morning news* the next day, there was majo on the front page kissing rosalynn carter's hand — the president's wife was visiting her husband's pentecostal sister, amy stapleton carter, who happened to live next to the underground commune. the news story under the photograph of the kiss said that somehow an unidentified man had slipped through the security guards, kissed her hand, and made a wish for world peace.

i didn't write the bears and tell them that i had been stopped in my tracks, but i stopped working on the promotion, and everything collapsed, and sun and wabun bear were disappointed, and finally i moved into the city and started working temporary jobs and selling used books and stopped trying to be a promoter — at least in *that* sense of the word.

but this short-lived history of me as a promoter is why, when rolling thunder decided to come to texas last winter, he wanted me to promote his tour. because after the bear thing, people saw me as a little bit of a promoter anyway; even if i had failed, they didn't care, because there weren't too many hippie promoters around for these kinds of new age/native american workshops and weekend seminars and such, which people were beginning to want to spend their money on.

well, i had wanted to meet rolling thunder ever since i'd read doug boyd's biography of him a couple of years before — he seemed very brave to me, single-handedly taking on the bulldozers clear-cutting the forests, and then he himself, as a powerful healer, intrigued me. but i didn't want to bring another majo down on me again. or one of his friends. so i wrote to rolling thunder's people, and i told them that workshop fees kept spiritual information away from poor people, who needed guidance at least as much as the upper-middle-class anglos who made up most of the audience for the national-circuit spiritual-advisor rounds, and that maybe the people with money could pay rolling thunder's way here if the workshops could be held free with requests for donations.

someone wrote a good letter back. it was thoughtful, explaining the needs of rolling thunder's people to support themselves. it let it be known that we weren't quite in agreement but could still be friends. so that was that. i was sorry i wasn't going to get to see rolling thunder, but i figured someone else would help him come.

sure enough, some little time later, a promoter named cathy lee called me, wanting to see if i could take part in the promotion of rolling thunder's visit to texas. he had found a sponsor in san antonio and was scheduled for workshops there but wanted to lecture for a night in austin while he was in the neighborhood. so we talked on the telephone, and i gave her this little philosophy that i had begun to put together, pretty much the way i had given it to rolling thunder's people. and we talked back and forth, and she could see my point of view, but she could also see rolling thunder's point of view, and there were many sides to the money thing. so then she said, well, if i couldn't help in the actual promotion, would i like to do an interview with rolling thunder, which she could arrange that i be paid for and which would appear in the local new age magazine and which would help in rolling thunder's promotion on later dates but which would actually appear after he had left?

so i rolled that one around and looked at it and wondered whether it fit into my ever-cooking-and-clarifying theories of economics and spirituality. now, it was true that i stood to make some money on an interview that rolling thunder wouldn't be paid for, but then, it was supposed to be advertising for his point of view. so we both had a little money consideration, a concern about support — no one's motives are pure. and i did want to be somewhat of a journalist/historian/writer, after all, and he wanted me to put his words on paper for him, to be broadcast to a larger audience. but there was also something else bigger than any of that, something called *inevitability* — the call to meet rolling thunder, the knock on my door not once but twice — and no one asking for any money for time spent with the well-known doctor; they were offering *me* payment instead. provided i worked/wrote. so i wrestled with these considerations for at least two or three seconds before i told cathy that, yes yes yes, i would be very happy to do an interview.

now, i do like to think of myself as a journalist at times, although every time i get around journalists i am reminded again that i am not one: they write in three hours what it takes me three years to think through. and i had never gone to a press conference before, not ever having reached that particular stage of promotion. but i figured there would be coffee and doughnuts there, since it was in the morning, which seemed pleasant to me, and i could find a corner somewhere and simply hang out and listen to rolling thunder answer questions. and i carried a boxful of pecans that had been sent down from my grandfather's home in east texas, picked by neighbors, to give to the visiting medicine man. on top of the box i wrote, *welcome home* — that being the greeting used at rainbow family gatherings to welcome new-comers to the woods. and at eight in the morning, walking along nueces street with a large box of pecans, with the grackles whis-tling and the sky clear and my hawk-feathered hat on, going to meet rolling thunder, i really felt pretty splendid.

but when i got to the ni-wo-di-hi art gallery, no press was there for the conference except a young woman cub from the *daily texan* and a contingent of photographers and technicians laying out mike cords and trying out angles. i handed the box to somebody, but there weren't enough people for me to hide myself among, so i ducked into the bathroom, and when i came out and there still wasn't anybody else showing up, i decided to duck out the back door. because i really wasn't prepared to ask rolling thunder any-thing. but cathy lee caught me in the hallway. she put a tape recorder firmly in my hand, and she said that since no one else was coming i should ask rolling thunder some questions.

so the *daily texan* woman and i were nervous and silent, but he sat down, lit up his pipe, and began to talk, giving us more than forty minutes straight on the tape recorder with very little prompting. and he was a wonderful, dignified-looking man with feathers sticking out of his cap and outdoor skin and with the young men of his tribe bustling around him.

but in that gallery everything was *high-waxed* and *framed,* very *inside* and *walled* — a strange place, it seemed to me, to try talking with an indian spiritual man. so i thought, *well,* my *inter-view with him sure wouldn't be like that.* i would ask cathy lee if we could have a walk by town lake instead, while we were talking. and i began to imagine us having a very good talk together. i figured i could show him barton springs, the heart of austin, and maybe even walk him to the low-water bridge and show him balcones fault, a superspiritual crack in the earth. then he could talk about the power of the earth and give austin people some good words about his impressions, coming to the heart of their city, healing power for the spring water. and in my mind's eye i could see the young men of his tribe, whom he was traveling with, walking and talking with some members of my own family, a little band of us meeting together underneath the treaty oak on fifth street, where warring tribes have met to work out peace terms together for hundreds of years, and all of us starting out from there.

and so that is what cathy lee arranged for me. she said rolling thunder liked the idea of the walk, and she said that maybe she herself would come along because it was such a fine idea.

on the day of the interview i drove over to the oak tree with my husband and brad, a young man who was living with us, and minki the dog. under the tree when we drove up were four of our friends with ryan the baby. there was also a car full of men with hats and feathers pulled up to the curb, but they were making no signs of getting out, and there was a young, long-haired man i recognized as one of rolling thunder's group, who was talking to my friends. when he saw my car, he walked over and stuck his head in the window.

it's too noisy for rolling thunder here, he said; we're going back out to the manor house where we're staying. and anyway, what's the purpose of all of these people being here with you?

i said, they're my friends; they wanted to meet rolling thunder.

he said, well, rolling thunder wasn't expecting any of this. it is entirely against protocol. you asked for the interview; you didn't say anything about bringing anyone.

well, i hadn't asked for the interview — i thought *they* had wanted the interview — and i hadn't heard the word *protocol* since a military-based childhood home built on the science of war and patriarchy cast me out to look for freer places, but that seemed beside the point.

so i said, well, i saw that rolling thunder had his tribe along with him, probably because it made him feel good to have them around, so i wanted to have some of my tribe with me.

that didn't make the young man very happy. he walked back to the rolling thunder car to consult. my husband jumped out of our car with the dog and said he would meet me back home later. my friends and baby ryan were looking awkward under the treaty oak. i got out of the car and the young man met me in the middle of the street. rolling thunder says you can bring your husband and the other one in your car, but that's all, he said. but he's not feeling good about this whole thing.

well, i'm sorry, i said. i didn't mean to do anything wrong.

but the young man didn't let go of it so lightly. the thing is, he said, this is the problem we have with white people, always making assumptions. you should have arranged having all these people with you beforehand. you should have asked if it was all right for them to come.

so i apologized again. look, i said, i'm just a fool. i'm just ignorant. i didn't think of the right way to do things.

well, he said, you need to learn to call before bringing people with you; that's the correct way.

that pressed me one time too many. i didn't say anything about the six other people accompanying rolling thunder in his car. instead, i said, look, let's not say it's the correct way; let's just say it's one way of doing things, because if you were to be invited to my place (and i waved at the noisy block where i had invited rolling thunder to meet) and you brought five friends, i would say you were all welcome.

so there we were, squared off at each other no more than ten paces away from the old treaty oak.

well, he finally said, let's not have any more trouble than we already have. like i said, you and the people in your car can come out to the manor house for the interview.

ok, i told him, i don't know where that is, so we'll follow you. i got back in the car and waved to my husband and friends, who were still standing at a distance trying not to be so noticeable, rolling thunder's car took off, and brad took off after it. pretty soon we were on the freeway heading north, rolling thunder's car being driven as if coyote himself were at the wheel, weaving in and out of the traffic ahead of us so fast that brad had to keep his foot all the way down on the accelerator and jump across lanes just to keep them in sight. and there were expired plates, an expired inspection sticker, and no insurance on our car — typical protocol for my particular tribe. but i figured if the police stopped us or if we crashed through the railing, that would be part of the rolling thunder story, too, just like the clouds rolling in from the east and the wind coming up on what had seemed in the morning a warm and sunny day.

but a sign finally appeared on the road — manor: 3 miles — and we were out in the country, driving past horse stables to a small farmhouse where rolling thunder and his people were already getting out of their car and going inside. it was certainly quieter and more peaceful than downtown austin by the treaty oak; even on the hike-and-bike trail, there would have been lots of traffic noise. and i was suddenly ashamed that i had even imagined that barton springs would have been a good place to talk with rolling thunder. so while i was heading for the front door with brad, i was thinking that, after all, we were getting a little time in the country away from the city; at least we were getting that. even if rolling thunder wouldn't talk to us. but before i could get to the house, the same young man who'd talked to me before stopped me again.

one more thing i need to ask you, he said. are you on your moon?

now, cathy lee had already talked to me about this, making sure not to schedule the interview during my menstrual period on instructions from rolling thunder, who'd told her that menstruating women had strange and static vibes that weren't any good for men to be around.

so i told the young man that it was all right, i had already been asked that question.

so you *did* get that part of the protocol, the young man said.

yes, i said, feeling my face sting red in spite of myself. i did get that.

so he let me go on into the house. rolling thunder was standing in the living room. he looked a little tired, but he smiled at us when we came in.

i'm sorry, i said, that i arranged everything wrong for you. i guess i really didn't know what i was doing. and the country is better; it's a lot quieter out here, for sure.

oh, that's all right, he said. i think it'll be a good day anyway.

the rest of the people had disappeared. so the four of us sat down, two strange pairs: rolling thunder wearing his feathered hat and arranging his medicine pouch beside him, me taking off my own hawk-feathered hat as a gesture of respect for the house and laying my own pouch on the floor, mine full of pens, pads, and a tape recorder; his young man with long black hair seated near the door, my young friend brad with his long blond pony-tail and wire-rimmed glasses sitting on the couch beside me.

i said, well, i hear you've been traveling around a lot.

he nodded. two months of travel, then one month at home; it's been like that.

where have you been traveling? i asked.

but he said, now, wait a minute, you interrupted me. i wasn't finished with what i was about to say. you see, i talk slow; the people in new york — my gosh, i can't keep up with them. so you need to slow down a little bit, get into a natural rhythm. people these days are wound up tight and don't know how to slow down.

then he started talking about chemicals in the air and in the food we eat making us half crazy, and how white people ate all the wrong things. and you can't survive on bean sprouts and peanut butter, he said, particularly if you live north like we do. you have to eat meat if you're going to do any work. and brown bread has more chemicals in it than white bread.

well, the dietary talk was pretty interesting, but he was telling me exactly the same things he had said at the press conference, when i'd seen him the first time, using the same sentences and the same words almost verbatim. but since he had corrected me at the beginning of the talk about interruption, i just sat there, nodding my head, not saying anything, while he continued to talk. i was even afraid to get my tape recorder out of my pack! i was afraid to get my notebook and pencil out! i simply sat on the edge of the couch nodding, trying to look as though i were hearing all of this for the first time, while he continued to talk in a monologue for about ten minutes. finally, apparently aware that as an interviewer i didn't seem to be doing anything, he said, well, you better get your tape recorder out. then get your notebook out and write down what i've just told you.

i felt relieved that i had been given permission to set up, and so i did, and he continued talking. white people are going to have to change their ways, he said, and the first thing they're going to have to change is how they treat each other. like old people, he said. when i was in new york, i saw old ladies carrying bags with all their belongings out on the streets. and so many in unemployment lines, and so many in nursing homes and prisons. something's out of balance, and the balance is going to have to be restored. white people could have learned a lot from indians when they first came here about living in balance, because we didn't have any of those kinds of problems ourselves.

it was a good message, a strong message, so i tried to look as though it were all new to me, even though we had sat across from each other no more than two days before while he told it to me and the *daily texan* reporter in the same words. but i didn't try to interrupt him in any way. i didn't even look at the list of questions on my lap that i wanted to ask him. and two flies started buzzing around my head out of nowhere, the first flies i had seen since last year, landing on my nose and my mouth, and i tried to wave them away, but they continued to buzz, and rolling thunder continued talking.

he said he was a cherokee and that his people had originally come from atlantis — part of them went to egypt and part came to this western world, and they were a very wise race and had mysterious ways of building the pyramids.

plus some slave labor, i said without thinking, as if some disagreeable female demon left over from my last moon had just taken over my tongue to dip a piece of loudmouth-anglo pseudo-history like a crossways oar into this river-like monologue we were paddling down.

no, they did *not* use slaves, he said. he glared at me, suddenly furious, pulled up his chair. don't *you* try to tell *me* indian history, he said. that's the problem with you white people, trying to say what our history is. his eyes were flashing, as if he would have liked to take me apart on the spot.

it's just that i worry a little bit, i said (determined to own this voice as a part of myself, for better or worse, since it had already spoken out), about empire building, regardless of where it's coming from. what kind of human sacrifice it took.

well, rolling thunder wasn't named that for nothing. he let me know for a good several minutes that he was displeased with my presence and my approach. he said i had no respect, and that that was the trouble with white people. he said white people had the inquisition — why didn't we talk about that? he asked me if i wanted to be just a writer or if i wanted to be a *great* writer, because if i wanted to be a great writer, then i had better stop looking at the bad and start looking at the good. and he said that usually people who were going to interview him were given a list of questions that they could ask him, and that when it came to me, someone had really goofed. and he said that he didn't like people making jokes about being ignorant either, and that when people didn't have the right kind of attitude he usually just threw them out. and he continued glaring at me like he was going to do just that.

well, i figured the interview couldn't get any worse, and that to die from rolling thunder's thunder would be as good a way to go as any, and besides, i had heard this tone of voice before from my own military father, except that he used to tell me my problem was being a stubborn female rather than being born white. so i said, well, the only reason i brought all this up is because in this part of the country, when you talk about going back to the old indian ways of doing things, you might be talking about cherokee or you might be talking about aztec, and it's probably good to sort out one kind of indian ways from another.

there was a little silence. he looked at me, then looked out the window.

they went down, he finally said (apparently referring to the aztecs), because they abused their own power. they brought themselves down. look, when i want to feel better, i can just look out this window, look at the trees, the grass. that settles my mind.

he looked out the window some more. i did, too, following his example. the thunder stopped and the atmosphere began to settle.

well, i said, would you talk a little about how you came to be a medicine man?

no, he said, i don't like to talk about that.

too personal? i asked.

he nodded. he was quiet for a little bit. he was tired of the interview and tired of me.

well, maybe i will say something about it, he said finally, mustering up the powerful forces of his own good humor. i knew that i was going to be a medicine man from the beginning, he said, except that's not what we call it. we don't say medicine man. and i had to go through seven trials in order to become one —

just then, the door burst open, and a tall man stumbled in, eyes glazed over as though he didn't know where he was, or like those of a drug-sick man looking for strong medicine. rolling thunder's young man, who had been sitting quietly throughout the interview, got up and started pushing him back outside. you don't belong here, he said. the tall man fell into him, then into the couch, leaning this way and that, while the young man got him out the front door again.

well, the vibes don't seem to be right today, rolling thunder said.

yes, i said, i think you're right.

the flies were still buzzing around my head. i can't even get these flies to stop bothering me, i told him, quite aware that the only flies were on my side of the room. i looked down at the tape recorder and saw that the tape was tangled up in the sprocket and wads of tape were coming out the top.

well, rolling thunder, i said, is there *anything* white people have to contribute?

well, yes, he said. but then he thought a long time before he said anything else. well, he said finally, money is all right; it depends. people say that money is the root of all evil, but it depends where it comes from and where it's going.

and that is all he said: white people and money, back to that again.

so brad gave him the tobacco he had brought for a present, and i packed up my bag, the tape recorder's tape frazzled and no good, the only part that had gotten recorded being that first part, which i had already recorded at the press conference, the message he wanted everyone to hear, the words he considered important enough to repeat and repeat: that white people are out of balance, that they need to change, and that the first thing to change is how to get along with each other.

well, i told him, once i had gotten my bag back together as well as i could, this has been a special afternoon for me because i've always wanted to meet rolling thunder.

he smiled broadly, then began asking me some questions,

friendly and kind, about what feather it was in my hat and whether or not i was part of the rainbow family. i told him i knew them — i wasn't exactly a part, but related. i told him that i was also supposed to be a little cherokee.

oh, yes, he said with a grin. i remember when i was young the braves coming off the reservations, making raids down here. we said that if the cherokees couldn't take the land with war, they would take it by love.

indian raids.

i kept smiling and nodding, but whatever gulf there was between us had just widened, my head suddenly alive with all the stories i had ever heard from my own east-texas blood relatives about settling wild country and witnessing massacre and death administered by indian raiders. rape and kidnapping! but after he recalled that little piece of history, i looked at him briefly square in the face, and even with the indian-raid horror stories superimposed around himself like an antique robe, he was a beautiful, old, wise- and strong-looking man to me. he was as smiling and as good-humored as my own east-texas grandfather had been, who had also found his peace in looking out at the land. rolling thunder puffed on his pipe and had a mild smoker's cough from smoking his pipe with so many strange people on this tour he'd been doing of the white man's land for months. *even though he charged money, i liked him.*

he asked me some other questions, said some polite leaving things, but i could answer only in monosyllables as i started moving toward the front door, off balance and dizzy, not much better than the crazed man who had been shown out just a few minutes before. there was no new age, and i was no promoter, nor any interviewer either. i was the oldest granddaughter and rolling thunder was the youngest grandfather, but there were still some generations to reach across, and we were male versus female, we were anglo versus indian — we weren't friends; we were almost enemies. we were the heads of two parallel histories coming up on different banks of the same river — the river versus — strange waters we didn't yet know how to cross.

July 1984

The White Man's Vision-Quest Journal

Gloria Dyc

JULY 30

One of Plenty Coup's "boys" (such a diminutive term!) picked me up at the Pierre, South Dakota, airport, a minuscule patch of cement amidst the rolling plains. I was disappointed that Plenty Coup didn't come himself; I have so many questions about the vision quest, the sun dance. And I wanted to share our past-life-regression work and the news that I was an Indian in a previous incarnation!

Bernie, my scruffy and slightly ripe chauffeur, guided me to a rusted-out vw and asked what I was "into." I explained my practice as a psychotherapist, my attempts to bridge spiritual and psychological concerns. I told him about my seven years as a

---

Fiction

---

meditator in the Vipassana tradition, how I thought this discipline would help me in my vision quest. I asked him if he thought the various shamanic and meditative traditions would converge, as they seem to be doing on the Coast and in Santa Fe. Bernie said, "Gee, I don't know, man. You got any money for gas?"

Bernie was apparently a meat-eater, if you know what I mean, and I had to roll down the window. He noticed this and apologized, explaining that he had been cutting wood for the sweat lodges all day.

Bernie told me that he met Plenty Coup twenty years ago. At that time Bernie was living on the streets in Denver, getting into crystal meth and petty theft. Plenty Coup adopted Bernie, an orphan, and Bernie had been following his "uncle" around ever since, helping out and keeping off the hard stuff.

THE SHADOWS DEEPENED on the rolling plains as we drove south to the Rosebud Reservation, past vast stretches of land without signs, gas stations, shopping malls.

Then Bernie asked if I would mind if he lit a joint. I said, "No problem." But I couldn't deny the feelings of anxiety and confusion. (Can you imagine, Sarah, someone lighting up a joint on the way to one of our retreats at Mandala?)

I began to have some very negative feelings, and I wondered if Bernie would pick up on them.

(Then I thought of you, Sarah, standing up and speaking out at our Vision Committee meetings, encouraging others to spread some of the light to grass-roots people: "Look at us; we're all alike. There are people out there starving for real food, for spiritual food. Buddha didn't step a foot on the path until he was exposed to sickness and death.")

Bernie must have been picking up on my thought-energy because he said, "So, Uncle tells me you're going to donate a beef to the camp. That's good; we're short this year, and there's a whole lot of hungry people to feed."

JULY 31, 2 A.M.

It was dark by the time we got to camp. I felt as though I were entering another time zone. There were campfires and tepees set up. (Just as you had envisioned, Sarah, when you predicted that I would be traveling to Dakota!)

Plenty Coup had about four tepees set up in his camp. He was sitting in a director's chair — his "Buddha" belly protruding from his striped knit shirt — giving orders, joking, telling stories. (Remember how we discussed his aura the first time he came to lecture on shamanic healing at the Mandala — you compared it to fire?)

Plenty Coup greeted me as I approached the circle of people around him. "You got a tent, John?" he asked. I told him I did. "That's *good*, 'cause otherwise I'd have to stick you in one of these tepees with one of my wives, and I don't know if you could *handle* that." Everyone laughed.

People continued to pull into camp for the rest of the evening. They came to greet Plenty Coup first; he's the central nervous system.

Later Plenty Coup took me over to the sun-dance circle, and he told me that he had found the perfect tree for the center months ago. Then he gestured toward the dark hills to the west of camp. "We're going to take you up there, tomorrow night or the next. The women we usually put over there." He nodded in the opposite direction.

Plenty Coup told a number of funny stories, but I wanted to record this one: "We were in the sweat lodge last night, you know, and we had maybe sixty rocks, and it was *hot*. When it's like that, you know, you're supposed to make your prayers short. But this one guy, he went on maybe twenty minutes, a half-hour. He covered *everything*. He started with every atom, then he prayed for the blades of grass, the insects, the reptiles. The four-leggeds. The winged ones. Then he started praying for people. His children, his family, his clan, his state, his nation. Then *all* the nations. He had to name each country, each county, each city. He prayed for the Chinese, he prayed for the Arabs, the people down in Africa. At least I learned some geography, you know. I never stepped foot in a school, and I'm proud of that. But meanwhile, you can hear all these men *groaning* from the heat. The door stays closed until the prayer is done; that's my way. You can smell the hair getting *singed* in there, it's so *hot*. Finally, there was a white man in there, and he couldn't take no more. He jumped over those red-hot rocks and crawled through the door. We found him later, curled up, all *red*. He looked like a newborn *mouse*."

July 31

Woke to great activity in the camp. The "boys" were coming back in trucks loaded with pine boughs to make a shade for the sun-dance arbor. The women were cooking around an open fire. I thought about helping out at the arbor, but I wanted to conserve my strength for my fast, so I decided to help out with the cooking. I introduced myself to a woman named Nina, an Apache from Arizona.

"You want to help? Here, flip the pancakes," she said, and handed me a spatula. Some old men standing off to the side laughed. They continued to drink their coffee, talking away in Lakota. (Most of them had teeth missing. I thought they were guaranteed dental care under the entitlements!)

One of the workers came over and poured himself a cup of coffee. He asked for sugar. One of the older men said, "Ask the *white* man for sugar. The white man has all the sugar." Everyone laughed.

My name has become the White Man. It doesn't matter how many times I have introduced myself — I am still the White Man.

Nina told me that men don't usually help with the cooking, but she was grateful for my help. I told her about my plan to go up on the hill. She asked me if I had ever fasted before. I told her about my monthly fasting and cleansing.

I asked her if she had fasted. She gave me a cryptic smile. "Oh yeah, we fast all the time," she said. "At the end of every month, when the money runs out."

I flipped pancakes until I was saturated with grease and smoke.

Plenty Coup approached and nodded to me and said something in Lakota to the old men, and they all hooted with laughter.

No one seems to be able to make decisions or take any action without consulting Plenty Coup. Money is short; time is wasted. (They could benefit from a little Western-style management!)

Example: Bernie came running up, saying, "Uncle, we need another truck to haul wood, but we have to pay this guy twenty-five dollars."

"Will he take my *credit card*, do you think?" Plenty Coup asked; then he looked over at me. I found myself reaching into my pocket. Later, Nina approached me for a donation for camp groceries. No problem. Then Bernie approached me for money for gas. I seem to be the only one around with money.

I know prosperity is meant to be shared, but I started to feel uneasy, confused. I went back to my tent for the afternoon to clear my mind.

I am troubled to be feeling troubled.

Plenty Coup told me that he was traveling to Pine Ridge to visit another sun dance, and that he would be back before sunset so we could begin to work on some of the preparations for my vision quest.

AUGUST 1, AFTER MIDNIGHT

I had been waiting for Plenty Coup to return to camp for most of the evening, but as the sun set about 9:30, I had to let go of the idea of meeting with him.

This evening I walked off into the woods and sat for an hour under one of the magnificent ponderosa pines, and I felt restored. Back in camp, though, some of the anger and anxiety and disappointment resurfaced. (I wish you were here, Sarah, to help me get at the core issues.)

The litter and garbage around the camp really bother me. Garbage bags overflow with cans, plastic cups, and plates. Garbage collects in the bushes, drawing flies. The words *Native American* and *ecology* have always been synonymous in my mind. So I spent some time going around camp, filling sacks and raking the ground. I am hoping that others will observe my actions and imitate me.

I feel like a freak at times. I have tried to talk philosophy with some, to explain our work at the Mandala. People seem indifferent, or slightly amused. (Maybe their view is just too *parochial?*)

I thought Plenty Coup would show some consideration and carry through on his commitment to me. I know there are so many others who need his guidance, but I expect a man to keep his word.

Plenty Coup finally came back to camp about ten, and I joined him and the others around the cooking area, determined to show some grace. The women fried meat and bread. (Their diet is heavy in starches and fat. No wonder the rate of diabetes and heart disease is so high. Oh, what I wouldn't do for a big bowl of brown rice!)

"Eat! Eat!" Plenty Coup said when he saw me. "We traditional men have *these*," he said and patted his belly. A few minutes later he took me aside and explained that he hadn't expected to be gone so late, but we would find time the next day.

Plenty Coup described the sun dance in Pine Ridge and ended up on a tirade: "They had maybe thirty dancers in Pine Ridge. Several white men. That's all I need — white men who want to dance. That one should be closed down. Lots of things going wrong. Then on the way back we ran into a relative of mine. Great-grandson to a chief, and he's *shitfaced*. They're going to get us when we're down. The state is going to step in here, take over. They'll want to tax us. We can't pay — they'll take what's left of our land. We can't be a *tribe* without land. This way here is the natural way. Communist way. Lakota way. Everybody works. If you work, you eat. I don't know. I'm tired of all these problems. My people are pitiful. Maybe I'll get a job with Honeywell or something. Get some credit cards. Maybe it's my turn to be rich."

Plenty Coup is a very angry man.

AUGUST 1, EVENING

I awoke at dawn to the sound of coyotes in the hills. The White River runs on the edge of camp, and I could hear it gurgle. There was an aura of rose and gold light over the hills. Plenty Coup was sitting in his director's chair in front of his tepee; his wife was braiding his hair. He had all his regalia propped up in front of his tepee — his war bonnet of eagle feathers, his sacred staff, the great eagle wing he uses in the sun dance. I felt as though I had slipped back in time to another era, when the Lakota still lived free on the plains.

Plenty Coup called me over and pointed out the morning star. His wife served us coffee. Then he began to talk about the loss of his mother, which happened only a year ago. He is the only son in the family. His mother had been sick for a long time, he said, with the "white man's" cancer, but the Indian medicine kept her going well into her nineties. When she passed on, they dressed her body in the traditional style and took her out in a horse-drawn wagon to be buried on her own land, and they sang the old songs. As Plenty Coup talked, his eyes filled with tears, and I saw his vulnerability for the first time. He can be so macho, you know, but that is a facade. "My people are pitiful," he said, "but we're going to *survive*."

I spent the morning watching the camp come to life. The same woman (or so it seems) rises each morning to make the fire in the main cooking area; the same few men collect wood for the fires. Most of the others rise later, when they can warm themselves around the fire. There's one woman who wanders around, a vacant smile on her face, not contributing much, claiming to be from some lost Eastern tribe.

Nina asked me if I knew how to get a fire started, and I shook my head. "You'd better learn; you never know when we might have a nuclear war, and we'll have to go back to the old ways." I laughed and told her I didn't think we'd have to worry about cooking fires in the event of a nuclear war.

"My people can survive *anything*," she said, with some hostility and defiance.

(Still can't get away from being the White Man.)

I began my fast today, even though Plenty Coup said I wouldn't need to fast until after I got out of the sweat lodge.

After breakfast I went with a group of men to collect rocks for the sweats, which are held all evening. They showed me which rocks were best — which would retain the heat, which would crumble. We had to use a crowbar to pry some of them out of the hillsides. I worked along slowly, wanting to conserve my strength.

They spoke Lakota most of the time, but finally one turned to me and asked in English, "So how did you become interested in our ways?" I told them about the Mandala Center, how Plenty Coup came out and conducted a sweat lodge for my clients. I told them I had been wanting to go on a vision quest for years and hoped to conduct a workshop on shamanic healing for the institute.

They were all quiet, looking at the rocks. Then someone told a joke in Lakota, and they all laughed.

When we got back, Plenty Coup said, "So, you got your ties together?"

Ties! I didn't know anything about ties! He told me I needed so many *hundred* tobacco ties — offerings of tobacco wrapped in tiny pieces of cloth — for the spirits.

Plenty Coup told me he'd take me to an aunt of his, and she'd help me make the ties. "What about a star quilt? You have a star quilt?" he asked. I didn't know anything about a star quilt! Plenty Coup told me that his niece made star quilts and that I needed to be wrapped in a brand new one when they put me up on the hill. "We have to do this *right*," Plenty Coup said. "Then afterward, you give that star quilt away to someone. That's the Lakota way. My aunt will help you out; she's a good woman. You should take her some *meat*. She's old now, and real pitiful."

Plenty Coup's aunt lives in a one-room log cabin in one of the nearby towns; the town must have grown up around the cabin. She is shrunken and deeply wrinkled and has only a few teeth left, but she is strong in spirit. There was a cot in the cabin and a wood-burning stove. She served us coffee that was so strong it was almost vile. There were paper bags everywhere: bags filled with clothing, odds and ends, herbs. Dried meat hung from a clothesline.

Plenty Coup explained the situation to her. "Good boy," she said to me. She took the package of meat from me and put it in the refrigerator, which was almost bare. She burned sage and sweet grass and then went about making the ties, ripping and

cutting brightly colored material, filling tiny squares with tobacco, wrapping them in string, with hands that were adept,
though arthritic. As she talked, Plenty Coup grunted his approval
or offered exclamations.

Then, quite suddenly, Plenty Coup announced, "A storm is
coming in from the west tonight, John. I don't think you'll want
to be up there on the hill tonight." His aunt looked up at me and
laughed. I laughed, too; I thought this was a joke. The skies were
perfectly clear — no sign of a storm.

But as the evening went on, it became clear that Plenty Coup
had no intention of putting me up on the hill.

So here I am. My ties are ready. I have my star quilt (for
which I paid three hundred dollars). I checked the skies again —
there are all kinds of stars.

My expectations were so high today; I feel deflated, and anxious, and angry. Are they testing me in some kind of way? Have
I become the brunt of some joke?

And how am I supposed to go up on the hill in this state
of mind?

AUGUST 2

So, I was wrong! At about three in the morning I was startled by
a terrible clap of thunder. There was an incredible storm — as
Plenty Coup had predicted! Lightning struck close to camp. The
frame of my tent swayed in the wind; water seeped into the bottom. (No, I would *not* want to have been up on the hill last
night.) By morning, the skies had cleared, and the temperature
began to rise. It was one hundred degrees today.

Finally, I'll be going up on the hill at sunset. Tomorrow is
the last day of purification before the sun dance. Plenty Coup
said he would pick me up before sunset tomorrow. "Don't forget
me!" I said. "If this heat continues, I'll be pretty dehydrated in
twenty-four hours."

Plenty Coup smiled and said, "When I was only nine, I went
on my first vision quest. In those days they still used the vision
pit. I was in a pit, *in the earth*, for four days and four nights. No
food, no water, nothing. Nine years old. You got to do it right to

get a blessing. You can't cheat. No canteen of water. No sneaking granola bars up there, John."

I asked him if I could take my journal.

"As long as you don't eat it."

I went through the purification ceremony. Plenty Coup lit the fire for the sweat lodge with some flints passed down from his grandparents. He lined up nine sacred stones and prayed and talked before striking one of the flints. The fire came into being — a wonder.

Plenty Coup said we would have only one round, rather than open the door of the sweat lodge several times, since I was going up on the hill. A few men crawled into the lodge with us, and it was so hot before the door was closed, I wasn't sure if I could endure it. The rocks were red with heat. As Plenty Coup sang and poured the water, I could hear the others groaning from the heat. I shut my eyes, for fear they would melt down my face like wax. When it was over, I found red, scalded spots on my cheeks, ears, and knees.

I gave Nina several hundred dollars for a beef; she and several other women have agreed to prepare the soup for a feast tomorrow, when I return to camp.

AUGUST 3, DAWN

The dawn has come with such gentleness. When was the last time I experienced dawn? The coyotes began to cry in the pale light.

I haven't slept, though I almost lost consciousness a few times, even while in a sitting position, my head falling to the side like a great weight.

By turns, I feel exalted and sorrowful. I may have felt serenity in the first few hours; now I feel close to madness.

The sky appears (unfortunately) to be clear. My mouth is dry.

By the time we drove up last night, it was completely dark. Plenty Coup knows the land well and managed to find this finger-shaped butte, which juts out over the pines. Bernie drove us up in the pickup. "This is where my relatives live," Plenty Coup said. Bernie laughed nervously. "Don't try to spook me, Uncle."

Many others have fasted on this spot. Plenty Coup lent me a

buffalo robe, so I can kneel or sit comfortably. He arranged the tobacco offerings in a small cedar tree, and prayed in the four directions, lifting his voice with such force it must have reached the camp below.

I felt overwhelmed and cried there in the darkness behind him. He shook my hand and proceeded back up the slope to the pickup, using a flashlight as a guide.

Far below I could see a fire in the camp. And then a complex, dramatic symphony unfolded. There was thunder and lightning in the west. The wind moved through the towering, dark ponderosas, and it all seemed orchestrated.

I prayed, I cried, I gave thanks.

It rained lightly. I turned my face up and caught the drops in my mouth. At four in the morning (by my estimate), I lost consciousness. Then I heard a rustling, like footsteps in the grass, and the snap of a pine bough. I remembered what Plenty Coup had said about his relatives, and my heart raced.

I didn't anticipate the fear. I have read so many accounts of medicine men and the spirits they encounter; suddenly I prayed to be spared that experience.

AROUND NOON

The sun is high in the sky now. The coolness of the morning has evaporated. There is no shade here. My concentration is ebbing. The periods I spend in prayer and meditation are shorter. Now I am becoming preoccupied with time. I keep thinking about the number of hours I will spend here before Plenty Coup picks me up.

I am worried that Plenty Coup will forget about me.

A while ago, an eagle flew overhead. I looked up at him. He looked at me. He circled a few times and left. This was a blessing. I felt energized, uplifted for a while.

Then anxiety and extreme discomfort returned. I berate myself for being so weak, so distracted. The anxiety intensifies.

I can't summon any saliva. My mouth no longer seems to be made of the familiar, wet membrane, but of a substance dry as leather.

Sweat is trickling down my back; I am attracting insects.

Is it noon? One or two, perhaps? I can hear the sound of a chain saw coming from the camp. I hear the laughter of children. Is anyone back there thinking of me?

AROUND FOUR

I am sucking on grass for the moisture. I cry and lick the tears as they come down. The sweat on my skin is sticky, like pitch from a pine.

I am having a vision of water.

Water in all its forms: Rivers of water — flowing, flowing. Diving into a lake of water. Water as it comes out of a faucet. A tall glass of iced tea, the glass opaque from the cold.

I don't care too much for Coke, but I'm thinking of it now. And lemonade — made from fresh lemons.

I try to pray, but I lose my concentration so easily. Then I break down in tears, appalled at my own weakness. Then I lick my tears, like a thirsty, trapped animal — devoid of any spirituality.

We take water for granted! It flows out of our faucets; we buy bottled, gourmet water. In Africa, women have to walk miles to bring home a jug of drinking water. Even back in camp, we've had to work for water — haul it, heat it, conserve it.

Food — we can live without it, for a while. Not water. We are water — 80 percent, I believe.

I think of our beautiful oceans. Oil spills and garbage washing up on shore.

I remember Plenty Coup talking about water rights — how the tribes would have to fight for water next, just to survive.

SUNSET

The sun will be setting soon. I worry that Plenty Coup has forgotten about me with all the activities in camp.

About an hour ago, I heard a pickup winding this way, its engine straining. I felt such relief! I even laughed at myself and thought: *This wasn't that bad.* Then I heard a chain saw; someone was cutting wood.

Desperation returned, and anger. I wonder if this is deliberate, if they are testing me down in camp to see how strong I am.

I pray for patience, but I can't hold out much longer. If they don't come soon, I will have to walk down myself. I *have* to have water.

The thought of walking back to camp is humiliating. But if it gets too dark, I may not be able to find my way.

My ego is taking on a life of its own. It walks around in its own body, grotesque and ludicrous. I hold back and view it with detachment, as the eagle must have viewed me.

AUGUST 4

Plenty Coup didn't forget me, though it was dark by the time they picked me up. I was confused and weak and could think of nothing except my first drink of water. I went through the purification ceremony again, and it seemed so unbearable I cried like a child.

The women in camp had made some sort of instant, sweet drink. I drank glass after glass, gulping and panting. Then I started on water, consuming glass after glass. "Thank you for doing this," Nina said, and shook my hand. They were passing out soup and fry bread in the camp.

I went down to the river and stretched out in the water. I stayed there, light as a hollow log. The stars were out.

Later, when the activities in camp had subsided, I saw Plenty Coup sitting on his director's chair. He was alone, and I thought it would be a good time to talk.

"Hau!" he said. "Everybody's got a problem. We're waiting for these people from Oklahoma. They're going to contribute a beef for tomorrow to help feed the people. Now they've got car trouble, and they're stuck in Nebraska. So how was it, John?"

I told him it was *hard*, much harder than I'd expected. He laughed. I told him about the eagle, my obsessive thoughts about water. He laughed again.

We sat there in silence for a while. Maybe it wasn't the right time, but I felt that I needed to be honest. I told him I had felt anger, waiting up there. I reminded him that we had agreed that I would be picked up before sunset, and I could only interpret his tardiness as a lack of concern.

"John, there was a lot going on in camp," Plenty Coup said, almost without apology.

"That may be so, but when you're up there, it's the only thing happening," I said.

This put Plenty Coup on the defensive. "We don't go by the white man's clock; we go by natural clocks, the seasons, the rhythms of camp. Everybody has needs."

"Maybe if you took advantage of our Western system of management, you would have more time," I said. Plenty Coup was becoming enraged, but I didn't really pick up on that until it was too late. "As it is, I'll be going home without a penny in my pocket, and I can't help feeling a little used. . . ."

Plenty Coup turned away, and then spun around and lunged at me, pushing me to the ground. I was stunned.

"*Used!*" he bellowed. "Maybe *you're* using *us*. You people come here. You want to learn our ways, just like that! "Oh, I think I want to be a shaman! Be sure to pick me up at sunset!" Our medicine people go up on the hill for four days and four nights. And most of the time the spirits don't come until the *very end*. You took all our land, and now you want our ceremonies!"

I could sense that people in camp were listening, but keeping a respectful distance. He talked about being at Wounded Knee, about being locked up in jail for a year. He talked about the 1868 treaty, how some of the chiefs were bought off with whiskey. He talked about all the cousins and sisters and brothers he had buried.

I got up from the ground when I felt he was winding down.

"This is all we got, John," he said, gesturing to the sun-dance arbor, the sweat-lodge area. His eyes filled with tears. "This is all we got left." Then he walked off into the darkness in the direction of the arbor.

I thought of following him, trying to resolve the situation; but at the same time, I respected his need to be alone.

A couple of camp dogs followed me back to my tent, and I felt like one of them. I had no one to talk to. I felt humiliated, confused, angry, exposed.

And I can understand his point of view, his anger and exasperation. I can imagine how ludicrous the idea of a shamanic workshop must be from their perspective.

(I was not expecting this kind of vision, Sarah. I wanted you to be here.)

The sun dance began today. The dancers were up before dawn, going in and out of the sweat lodges. I was exhausted but excited, and I summoned my strength so I could see the entry into the arbor at dawn. All the male dancers were lined up in their ankle-length red skirts, holding their eagle-wing fans. The women wore long cotton dresses of various colors, crowns made from sage.

Plenty Coup looked magnificent. He wore a deerskin breechcloth. His hair was loose; the head of an eagle was attached to one of his wrists. He carried a cane with a scarf and an eagle wing attached to it.

Before they entered the arbor, he talked about the significance of the dance. "You are going to fly," he promised them. "You're going to be between the earth and the sky."

The dancers made shrill sounds on their eagle-bone whistles as they made their way to the entrance of the arbor.

All day I've listened to the songs and the drumming. There are some fifty different songs, but I have to listen very closely to distinguish among them, since I don't know the language.

By midday I could see the lips of the dancers were dry and cracked. It appeared painful for some to walk over the grass after so many hours of dancing. One woman dropped to her knees, overcome by the heat. They are not supposed to have food or water for four days, though most of them have soup and water at night. Plenty Coup said he would like to do it the old way, but the people are weak.

And as I looked at some of the dancers, I saw that they had been transformed, that they had passed beyond the realm of suffering, and by the look in their transfixed, glazed eyes, the slight, awed smile, I could see they had a vision of a world beyond this one. They were *flying*. They were dancing between the earth and the sky.

Bernie will drive me back to Pierre in the morning so I can get back to the Coast; I hope he bathes first, or at least takes a swim in the river.

August 1991

THE END OF A SIXTIES DREAM?

AN INTERVIEW WITH STEPHEN GASKIN

Michael Thurman

O*n Columbus Day 1970, more than two hundred San Fran-cisco hippies set off in a caravan of remodeled school buses on a cross-country pilgrimage. In the lead bus was Stephen Gaskin. A former Marine and college teacher, he'd started experi-menting with psychedelics in the 1960s and had begun holding Monday-night discussion groups at San Francisco State College on drugs, religion, honesty, and peaceful cultural revolution. (The talks are included in Gaskin's first book,* Monday Night Class.*) Within several years, the group grew from a few friends to more than a thou-sand listeners, and by 1970, Gaskin's most dedicated followers were ready to leave their environment of psychedelic serendipity to start a community of their own, based on self-reliance, voluntary simplic-ity, hard work, and faith in God.*

In Tennessee, they bought seventeen hundred acres that became known simply as the Farm. Members of the Farm, who eventually numbered fifteen hundred, signed over all their worldly goods upon joining. The Farm grew into one of the nation's most ambitious and successful communes, with its own schools, clinic, soybean dairy, bakery, and work crews for carpentry, farming, and so on.

It must have been quite a shock to the neighboring families to find those psychedelic school buses, looking like the ships of an extraterrestrial invasion force, in their backyard. But the Farm soon established bonds of friendship with its neighbors, often sharing information, farm equipment, and manpower. Gaskin recalls going to the local bank to secure the initial loan to buy the land. "The bank guy said, 'It's not just that you're these out-of-town hippies we've never seen before; it's also that it's the largest loan that anyone's ever asked for from this bank!' Since then, two of those banks have folded, and we haven't. No one would have predicted that."

Over the last fourteen years, 150,000 visitors have passed through the Farm's front gate, many simply to share in the vision of an alternative way of life; the Farm's midwives have delivered 1,250 babies, by natural childbirth, for Farm members and others; and an international relief project called Plenty has gotten its start on the Farm.

The Farm has also paid its dues. There have been hard winters, a hepatitis summer, and a bust for growing pot, for which Gaskin and others spent a year in jail. The community has evolved spiritually, culturally, and technologically — from tents and oil lamps to computers and solar panels.

It was 1980 when I was last on the Farm. When I returned in 1985 for this interview, I expected to see the same old Farm, more or less. From the minute I arrived at the gatehouse, however, I could sense something was different. For starters, I was the only visitor. Before, I had seen throngs of curious sightseers, veteran pilgrims, and hippie children along the road; now there was only the sound of my car engine. The fields that usually would have been plowed up for spring planting were idle, full of grass and wildflowers. The canning-and-freezing building, the radio station, the community kitchen, and the huge tractor barn were all vacant and boarded up. The Farm looked more like a psychedelic ghost town than like the busy community I had previously seen. Some of the houses I had helped work on during past visits were now empty, many with windows broken; others had been torn down. The children I saw were not tie-dyed replicas of their parents but kids wearing designer jeans and jogging shoes. Their hair was cut shorter, and video games had them in a frenzy. In short, they seemed a lot like normal middle-class kids, except more mature, self-confident, and likable.

As I talked with a few of the members, I kept hearing phrases like "before the change" and "the new system." Gradually, I began to piece together what had happened. The Farm's only substantial source of income was the construction crew, which did outside work, and when interest rates went way up, business began to slump, and the burden of providing for the whole community became too great. There was a danger of having to sell off part of their land to survive.

Besides changes in the economy, there were internal tensions. Many people felt they weren't accomplishing anything, or that the standard of living was too low, or that they were being treated un-fairly. Other families, now in their midthirties, were growing tired of living with dozens of other people and wanted a more traditional nuclear-family structure.

After selling their soy-foods company, Farm Foods, to pay off their land mortgage, the board voted to change the basic financial structure of the Farm; total collectivism had become economic sui-cide. Now each family had to provide its own income and pay a weekly tax that went to support the few remaining services, such as the school. But many families were unable to find outside jobs, and others disagreed politically with the changes. People began leaving, and the Farm's population dropped steadily from fifteen hundred to the present three hundred adults and children.

The last time I had sat at Sunday Morning Service, there had been a meadow full of people, and Gaskin had needed to use a micro-phone. On this Sunday morning, we numbered fifteen and sat in a small circle, exchanging ideas. But the feeling I came away with was one of determination, not defeat. These were not starry-eyed hippies looking for a free ride, but seasoned homesteaders. Most of them had arrived with the original caravan; they had seen it all come and go. They now see their practical knowledge as an untapped resource and are considering using the Farm as a conference site, sort of an Esalen for community training projects.

Gaskin is something of a sixties icon, a man as hard to pin down as his community. He refers to himself as "a nonviolent social revolutionary," which includes being an author, a teacher, an activist, and parish priest to his community. His books include The Caravan, Mind at Play *(both out of print),* Rendered Infamous *(Greenwood Publishing Group),* Amazing Dope Tales *(Ronin), and* This Season's People *(Book Publishing Co.).*

Is he a leader? Although he shuns this label, he has been criticized for being too egocentric and demanding. Since he is the Farm's founder, no one can dismiss his strong influence on internal affairs, yet the Farm has always been governed by an elected board that may or may not agree with him. Is he enlightened? Gaskin sees that as more of "a process, and the enlightened person can think an ordinary thought and be ordinary, or the ordinary person can think an enlightened thought and be enlightened. It depends on your contract with yourself. I made a contract with myself about a way I was going to be the rest of my life, and I haven't broken that contract." For the most part, he is a self-taught avatar, a spiritual maverick who draws his teachings from Zen, Judaism, Christianity, and peyote rituals. Rather than leading the community through a daily set of yogic calisthenics, he believes in sticking to the day-to-day practical business of the Farm. "What we are trying to do is take our yogas out of real life, not to manufacture things to use up our senses with when we have this humongous amount of work to do."

Is he controversial? Definitely. And he shows no signs of changing any time soon. He sees no distinction between spiritual practice and political action, and in the past he has aligned himself with the antinuclear movement and Native American land rights, often traveling around the country to speak on their behalf. As the founder of Plenty, he has worked for peasant reforms in Central America, which earned him the first Right Livelihood Award, sort of a New Age Nobel prize.

Though his waistline and hairline are beginning to reflect his age, he says he will always remain a hippie at heart. His face is more wrinkled, but his eyes still sparkle with a mixture of mischief and inner wisdom.

Gaskin wasn't as eager to talk about metaphysics as about his family's new cottage industry, The Practicing Midwife, *a magazine begun by his wife, Ina May. It deals with all facets of natural birthing, particularly the political issues, and has turned the upstairs of their home into a computerized beehive. Stephen was particularly proud to show me his latest personal project: installing their first indoor bathroom. He joked that "when it's done, we're going to celebrate by having a pot party." Computers? Fancy indoor bathrooms? Could this be the same man who once stated that he "fought every inch of electrical wire that the Farm wanted to put up"? It simply reveals someone who has never been afraid of change, either personally or culturally. He seeks constantly to stay at the vanguard of his generation, yet always with integrity and a moral consistency that he traces all the way back to the spiritual truths he encountered on Haight Street.*

Those interested in learning more about the Farm, or Plenty, its international aid program, can write to Vickie Montagne, 34 The Farm, Summertown, TN 38483, www.thefarm.org.

Thurman: What was your religious background?

Gaskin: Not much. American Christmas card. My mother was half agnostic, and my father won't talk about it at all. There was no pressure on me to be religious. I think that's why it was so fresh to me and why it turned me on. I've always thought organized religions are a bore. They're just absurd. What I have come to respect is the continuing, unending stream of people who have a spark or flash of enlightenment that changes their view so much they usually have to tell about it. They leave records and tracks, and you follow them like an explorer, knowing that there are other people who have been through these changes. And if you have the good luck to read good books, you find out that, as far back as we have records, people have had similar questions, concerns, problems, and flashes of enlightenment. So you find yourself in the mainstream of humankind.

Thurman: How was it that you began to fill the role of spiritual teacher?

Gaskin: I was about ten years older than most of the hippies, so I found myself getting loaded with a lot of responsibilities. I had also been in the Marine Corps; I had been to Korea; I'd carried dead and wounded. I couldn't listen to any bullshit about military takeovers. Also, I was just coming into my own at that time and realizing that I had to get rid of the part of life that was just trimming. See, I believed, and I still believe, in our hippie realization of this cycle we're in. And although it caused a lot of us to become interested in the Eastern religions, because they had such interesting cosmologies and psychologies, I have always tried to keep faith with the fact that we had a realization that was real and precious and common to our generation. I don't fit into any of those Eastern religions. I haven't been to any swami school; no Zen master has given me any stripes.

Thurman: If you hadn't been a teacher, what else might you have been? What other direction might you have gone in?

Gaskin: Well, I don't make them mutually exclusive. I am a nonviolent social revolutionary, and I think that anybody who follows religious disciplines ought to be one, too. And people who say they're religious but aren't social revolutionaries are fooling themselves, because if religion is to have any validity whatsoever, the gap between the real and the ideal cannot be so unbridgeable that it's all pie in the sky. There's got to be a chance for justice on earth.

Thurman: What has happened to the enormous energy that was generated in the sixties?

Gaskin: Some has fallen on stony ground, some of the seeds have been eaten by birds, and some of it has fallen places where it has returned a hundredfold. That's what happened to our revolution. It's still there. Although there's not a visible hippie tribe on the streets or roads or in the cities, as there used to be, it doesn't really matter, because we permeated the culture. The popular music of this country is now rock-and-roll, you know. You hear people on *Hee-Haw* talking hippie talk that was invented on Haight Street. They say, "Dig this." Everybody does. We permeated the culture, and that's what it was supposed to do.

The other thing, to me, is what I talked about earlier — not to forget that this was a homegrown set of realizations that hundreds of thousands of people had, and not to try to tie it down too much. I know and am acquainted with about half the religious teachers you hear about on the circuit. That's all fine, but they didn't bring those ideas to this country. Those ideas grew out of this soil like mushrooms, right out of the cultural conditions we'd gotten ourselves into.

Thurman: As a way of coping with, or changing, those conditions?

Gaskin: You've got to be a rich country to have hippies. They're an interesting class. They're a free, privileged scholar class that can study what they want. They're like young princelings. That's why the only other places to have produced hippies are countries like Germany, because they're rich enough, too. It's really kind of an upscale movement, in a way, except for when it broke through. And when it broke through was when it was the most revolutionary, and when it really scared the Establishment, because hippies bond across cultural, religious, and class lines. That's terrifying to the Establishment. That tears down everything that they've stood for for five thousand years. The communists want their dynasties, too; everybody wants their dynasties. They don't want people to be able to just step out of one culture into another. They can't stand Ford heirs chanting, "Hare Krishna." People who were supposed to have lived off the wealth of old Henry Ford going out and sitting cross-legged and wearing ponytails? It ain't hardly civilized!

Thurman: Was some of that energy maybe too optimistic? Expecting something to happen too soon?

Gaskin: Things did happen. We got out of Vietnam. We made it so that you couldn't run a racist society separate from the rest of the United States, so that the Constitution reached down into corners of Alabama and Mississippi. We got rid of a president who was a tyrant. We brought new forms of education to other countries through the Peace Corps. There was a tremendous cultural flowering that took place. All flowers eventually curl up. But the significance of the flower is in the seed. And the seeds were planted.

You could say we sent out a call. I guess I'm one of the guys designated to stay by the phone and answer the calls as they come back. And I get those calls. I get letters from people who were babies when I was a hippie; who, as soon as they grew up and looked at the world, said, "Oh, I wanted to be one of those. Are there any left? Is it too late?" They say, "Did I miss it?" And I say, "No, it's part of an ongoing thing, a cultural thread that should continue forever." It helps keep societies from going crazy. Societies are too powerful by definition. Everybody has to be an anarchist to some degree. The amount of police power that societies are developing is terrifying. There always have to be people who, not from malice or greed or evil, but for the best social reasons, remain outlaws.

Thurman: How hard has it been for you not to set yourself up as a guru figure? Is it tempting?

Gaskin: Not to me. Not the trappings. I might fall into excesses, but I'm too much of a down-home kind of person to get into ceremony and pomp. I was amused when I saw fifteen hundred hippies drag a Juggernaut with Bhaktivedanta on it through Golden Gate Park. I thought that was funny at the time, and I still do. I think my temptations are different. When you're in a position of having a lot of responsibility and can't lay enough of it off onto others, you tend to act fast. You shoot from the hip and are brusque because you've got so much loaded on you. You want to get it over with quickly. It's the same condition that doctors find themselves in. Because, to a doctor, you're just one of a thousand appendectomies. But to you, it is maybe the only hospital trip you'll ever have, and it's a hell of a big deal. The doctor isn't going to give you as much time and importance about it as you really think he or she should. I can fall into that, and I have to be aware of it.

Thurman: So there will never be any Stephen Gaskin School of Meditation?

Gaskin: No. People have hung stuff like that on me, you know, but I haven't liked it. And, in spite of what people have said, I don't want it. The old religions are full of hierarchy. Maharishi, for instance, now gives initiations by videotape. That implies a great deal of hierarchy, if one guy's blessing is so good that you have to get it by tape instead of in person.

People have spoken about my "followers." Well, I don't have followers; I have friends.

Thurman: Is it necessary for a community to have a central figure or leader?

Gaskin: I don't know if it is or not. There are millions of villages all over the world that don't have one. Although somebody usually seems to evolve to the position of elder or something like that.

Thurman: How is the Farm different from some of the sixties cults that were established?

Gaskin: I remember a long time ago, when we were still on coal-oil light, seeing two paths: One of them went to a place where we became almost a quietist village, where all we did was polish our craft, do our meditation, and study our thing. That looked like a dead end to me; it had no freedom in it, and we would begin to be stagnant if we went that way. The other path was to grow, and influence and interact with the world.

Thurman: What mistakes have been made by other communities that have failed?

Gaskin: I don't think it's a question of mistakes. Mistakes imply that you should have known it already. I think there are normal evolutions, and that if you enter into community without the awareness of those evolutions, you'll evolve yourself out of being a community, if you're not careful. In Israel they talk about the kibbutz and the *moshav*. The *moshav* is like a kibbutz that has evolved away from collectivity and toward the families having more individual control of their money. I think we've become somewhat of a *moshav*.

Thurman: Let's talk about the changes the Farm has been going through. When did that start happening?

Gaskin: I think our peak was about four or five years ago, with about fifteen hundred people on the property; there were some twelve hundred residents and the rest were passing through, here to have a baby, study soy foods, and so on. The last of the Carter administration and the beginning of the Reagan administration were very hard on us. We had a 105-man construction crew, and when the prime interest rate went to 21 percent, nobody would build a doghouse. It just stopped us: imagine 105 jobs gone. We always used to try to rotate the people who had to go off the Farm and work. But we didn't really do that as much as we had intended because people got to be journeymen in their craft. They didn't want to rotate. And they wanted to keep more of the money they made. So we hit a kind of plateau. We were an absolute collective on an honor system, which meant that the community coffers leaked like a sieve. People had checkbooks who had no reason in the world to have them, and they were running funds out that somebody else was working by the hour to earn. We didn't have enough accountability. We didn't have enough skill in accounting, much less a philosophy of it.

Thurman: Were you overpopulated?

Gaskin: In the sense that we had too many people here who were not here for the community, yes. We had dedicated people, guys who had been carpenters for several years, who really wanted to be part of the community. They got burned out because there were day-trippers soaking up the funds.

The question of self-government is something we've worked on longer than almost anything else. I don't know how many different kinds of boards and committees we've tried. And what seemed to happen to them is that they'd all lose their clout in about a year or two. They'd make a few moves, and then people would just quit listening to them.

Eventually, the idea came out that the state requires us, as a nonprofit corporation, to have a board of directors. So we said, OK, let's take *that* board of directors and make them the governors of the Farm, and they will be responsible to the state about being the type of nonprofit corporation we say we are. And that will put people's tails on the line about maintaining the standards.

So we elected a nine-member board. And because of the kind of troubles we were in, the party that got the most seats was made up of business-minded guys who said, "We are sloppy; we have got to be accountable." So we elected essentially a business board — I think seven out of nine. We got smarter about business, and things tightened up real fast, but there were side effects.

Some people had to leave because they couldn't earn as much as was required under the new system.

Thurman: You mentioned a two-party system on the Farm. How does that break down?

Gaskin: The Farm industries and businesses couldn't get outside loans, and they wanted to separate themselves from the community enough to have financial independence, like the *moshav*. So they bought the businesses from the Farm. For instance, the Farm owns only about 25 percent of Farm Foods right now. The employees and investors own the rest of it. So they are making more money than other people on the Farm who are working by the hour at the trailer factory, for example. We're beginning to have two social classes here.

The other party is more the folks who are interested in living on the land, who are committed to the idea of community as an ideal. They are interested in the Farm becoming more of a conference center and maintaining our position by a transmission of our learning.

Thurman: Your population is so much smaller now. Was there a mass exodus at one point?

Gaskin: It happened in waves, and over different issues. It ranged from folks who came here for a formal spiritual community, which we aren't, to folks who thought they were going to get rich and have a swimming pool and live better than this, and decided to go out and do it by themselves. And everything in between. So people left for many different reasons. There are people out there who are my close friends who write me really lovely letters, and then there are folks out there who are mad at me and think I caused them to waste a couple of years of their time.

Thurman: So there were bad feelings?

Gaskin: It was like a divorce. There were people who had been here for years who, other people felt, had contributed nothing and should be on their way. And there were other people who felt they had contributed a lot and so should get a bunch back when they left. People who are on our boards now are confronting all those problems and trying to solve them in the fairest ways we know. The nucleus of people who are here now are pretty sure we want to stay here. If we didn't do this, we'd probably start something new somewhere else, and if that's the case, why not continue with this, which is already going?

Thurman: Did it look as if the Farm might fold at one point?

Gaskin: It was scary for a little while when we were counting noses to find out if there was going to be a skeleton crew left who wanted to keep doing it.

It's interesting that, after we started having television (during Watergate), the kids began seeing TV ads and began caring how they looked, wanting to look the way kids looked on TV. So some people who, for themselves, would go with long hair, bluejeans, and outhouses, for their kids had to have beauty-parlor hairdos and designer jeans — because they felt that they weren't being good to their kids if they didn't give them all of that, although they had given it up themselves.

At first, I took very seriously a lot of the shenanigans that went down here, much more than I needed to. Then I saw that you can't stand in the way of a cultural movement and mass generational changes.

Thurman: Were people overly optimistic that the Farm would continue the same way that it was, as a collective, indefinitely?

Gaskin: Those of us who are seriously committed to community wanted it to continue. But there's no sense in being collective out of principle if it's not working. So what we're doing now is reassessing the things that we do want to be collective about. I don't want to be part of anything that's so institutionalized that you spend more time being mad at the institution than being helped by it.

Thurman: What do you see as the importance of the Farm?

Gaskin: There have been four thousand residents of the Farm over the years, and the vast majority of those people are pacifists, conscientious about what they eat, caring about what's going on in the Third World and other cultures. Very few of those people turned out to be Republican congressmen. Most of them are still pretty idealistic folks.

We really helped the soy industry get off the ground in the

United States. At one time our tempeh shop supplied the spores to all the other shops in the country. We had been living on soybeans for years by the time other people were playing with it to find out if it was cool or not. I feel we have helped out the movement all the way along the line.

There's no such thing as a free lunch, and there are a lot of folks who worked real hard for the folks who had their good free experience here. So now the experience is going to cost a little bit, and we'll try to be as sweet about it as we can.

Thurman: What type of economic system does the Farm have now?

Gaskin: We pay taxes now, and we're more of a cooperative, whereas before we were a collective. The main difference is that all our money used to be handled centrally. Now everybody earns their own money.

Thurman: So you could have some people on the Farm making more money than others.

Gaskin: We already have that. We have some people who have a hard time paying the taxes we have to pay right now, and for other people it's not a burden.

Thurman: How is that going to change the fabric of the Farm?

Gaskin: Hopefully, we'll raise the standard across the board to where those inequities are not so obvious. I'm not going to begrudge my friend Bernie the right to drive a new car just because my car is fourteen years old.

Thurman: Would you say that in the past you weren't choosy enough about whom you allowed to come live here, and that now you're tightening up on that a little more?

Gaskin: No, I couldn't say that, although at one point we had people who were coming here just because they were in love with the idea of us, but did not seriously understand what it took to keep the idea going.

Thurman: They were looking for paradise?

Gaskin: Yeah. And there were other people who knew that they were working harder than what they were getting back for it. It's OK to *choose* to put out more than you're getting back for the purpose of doing a good thing. But to fall into it because other people don't pay attention makes you mad after a while, and some folks got mad.

The mistake that we made — and that, I think, a lot of people make — is estimating our income by taking the number of people in the community and multiplying it by the amount of money that each could make if they all had a job. Because it isn't going to work that way. There's going to be a lot of people who aren't going to make it, or are going to get sick, or something like that. We never made anywhere near the amount of money that we had projected.

Some of the folks who didn't deserve to get hurt by the change of systems got hurt by it. They were working at what we call "public works," like farming, sanitation, teaching, and medical services. Those people didn't have a way to make a living and weren't trying to make a living. They were trying to serve the Farm. A lot of those people had to leave.

Thurman: Now that the Farm is getting more practical, is there a danger of losing your spiritual perspective?

Gaskin: There's always a danger of losing the spiritual. But one great blessing to the whole Farm is that we are host to Plenty International, an international overseas-aid group that has already grown another organization larger than itself in Canada. That helps us to keep things in perspective.

Thurman: You used not to be involved in the local politics here, but lately that has changed. Can you talk about that?

Gaskin: We didn't want to upset the balance of power when we came in here. But now we have two members on the county council. And we seriously helped put our congressman in office by producing a computerized mailing list for him. A week after the election, the local news said that one of the best-kept secrets in the election was Congressman Bart Gordon's computer crew from the Farm. And those guys got an offer to go to D.C. and be on his staff, too!

Thurman: What changed your attitudes about local politics?

Gaskin: When we first came here, all the people who were running things were in their fifties and sixties. We were the kids, and we asked for sanctuary from the older folks. It was very kind and good of them to give it to us. That's one of the reasons we have such a strong love of the local Tennessee people — we felt as if they gave us sanctuary here. And then the older folks died off, and the baby boomers are now in charge. So we got to be part of the action. When we had our first election, it wasn't us against the Tennesseeans; it was us and one wing of the Democrats against the other wing.

Thurman: Personally, how has your life here changed? What kinds of things do you do now?

Gaskin: I get to do more of what I want to do now.

Thurman: What do you do less of?

Gaskin: Stand around and jawbone with people. I love people, I'm a very gregarious dude, and I can rap and rap. But I really like having the chance now to clean up my yard to the point where my father won't be ashamed of it every time he comes to see me. And having the time to pull off some of my projects here at home. This bathroom project is significant. I'm really interested in people taking care of themselves and letting me off the hook so that I don't have to take the blame for everything that happens around here.

Thurman: You've caught a lot of criticism from the feminist movement about the more traditional male and female roles here on the Farm. Has that been justified, to some degree?

Gaskin: I haven't felt confident enough for several years now to have any opinions about what women ought to do. As a wild, free hippie who had all my social mores get blown to hell by a couple of hundred acid trips, and then grew them back from civilization over the next decade or so, I can't defend all the places I've been, other than to say, "Yeah, wasn't that a trip?" But I would like it to be known that, for instance, pornography causes violence against women, and I am disgusted by portrayals of women in subordinate/slave/bondage roles. I have daughters, and I would be revolted if someone wanted to treat them in that fashion. I am also the general manager of Ina May's *Practicing Midwife* magazine, which is a feminist cause.

Thurman: Some of the criticism has been from people

who have come here and seen women in traditional roles, such as child care and cooking, and have felt that the Farm was holding women back.

Gaskin: I think that was an idea that was having its time, and everything was being checked out against that paradigm. But there have always been women involved with management on the Farm. When we suddenly had the question put upon us of how to make a living here, in some of our families the man got a job, and in other families the woman got a job. And whoever stayed at home helped take care of business because it made the whole thing work, whether they were the mommy or the daddy. And I respect all that.

Thurman: Let's talk about Plenty. How did it get started?

Gaskin: Our first operation was to send fifty thousand bushels of grain to Spanish Honduras after a hurricane took out their crops. We also did some local tornado relief in Memphis in 1974. We were already an established aid organization by the time of the earthquake in Guatemala. Plenty was our response to the question of whether or not we were going to be a quietist community. We said, No, we're going to build an arm that reaches out to other folks.

I used to travel on the dollar's favorable exchange rate. I would go to a poor country and live a long time on not much money. Now I feel that's immoral. We wanted to make a way that we could go to those same countries with good karma, meet the people, and come in as helpers instead of as tourists.

Thurman: Plenty is the only thing for which you've ever asked donations, unlike other communities. Why is that?

Gaskin: Because I thought asking money to support the Farm itself was wrong. I was trying to teach that the day of the begging monk is over, that a monk ought to pull his weight if he wants to help the world. A monk these days should have a job.

Thurman: What kinds of things does Plenty do that were not being done?

Gaskin: The concept of foreign aid prior to us was *big*. They did a $200 million railroad project in some African country, and when the guys who built the project went back home, the whole thing collapsed. And they built big, Canadian-style wheat farms in Africa. But you've got to have Canadian-style farmers to run them, too. So now the people who are studying aid groups are saying that grass-roots assistance — building from the ground up, helping the peasantry get stronger — helps the whole country. And our kind of aid is becoming really relevant. In Guatemala we've put in twenty-seven miles of water pipeline. After the earthquake, in concert with other groups, we constructed a prefab-house factory that built twelve hundred houses, as well as rebuilding twelve schools. We've done soy-foods projects, reintroducing seeds and beans that are indigenous to each country. We have two hundred farmers growing soybeans in Guatemala now, and about one thousand families using a soy dairy there.

Thurman: How many countries have you done work in?

Gaskin: We had a crew of four in Bangladesh for a while; we've got people working on a soy dairy in Sri Lanka. In the Caribbean, we've been to St. Lucia, Jamaica, St. Vincent, Dominica, and Antigua. We've been to Haiti, and we want to do an ongoing project there, but it's hard to make it happen because Baby Doc still rules the roost. Plenty is also interested in Grenada, and we'd like to go to Mexico; we're currently negotiating with the Mexican government to see what we can do for them. We're saying, "How can we help?"

We're interested in developing businesses at a small, grass-roots level, because these peasants don't even have a mom-and-pop grocery store. We make a distinction between rank capitalism and the petite bourgeoisie. Small business is a good thing for the world. A lot of these businesses form a thick web that makes for a good civilization. If you haven't got a lot of small-scale capitalism going on, then you've got multinational corporations and slaves. We don't buy slaves anymore in this country — we just rent them.

Thurman: So you're promoting self-sufficiency in these countries?

Gaskin: Exactly. We don't want them to rely on us. And if the people there would like to set up a collective to receive the goods we give them, that's fine, but we don't demand it. We don't care if somebody makes some money on a project we set up, because we figure that any cash flow has got to help a poor country.

Thurman: The more I look at the culture these days, the more it looks like the fifties all over again. Do you feel we're headed for another sixties-type phenomenon within the next decade?

Gaskin: Yeah, we're getting nostalgic for the action that we used to have. We're getting so bland now, and I really pray that we get to see another burst of energy. When the sixties happened, I didn't know there were such things, and it lifted me up and blew my mind and informed my consciousness in a way that was a million times heavier and more interesting than anything I'd experienced before. I think it did that for many people. And now, knowing that such a thing can happen, I can just sit here and wait for it to happen again — like "Yeah, here it comes!"

I believe that it is cyclic, and I think the reason that Reagan is getting such good play right now is that we were a little extreme in the sixties. And now we're being paid back for fucking in the streets. Next time, I'm not interested as much in that as I am in making real, solid social changes that last for decades.

August 1985

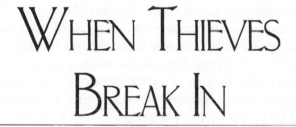

WHEN THIEVES BREAK IN

Stephen T. Butterfield

Last year I was robbed twice.

My house is high up in a river valley, surrounded by hills and trees. On clear fall nights I get a full view of the stars. All the many voices of the mountain wind sing around the dormer windows that I built. It is a little magic cottage, not quite wilderness, but a good place to write, paint, meditate, and live. I pay the bills with my own work, and when I die, maybe a bit of the magic will be there for someone else to enjoy.

Such peaceful, isolated, rural houses, obscured by woods, miles from the nearest police, are sitting ducks for thieves. My neighborhood had ten burglaries in a single summer. Before robbing an area, burglars often take pictures of the houses, watch the residents, and make hang-up calls; they might pose as door-to-door evangelists. They hit most often in the early morning. On one job, they bypassed the locks and cut right through the wall with a chain saw. They take guns, stereos, cameras, VCRs, tools. They will even pull out sump pumps and cut up plumbing for the copper. Sometimes they defecate on rugs and smear their excrement on the wallpaper.

If the mark shows up unexpectedly, what began as a burglary may explode into homicide. The owner of a local marble company interrupted a robbery at his place of business, and the thieves beat him to death with crowbars. In the same town, while robbing the house of a handicapped man, burglars tied him to his wheelchair and smothered him with a pillowcase.

One day in June, I was building a front porch. My domestic partner, Linda, left for work, and I went into town to get supplies — an errand that took me less than forty-five minutes. When I came back, my kitchen door had been kicked open and my house ransacked. The missing goods were easy to carry and resell: stereo components, binoculars, watch, jackets, recording equipment. The burglars must have been watching from the woods, waiting for us to leave; a lookout hiding at the end of the road with a walkie-talkie could have given warning of my return.

Before this, I had felt vaguely sympathetic toward criminals and convicts, influenced by movies to regard them as twisted victims with perhaps heroic dimensions, getting a raw deal from the American gulag system, fighting back against an oppressive society in the only way they knew how, through direct action. These are the kinds of fatuous illusions most easily indulged in by naively sentimental "progressives" who have never been carjacked, mugged, raped, robbed, or beaten, or had a loved one robbed, attacked, or murdered by thugs. The best cure for such ideas is exactly what happened to me.

The state police came and made a report. That was all. House burglary is such a common occurrence that they scarcely bother to investigate unless it involves homicide. What, you don't have triple deadbolts? You don't have serial numbers recorded? You didn't videotape the burglars? Well, a fool and his possessions are soon parted.

But the loss of possessions is not what mattered; I have given away much more than their value. What mattered was being reduced to an object of prey in the one place you imagine is a refuge from attack by predators: your home. Be it ever so humble, home is where forced entry violates a lot more than property rights. Strangers had used the woods on my land to track my daily moves, to spy on my intimate moments, to note when I would be most vulnerable, and to hit when my back was turned. To them, I was less than a rabbit to a fox. They did not care that I could smell their sleazy presence for days afterward in my bedroom, and they would have laughed if they had known I did not even want to sleep in my bed anymore — which I had constructed with my own hands — thinking how they had walked on the bedclothes and torn up the bookcase going after the VCR like pigs rooting under a log.

I had good reason to believe they would come back. They now knew the inside of my house and the rest of its contents. A few nights later, I heard human footsteps in the woods. The next morning, I found things displaced: lumber askew and the shed door standing open.

To keep things in perspective, I reminded myself that robbery and pillage are the rule in history, not the exception. The Vikings lived on plunder for generations. If you were a Saxon farmer on the east coast of Britain in the ninth century, your worst nightmare was the sight of those long, curved-prow ships, lined on either side with round shields, rowing up the estuary toward your village. Plunderers are almost always young: the young more often have the energy for this lifestyle, and the stupidity to find it appealing. Their most common victims are the weak, the disabled, the solitary, and the old.

I began my recovery by telling all my neighbors up and down the road what had happened to me. Most of them I was meeting for the first time, although I have lived here for eight years. Several of them had also been robbed, some here, some elsewhere. Our isolation makes it easy. Before modern times, the whole of an English village would turn out at the cry of "Thief! Thief!" and join in pursuit. Today, when neighbors have lost that willingness to watch out for each other or, worse, when they don't even know each other's names, there is no longer a community, however close together they may live. Such a neighborhood, defended by nothing but a small cadre of overworked police who often have more contempt for victims than for criminals, is a robber's paradise.

To be fair to the police, I should note that they broke two rings of thieves here recently and solved a number of local murders. But, underfunded and underappreciated as they are, they cannot possibly stem the creeping disrespect for law that has permeated and bankrupted this country from the top down since the Nixon era; nor can they ever substitute for the mutual aid of neighbors and families.

Lest anyone suspect from my attitude that I am not spiritually evolved, let me demonstrate my awareness of the options that emanate from the higher planes. I could put a sign on my door saying: PROPERTY IS THEFT. WALK IN. I could sell what has not already been stolen and donate the proceeds to charitable causes. I could recall the precept in *The Cloud of Unknowing* that those who do me harm in this life are my full and special friends. Imitating the good bishop in Victor Hugo's *Les Misérables*, I could advertise in the paper that the guest to whom I "gave" my stereo forgot one of the speakers and is welcome to return for it at any time. I could see the burglary as a message to simplify, simplify: I own too much; I am too attached to my home and my things. Who needs this computer, these guitars, these bathroom fixtures, this bed, this water pump, this income? I could write poems thanking the burglars for enriching my path.

Instead, I bought two cans of mace and a 9mm semiautomatic pistol. That one act was worth all the thousands of hours I have ever spent on spiritual practices and all the spiritual books that take up space on my shelves. Any desire I had to appear highly evolved vanished in an instant. Along with it vanished the lingering guilt, the longing to be thought wise and compassionate, the hope for progress toward enlightenment that haunts all of us who have devoted years of our lives to the Path. At long last I did not give a rat's ass for the Path, or for the opinions of so-called spiritual masters either, many of whom are thieves themselves.

At neighborhood-watch meetings, I learned that most of my neighbors own guns, too. With a few exceptions, they are not hunters; they own guns for self-defense.

"I hang out on my front porch and shoot into the woods," said a Marine Corps veteran. "I shoot randomly at odd times of the day or night. Just to make people think I'm crazy. They say, 'You don't want to mess with that guy, because he shoots at anything. Stay away from him.' "

"We were going out last week," said one woman, "and I had to remember to hide this and lock that, and as I was taking all these precautions, I got furious that I should be forced to go through this. You can't live in peace in your own home."

"And if you try to defend yourself," another said, "you're the one on trial. Criminals have more rights than we do. We've got to change the laws in this country. If someone breaks into your home — I don't care if it's night or day — you should have the right to shoot him, period. No questions asked."

"Even when the police nail them," said another woman, "the courts turn them loose. There's no more room for them in the prisons. And if by some miracle they *do* time, they're back out on the streets in two years, doing the same damn thing."

"It's the recession. There are no jobs."

"But these professional burglars aren't poor. The guy they caught breaking into the condominiums up on the mountain was driving a brand-new BMW and wearing an eight-hundred-dollar suit."

"Is this a gun like Dirty Harry's?" asked the seventy-year-old woman who was watching my house so Linda and I could leave for the day. "I want to shoot a gun like Dirty Harry's. Will you show me how to use this?"

"Sure," I said, taking her out to the backyard. She laid down a tight pattern of holes that would have done credit to Ma Barker. "What do you mean, show you how? Why don't you show me?" I said. We laughed.

"My son is coming by in the afternoon," I said. "He'll be driving a rusty Toyota."

"Yeah, well, he better have some ID or I'll mace him."

I love these people.

The gun felt heavy and solid in my hand. Linda's fury outdid even mine. We blasted cans, boards, paper targets, and burglar silhouettes drawn onto scraps of sheetrock. I enjoyed the sudden recoil, the acrid smell of gunpowder, the burst of gypsum that accompanied each hit.

When practice was done, I tore down the weapon and cleaned it carefully with nitro solvent and gun oil, reflecting on the sobering facts that it could as easily be used on myself as on someone else, that most gunshot wounds inflicted by homeowners are against their own families, and that guns are extremely attractive items to burglars. Nevertheless, I decided to join the National Rifle Association. Taking weapons away from besieged citizens is like depriving outdoor cats of their claws. Hornets have stings, hawks have talons, porcupines have quills. Reality is always armed. The alternative is to run or fly. I can do neither.

Of course, owning a gun is one thing; shooting a live human is quite another. A night burglary on a garage three miles down the road tripped a security alarm, bringing the resident out of his house with a .357 Magnum. He was the owner of a bar and had already been robbed several times. He shot and killed one of the four intruders. They turned out to be local high-school boys looking for alcohol. The boy he killed was on the football team. The school had a big memorial service, and the bar owner was indicted and tried for manslaughter. The jury acquitted him, but the newspaper made him look like a child-killer, and his legal fees pushed him to the brink of ruin.

This is one of the sad ironies of self-defense: the intruder we are most likely to shoot is not the experienced thief, who is skilled at evading risk, but the adolescent kid, whose brains are on loan from a Nintendo cartridge and who regards burglary as a thrill akin to bungee-jumping.

If I waged war to defend my right to live in my cottage unmolested, what would happen to its magic? This question was an afterthought. At the time, there was no doubt in my mind, as I sat working in my study with the gun close at hand: any thug who sent his foot through my door while I was home would meet a hail of lead. I was looking forward to it.

When I was sixteen, a gang of boys bullied me without mercy for months. My friend Bill, who went to another school, was dating a girl claimed by one of the gang members, and they threatened to "waste" us both if he kept seeing her. Finally, they challenged us to a rumble behind an abandoned church. Pushed to the wall, I decided to take my father's advice about standing up to bullies: whether you win or lose, he said, they will leave you alone. I told the gang they were on; they could bring anybody they wanted, and we would do the same. In those days kids did not carry guns, but knowing there were only two of us against all of them, I expected to be killed nevertheless. We spent the hour before the fight on the toilet. Our plan was to run straight at them from the bushes, go for the biggest guys first, hit them hard and fast in the privates, and keep brawling until we got knocked out. I carried my wallet so the police would be able to identify my body.

We danced toward battle like wild angels, clutching home-made brass knuckles and punching shadows. I did not give a damn anymore whether I lived or died. For the first time in my life, I knew what it was like to lose all fear.

We clasped arms in the blood-brother shake: *This is it, buddy.* We touched our fists together and charged, screaming, into an empty lot: no gang, nobody.

On a side street, we found Bill's rival, the gang member who was dating the girl, walking toward her house alone. He said he knew nothing about any rumble and wasn't looking for a fight. Bill punched him in the face and broke his nose. It sounded like a pumpkin hitting cement. The kid dropped to his knees, crying. There was so much blood I thought we had killed him. Bully and victim had suddenly switched roles.

All these threads make up my image of fighting back: the humiliation of being victimized, the restorative joy of fearless rage that sweeps away all hesitation and doubt, the committed bonding with allies; and finally the sullied victory, the needless overkill, the shame of having become what you set out to fight.

How many thousands of years have we enacted this ritual? No part of it can be separated from the rest. And yet one thing stands out: that gang never bothered me again.

It is possible, too, that I would choose to give my possessions to a needy predator, but such giving is infinitely more meaningful if you have the power to say no.

Our neighborhood-watch group made lists of motor vehicles that belonged in the area. When I saw a strange car parked by the side of the road, I'd pull up close behind it in my truck and write down the license number, description, date, time, and place. If someone was inside, I'd lean over and politely introduce myself. "You live around here?" I'd ask. "I'm sorry, I haven't met you. What's your name?" If it turned out to be a resident, I'd laugh and explain my mission. We'd swap burglar stories. We also watched for illegal trash dumpers — contractors from points farther south who preferred to dispose of their waste on Vermont back roads rather than comply with their own state laws. The sense of community, I learned, was alive in this place, after all. I had just never plugged into it before.

Watch work has many satisfying and funny moments. One morning, I unloaded a round metal cover from my truck and rolled it into my woodlot to cap my spring. A young man driving by in a TransAm stopped and followed me on foot. I closed my hand unobtrusively around the mace in my pocket.

"You live around here?" he asked.

"Yup," I said, "right here, to be exact."

"I live a couple houses up the hill with my parents."

"Pleased to meet you." We shook hands.

"When I saw somebody going into the woods, I thought I'd better stop and check it out."

"Good," I said, laughing. "Thank you. Keep doing it."

Once, I came up behind two men in a parked truck videotaping a farm. They looked like hunters, but why would hunters videotape a farm? And why was their license plate smeared with mud? I jotted down a description of their vehicle. Just then, they saw me and roared away. I was happy for a week.

This kind of group self-defense is what heals the loss, the insecurity, and the sense of powerlessness that attend being violated by burglary — not meditation and prayer, forgiveness, compassion, or the quest to conquer hate. Each step brought me new strength. Installing a burglar alarm, I learned more about electronics than I ever wanted to know. To put up a gate across my driveway, I dug down five feet, sank two heavy steel posts, and sealed them in concrete that I mixed myself. Being lung-disabled was just another obstacle to work around, like a big rock. While I was working, a neighbor's German shepherd came over to say hi. She sat in my lap, licked my beard, nosed my pockets, and lay down on my shovel. I couldn't stop laughing. She crooned happily and pawed my chest. That was it: a watchperson should have a watchdog.

I bought a ten-week-old shepherd puppy, my first dog, though I'm past fifty and have always loved cats. We named her Juno. She was adopted to guard my home, but it didn't take long before she upstaged her purpose. She had to be housebroken, fed, cared for, exercised and played with every day, introduced to the neighborhood, trained to come, sit, stay, lie down, and surrender harmful objects, and taught where she could and could not go, and what was OK to chew.

Juno can put on a tough act now that she weighs eighty pounds. Her tail sweeps the ground, her neck bears a thick tan ruff, and her ears stand straight up. At the sound of an approaching stranger, she leaps to her feet and bays like a wolf. I hold her collar to keep her from flying out at the UPS man while he hands my packages gingerly around a crack in the door. He probably doesn't suspect I'm restraining her from smothering him with kisses and beating him to death with her wagging tail.

I can only wonder how many centuries of bonding created this irresistible devotion between human and dog, but it swept me away as if we'd been born for each other. I had not suspected what passionate souls lay behind those wet noses and panting tongues. When she ran toward me at full tilt, her legs could not keep up with her round, bear-cub body, and she tumbled into my arms, wiggling and struggling to wrap her paws around my neck. If I sat in the kitchen, she wanted to be under my chair. If I sat in the living room, she wanted to be next to my feet.

When I left for work, she followed me along the fence and whimpered mournfully, wagging her whole butt and prancing her front paws back and forth, struggling to obey my order to stay. Coming home, I could see just her ears sticking straight up above the glass panel of the breezeway door, cocking this way and that to catch the sounds of my approach. As I came in, she threw herself at me in a madness of joy. I knelt down and she climbed on top of me, crying and licking my face. I was always laughing around her. I thought about her all day. I could hardly wait to get back to her in the afternoon.

Her favorite activity was walking — nothing special, just prancing along happily beside my legs as I plodded slowly through the woods, panting heavily. She made wide circles around me, nose to the ground, looking toward me for direction, listening to my wheezing breath with her head cocked sideways. For a game, she stole one of my gloves and teased me with it by staying just out of my range. When I fell once, she came running back and stood over me, anxiously sniffing my face until I got up. While I rested for a moment to catch my breath and admire the moss on the stones, she sniffed old stumps, dug between roots, ate some deer pellets, and lapped at a stream. Then she sat beside me until I started moving again. On the road, when cars passed, I called and held her close to me. After the first few times, she trotted over to me at the sound of a car and stayed beside my feet until it passed. We moved and rested in the same nonverbal current, like two parts of the same animal, learning each other's lessons and dreaming each other's dreams. We dreamed of furry things that live in holes, huge, cathedral-like pine corridors, tangled brush, forgotten paths, rank logs, and hidden pairs of eyes.

One night, waking from a sound sleep, I knew by the ominous silence that she was sick. And I knew by the pain in my heart that now I had something else to lose. I found her hiding in a corner, head down, as if ashamed that she had vomited. I cleaned up the mess and slept beside her, imagining the worst.

The next morning, I waited for news from the vet like a mother with a child in the hospital. It was nothing much: gastritis, probably from a change in diet. Taking her home, I put my arm around her, telling her how beautiful she was. She got carsick and threw up in my lap.

By winter I could no longer ignore the way she limped when running, stumbled easily, and sat whining and wagging her tail while the dog from across the road ran circles around her. I exercised her by throwing her ball so she could chase it and bring it back. One evening, she dragged her body toward me with the ball in her mouth, swaying sideways and hitching herself along like a cripple, determined to please me even on the point of collapse. I quit the game and brought her inside. She lay down gratefully, licking my hand.

It turned out that she had hip dysplasia, the bane of German shepherds. Early in this century, shepherd breeders got the bright idea to develop a sloping back in their dogs. They thought it gave a fashionable, attractive profile. They got the sloping back, all right, and along with it a gene for misshapen hip sockets that on x-rays look like melted plastic. Juno's hips were the worst my vet had ever seen. At eight months, she already had arthritis spurs that made getting on her feet in the evening a painful chore. They say dogs resemble their humans; it figures that I would end up with a handicapped dog. Linda drew a picture of Juno sitting on a skateboard, wearing dark glasses, with a cup of pencils by her side and a sign saying: GUARD DOG. WILL BARK FOR KIBBLES. PLEASE HELP.

Hip dysplasia is sometimes a cause for putting a dog away, but young dogs can be helped by detaching the thighbone from the hip, rounding the end, and supporting it with leg muscle. This operation will save Juno the use of her hind legs — for which I will pay, without regret, almost half what I lost in the burglary. So I'll end up spending on the dog the worldly goods that she was supposed to protect.

Sitting beside her and stroking her belly, I thought about all the abandoned, starved, tortured beings of the world, the animals in laboratory cages, the monkeys with their eyelids sewn shut because some clown wants to write a thesis on the effects of light deprivation; I thought about the children who get much less to eat than Juno, the workers who cannot afford the luxury of buying operations for their dogs because to do so would deprive their kids; I thought about the lands where humans would slaughter Juno without compunction and hang her up in the market for meat; I thought about my own rage, which had somehow given birth to love, leaving me even more vulnerable to worry and grief; I thought about the magic — it had not disappeared, only taken new forms; and I thought, yet again, about the Path, which I had abandoned, but which by no means had abandoned me.

When I got tired of all that thinking, I opened a can of chicken and fed it to Juno with a spoon. She liked that. It saved her the discomfort of getting up. She thumped her tail on the floor.

April 1993

Notes

Chris Bursk

1.

Fucking bitch. You mess with me
and I kill you.
You mine. At first I crunch up the paper,
the note Jesse was careless enough
to leave around
after class. It sits beside me now,
loosens its fist,
unfolds by itself.
I keep crumpling it up just to see it breathe,
open up again.

2.

Why do I keep coming back to this jail,
shouting, pounding my fist on the desk,
whistling, falling on the floor,
clutching my heart?
Look, if you're not going to pay attention, OK.
But at least be honest,
get up from your seat,
and right in my face give me the fuck-you finger.
Finally, a few students will look up;
even the couple in the corner
will stop fondling each other under the table.
Is this the only reason I'm here,
so Annette can brush against Bobby D.,
rub her leg against his?
This afternoon, when the women
couldn't come to class,
some of the men actually stuck around,
spent the hour talking seriously —
about jailhouse romance,

trying to define it: how one needs a woman
to look for across the cafeteria,
beyond the fence at yard-out, somebody
to smuggle notes to, poems
copied off greeting cards, folded
into small squares, and slipped
into cigarette packs.
It's not the women's fault,
Lorenzo said. *But none of us talk straight*
when they're here, none of us.
We've got too many Cadillacs, T-birds,
custom-made Mercedes
driving around this room.
We're too busy spending the thousands
we've got stashed
somewhere, for just the right woman.

3.

Lorenzo loves robbing clothing stores,
gazing into their mirrors
so he can see himself holding a gun
on all those smartly dressed men,
imagining himself stepping into the glass,
the perfect getaway.
This afternoon I tried to get him to say more
about his wrestling scholarship,
why he gave it up when his father died,
along with his collection of math books,
his beloved calculus.
How come he won't write about
when he was a boy?
But no, he'll show me instead page after page
from a book he wants me to help him finish,
about out-of-body travel,
each journey of his recorded,
each waking cry witnessed by cellmates

and guards,
date and time,
all the unexplained phenomena.

4.
What am I to do with Michael?
He's in detention again.
Today I had to unclench his hand
from his pillow,
those fingers too long, too powerful
for the skinny body that's an embarrassment
to him, seventeen, the prison's baby,
a boy whose soft features won't blunt
for years, probably, and then
much to his relief.
I rubbed his shoulders, lifted his face
so he couldn't pretend to
fall back asleep.
What do you fucking care?
I wanted to get hit.
I wanted some evidence to show the warden,
a bruise the size of a hubcap.
And when Michael pulled down his shorts,
wasn't he punishing me for my naiveté,
my advice? He'd rehearsed in his cell,
as I'd suggested, whispering the words
as if this time he might stay calm
and the guards would have to listen,
admire his logic. *Fuck off.*
I know you're just trying to help,
but fuck off, won't you?
He went on lacing and unlacing
one of his torn sneakers.
It's not my fault. It really isn't.
They shouldn't have done this to me. You
shouldn't have let them. Then he was pressing

into the dark of my shoulder, pounding his fists
against me, this righteous, sobbing anger,
this rage that he sometimes thinks is all he has,
all he can count on.

5.
As usual, Will was waiting for me
after class, and as usual,
he wouldn't say anything
till I coaxed him — it had to be my idea,
not his, to talk. Part of the story
he'd told the cops: Annie had been beaten up
then raped by her old boyfriend,
who'd claimed she had money of his.
She'd begged Will not to come over,
to give her a few days
before they saw each other again.
Will hadn't told the cops he'd been drinking
with friends, one of whom had a gun,
and they'd gone over to the ex-boyfriend's
apartment building,
knocked on the door, and when
it'd opened
Will had pushed his fist right away
into the man's stomach. His friends
had held the man
when he'd crumpled over, and Will
had had to lift his head
with one blow, before
with the other he could locate
the man's belly.
But when the woman began to scream,
when she called out the man's name
and Will realized this was the wrong person,
the wrong apartment,
still he didn't leave,

didn't stop shoving his fist
into the man's gut till his friends
drew him away,
and still he wouldn't be persuaded
from the parking lot;
he had the gun out,
and this was how the cops had found him.
What else could I do? Let the creep
get away with what he'd done? I still had to
find him, didn't I?
What would you do if your wife had been raped?
Would you sit down and draw up
a list of alternatives?

6.
When the class was asked to list their talents,
Sly put down "adjusting."
There's nowhere I can't adjust.

Yeah, Gaelinda had to laugh,
Nowhere, honey,
laughing her fuck-you laugh.

And when Dwayne confessed to how much
he liked being runner
for the infirmary, maybe he'd study
to be a nurse,
his voice suddenly boyish, serious
like a kid's when he's talking about baseball
and playing in the majors someday,
Gaelinda called over to him, *Homeboy,*
I know you too well.
I know you from the streets.
Today she wasn't going to let him be anything
but an inmate.
I'm good at fucking men over —

put that down on your list. Don't matter
where I am,
I do my dance.

When she was leaving, she balled up
the assignment,
made many sharp edges to it,
many flat surfaces buckling on each other.
The paper almost barked.
It was a small pack of dogs in my hand.
You think I got room for this shit in my cell?

Afterward, I stroked the wrinkles out.
She'd signed her name in flowery spiralings
and swirls, and over the *i*
there was a small heart,
and under the column Obstacles
to Developing Your Talents,
in lovely, looping capitals,
she'd written, NONE.

7.
Is this all we are to you — fuckups,
dope fiends, pimps,
a few murderers? Funny stories
to tell at a cocktail party? "Why, just today
I was chatting with my friend
the baby-raper"? Just who
are you going to show these poems to, these
mind games?
The warden? The shrink?

Was this what the rest of his life
would be like?
Stephen demanded to know. In a room
just big enough for a table, a broken

upright piano,
a stack of 1944 Unitarian hymnbooks,
to sob to a lame group like this,
to an even lamer volunteer,
a teacher who'll make a joke out of him later
to impress colleagues, coeds.

8.
My college students are writing in favor
of longer sentences, capital punishment.
It's not fair, they say, turning
from a film on child abuse. It's justice
they want. For the two prim girls
at the shelter,
the baby whose mother held it
over a bridge railing,
the tiny one who had to be freed from
her mother's tight grip.
And justice, too, for the boy
who came home to find his dad
on the back porch
hanging by his neck?
Who tried with a kitchen knife to saw
his father down,
who threw it at the neighbors
when they tried to lead him away? Justice
for this young man,
in jail now for a string of burglaries?
And justice for that old machinist
Mr. Mac, whose cough is a rusty blade
driven so deep into his body
he leans over as if stabbed? Two months ago
he killed a child;
the car leapt the curb
as if someone had wrenched the wheel
from his hands.

He told me he'd been drunk.
Can't you imagine how he must have felt,
this seventy-two-year-old man
who'd never missed a day of work,
facing one-to-five now, mandatory?
No, my students answer;
no, they've never done anything that wrong,
nothing a criminal might do. A man
ought to be made to pay
for what he's done. He ought to be
made to pay.

9.
It's not fair. That's what I hear in jail, too.
My class here wants justice also,
by which they don't mean they're innocent
of all charges against them,
but only that life's been punishing them
for something more
than what they did with a knife or a gun,
what they started to walk out of a store with.

10.
Michael tried to explain his mistake,
as he called it, how once he'd begun,
he had to keep stabbing.
As he talked, the guards came closer,
as if they could tell from the little trembling
of his hand, his head's tilt,
his voice rising to a harsh, thin whisper,
that he was about to start something
only they could stop.
Why did I have to make things difficult
for him — a counselor
who didn't even come in with Jesus,
only paper and pencil?

At least at home Mike could run away,
climb his railroad bridge and sit for hours
with hundreds of feet of air between him
and his stepfather, planning
what to say to keep the bastard's hands off.
What do I want from him?
Why have I given him this old watch?
Do you think I have to worry about time here?
It matters if I'm late?
Lifting its little coolness to his cheek,
its scratched bubble face,
he told me about his first watch,
the one he'd taken from his stepfather's bureau
as the man slept —
the first thing he'd ever stolen —
and he'd climbed the railroad bridge
and not till the top had he slipped it on.
Maybe he'd never go home,
maybe he'd stay high up there, sleeping
with the winds
and the ghosts of the trains. The watch's face
had glowed,
and he'd cupped his hands around it,
squinted till there were tiny sparks
radiating out.

II.
Maybe Stephen was right
when he told me the warden let me
come in here
to teach my course, to tutor,
to put out the inmate newspaper
just so he could look *liberal*, his jail *progressive*.
Don't I dress carefully
so the guards will know to let me out
one iron gate,

then another, then a third?
Each time I leave, I wonder if I haven't gone
into this jail
just so I can walk out of it,
whistling into a cool October, so much
sudden, bright air,
all these open spaces welcoming me,
offering no resistance.

12.
Finally, even Heavy agreed
to read aloud today,
this quiet, 280-pound passer of bad checks,
his bold, graceful handwriting
and the silk bandanna he washes out
each night
his only flourishes allowed here.
Tomorrow he is getting out.
What if he fucked things up again?
What could he find to talk about
with his sons?
He wanted to bring home a toy for each,
just like a normal father
returning from a business trip.
His voice kept falling in midsentence,
and I had to keep encouraging him
to speak up,
and once I saw Bobby D.'s lips moving,
as if someone had to finish
his homeboy's words;
we were all leaning forward — Sly,
Dwayne, Lorenzo, Stephen, Will, old Mr. Mac,
Baby Mike, Eddie, Annette,
Boom-Boom, Mary Ellen, even Gaelinda,
even the Ice Man, Jesse —
as if with each person's turn to read

we were standing at a door
slightly ajar, trying to follow the light
all the way in,
as far back as possible, and then imagining
everything else
that was left in the shadows.

13.
Out of sight of the prison,
I pull off the road, open my notebook,
and am writing,
my face so close to the words
I might be arguing with them, whispering.
Looking up from the page just now
I almost expected to find myself back in jail,
but when I gazed into the rearview mirror
I saw only my face, tinged yellow
and gaunt, the forehead unpardonably stupid.
How could the love I'd felt
just a few minutes ago
not still show? Listening so intently,
hadn't I, too, been transformed?
I'd thought, surely,
I'd have been able to bear this grace
past the first light, the next,
carry it into any traffic.

November 1987

Aliens In The Garden

James Carlos Blake

There were, Julio thought, two clear advantages to working in the fields rather than in the groves. For one thing, you did not have to climb a ladder to pick tomatoes, and thus there was much less risk of breaking your bones. Besides that, a full basket of tomatoes weighed less than half as much as a boxload of oranges, a difference for which your back was grateful at the end of a day in the fields. Also, he suddenly remembered, there was the matter of snakes. In the fields he had seen only one, a watery green whip said to be harmless unless you were a bug or a mouse; but the groves had been crawling with mean little rattle-tailed vipers.

He was working swiftly but with care not to bruise the ripe

fruit. He flipped another couple of tomatoes into his basket, then pushed it a few feet farther along the low row of plants. He paused to wipe sweat from his face, being careful not to get insecticide in his eyes, and considered the import of his reasoning. *So now I think of a ladder as a risk*, he thought; *of a hundred pounds of oranges as a great burden. Sweet Mother of God, I am thinking like an old man.* He felt a rush of anger, followed almost instantly by confusion because he could not say what, exactly, was making him so furious. Of course, simply being here in this field, breaking your back under a broiling sun, was sufficient reason for feeling bitter. But there was something else to this, something that had been creeping up on him for the last few days, something beyond the natural bitterness of a life of hard labor. He could not define it, but he had gotten to know very well what it felt like: Like large, sharp teeth. Like the teeth on the huge black dogs the chiefs brought with them to the fields whenever they feared trouble. Teeth like those, ripping deep in the hollows of his chest.

"Get your ass to work, God damn it!" Gene, the worst-tempered of the crew chiefs, had spotted him gazing dumbly into space. "Ain't no moffuggin' picnic!"

Although Julio did not know much English, *God damn* and *moffugger* he had come to understand quite well. One did not need to know English in order to comprehend Gene's commands. *God damn you, negrito*, he thought as he resumed picking. *God damn all of this.* He wished he were back in the groves, snakes or no snakes. In the groves you had shade — and so what if the air in an orange grove could get so thick and heavy it was hard work just to draw breath? At least you could work standing up, even if it *was* on a goddamn ladder. In the fields you worked in the open sun, sweating like a mule, crawling down the rows on your knees, your back bent and your spine cracking, breathing dust and insecticide fumes. The bug spray burned and stained your skin, and you carried the smell of it everywhere, even after you washed.

To hell with the risk on the ladders, Julio thought, *and with the hundred-pound weight of the orange boxes.* The groves were better than this. A man might have the bad fortune to have to work hard all his life, yes, but not on his knees. No man was meant to work on his knees.

Well . . . a priest, maybe, he thought, and tried to laugh.

HIS FIRST THREE weeks in this country had been spent in an enormous orange grove where he had arrived crammed into the back of a truck with sixteen other men. The truck journey had originated somewhere a few miles north of the Rio Grande and west of Langtry, Texas. They had followed a coyote, a smuggler, across the river in the middle of a moonless night, splashing and falling and choking on their pounding hearts with the fear of being captured by *la migra*, the American border patrol. They had walked and walked, shivering in the desert wind, their clothes soaked and stinking with river mud. Then, just before sunrise, the truck had found them, as planned. They were ordered into the back of it, and a rolling door was yanked down behind them and locked. Two days later, ravenous (they'd had plenty of water in the truck but no food) and reeking of the piss they'd had to sleep in after several of the waste cans had spilled or overflowed, they arrived in Florida (so lovely a name, *florida*: "full of flowers"), a world apart from the arid flatlands and barren mountains of his homeland in Coahuila. When Julio stumbled out of the truck and saw the endless rows of orange trees, he thought he had come to the garden of God.

They were housed in decrepit trailer homes set in a clearing in the grove, were fed from a mobile kitchen — a camper-backed pickup truck that came twice a day with its rations of beans and rice, tasteless white bread, bags of corn chips — and were worked every day from first light until dusk. They got paid every night according to the number of boxes they had filled with fruit, and even after the bosses had deducted from his wages their expenses for housing and feeding him, Julio was left with more money than he had ever earned for a week's work back home. He allowed himself a few dollars every week for beer and candy bars, and for gambling a little in card games, and the rest he kept in a tight little roll held with a rubber band and tucked in his underpants. Every night he fell asleep with his hand cupping the roll protectively.

Every day, as he went about his work — scaling a ladder set against a tree, plucking oranges and dropping them into the bag slung over his shoulder, descending the ladder and emptying the bag into a field box, remounting the ladder — every day he daydreamed of the glorious return he would make to his village in Coahuila. (He had promised his wife he would return in the spring, no later than midsummer — and perhaps he would, if he did not decide to stay here awhile longer and accumulate even more money. A promise to a wife was a sacred thing, yes, but always vulnerable to the unpredictable circumstances of a man's life. And who could predict for certain whether a man would be satisfied with the amount he had saved by midsummer, or would wish to add a little to it?) Whenever he finally did go back home, his roll of money would be as fat and heavy as an avocado. He would be hailed as a rich man, much respected, his family proud and widely envied. Such, at any rate, was the character of his dream.

But then one night, before he had been in the grove a month, the authorities showed up. Trucks and cars with blue lights flashing on their roofs. Lights everywhere — headlights, spotlights, flashlights. Loudspeakers blaring in Spanish: the police, *la migra*, *federales* of all kinds; everybody is under arrest, don't try to run away or you will have worse trouble. Julio would not have run, would have surrendered himself and his dream on the spot, so stricken with fear and despair was he, but everybody else was running, and so, infected with panic, he ran, too. He ran straight to a pickup truck in which a crew chief named Walt was driving away even as several other men clambered into its bed. Julio leaped aboard, and somehow — who could ever say how? — they escaped, five illegals and the gringo chief. All the others, it was assumed, were caught.

By the following afternoon, he was picking cucumbers in a field just outside a town called Immokalee, much farther south. In town, he rented a cot and a locker in a place named the Ross

Hotel, a name suggesting amenities far beyond the reality of that dimly lighted, onetime warehouse of unpainted concrete block. The single large room was simply a dormitory for transient laborers, a place even more malodorous than the truck in which Julio had ridden from Texas. Cockroaches skittered across the floor, and the walls were covered with crude sexual drawings and scrawlings in English and Spanish. "Put bars over the windows of this place," Julio heard someone say, "and it could pass for a well-furnished jail in Sonora."

HE HAD NOW been living in the Ross for nearly two months. Every morning before sunrise, he would walk to the farmers market and board a clattering bus jammed with several dozen men, and they would be driven out to some field or other to pick produce under the supervision of a handful of crew chiefs.

Today, as they had been for the past two weeks, Julio's crew was working in a tomato field. And today, as he had been too often lately, Julio was painfully hung over. In truth, this morning's bad head was one of the worst of his life. He had not even made it back to the Ross the night before. After staggering out of the Green Rose cantina at closing time in the singing company of his friends Francisco and Diego, he had bid them good night and headed toward the Ross. But somehow he'd ended up sleeping in the palmetto thickets alongside the main highway.

He had awakened with the first light of morning jabbing into his eyes, and a head like a rotten watermelon. Lying on his side, damp with dew, he was staring into the pink eyes and long, whiskered snout of a curious opossum standing two feet from his face. The ugly creature scurried into the bushes at Julio's startled groan. Through torturous effort he got to his feet, repulsed by the smell of himself and the rancid taste on his tongue. A few minutes later, he was trudging toward the fields, keeping to the shoulder of the road, when a passing truck stopped for him. The driver was a happy Chicano who spoke bad Spanish but was glad to drive Julio directly to the fields where his crew had already been at work for two hours. At first, the crew chief, Gene, a bad-humored black man with whom Julio had never gotten along, was not going to let him work. He berated Julio for a drunkard who had "another think coming" if he thought he could just show up any old time he wanted and be allowed to go to work. Finally, after cursing at him for nearly five minutes, he relented; the day's order was for fully ripe tomatoes, and Gene needed every picker he could get. "Get your sorry ass into the row over by Big Momma P." Julio snatched up an empty basket and hurried into the field.

The moment he bent to work, he knew the day was going to be a bad one. Although it was not yet midmorning, the sun was already fierce and pressing hard into his back and scalp. Because he had not returned to the Ross before coming to the fields, he had not gotten his cap — bright blue with KC stitched across its front in white letters — and would therefore have to work bareheaded. He did have a red bandanna in his pocket, and though he was sorely tempted to wear it over his head and tied under his chin, he didn't. That was the way the field women wore them. Instead, he simply rolled it and tied it around his head as a sweatband.

Every time a picker filled a basket with tomatoes, he carried it back to the end of the row to have the load inspected by a checker. If the checker approved the load, he gave the picker a ticket bearing a cash value equivalent to wages. At the end of the day, the picker traded his tickets for their total worth in cash, his day's pay. Today each ticket was worth thirty-five cents. Once the checker OK'd the load, he called for a toter — a carrier — to take the basket of tomatoes and deposit it into a truck.

Last week the call had been for green tomatoes, the hardiest kind and hence the easiest to pick and the simplest sort of load to check. The pickers had been able to work fast, and the checkers had hardly glanced at the loads before issuing tickets for them. The toters had dumped the greens into the trucks as casually as baseballs.

All this week, however, the order was for red ripes, so the work was, naturally, going a lot more slowly. The pickers had to be mindful to pick only ripe fruit, and to handle it gently. The checkers had to inspect the loads more closely to ensure that nobody was trying to get by with hiding greens and pinks under a top layer of reds, a practice not unknown among pickers trying to fill baskets as fast — and as easily — as they could. The toters, of course, had to be careful on the trucks. But the crew chiefs remained under pressure from the growers to keep the process moving as rapidly as possible, and they stalked relentlessly from one end of the field to the other, cursing, commanding everybody to work faster, God damn it, work harder. The call for red ripes always put everyone in a meaner temper than usual.

Julio was gagging on insecticide fumes, but he fought down the impulse to be sick and forced himself to concentrate on his work. The bugs had been particularly bad this season, and the field had been sprayed every day. The plant leaves and the fruit gleamed with the oily black poison. Before Julio had filled his first basket of the day, his hands were encrusted with bug spray. Long before the end of the day, his arms would be smeared to the elbows; the seat and thighs of his pants — where he continually wiped his hands — would be stained even darker than they already were.

The air was full of dust as well as the stink of insecticide, and as the smear of crushed, discarded tomatoes grew along the rows, so did the swarms of huge green flies attracted by them. A worker who was not careful when breathing through his mouth could find himself suddenly choking on one.

Julio hurt all over. Every heartbeat thumped against the top of his skull. His eyes felt too large for their sockets. The insecticide seared the open cuts on his hands. Sweat seeped out from under his headband and rolled into his eyes, and when he wiped it away with the back of his hand, his eyelids were left burning. His tongue was an oil-caked rag.

A hundred yards away, set on a crate in the center of the field, was a fifty-gallon drum of water covered with a wooden board. Throughout the day, workers would go to the drum and dip water with a pint-sized can attached to the drum with a wire. No worker ever drank fast enough to escape the crew chief's angry orders to quit goofing off and get back to work.

Despite his present agony, however, Julio would not go for water, not yet. He had to earn that first drink of the day. It was a game he played with himself. The game had but one rule — the only rule out here not imposed on him by the chiefs: he could not have a drink until he had worked his way past at least one of the pickers in another row. He had imposed this rule on himself on his first day in the fields, when he'd noted that many of the other pickers were obvious *borrachones* — habitual drunkards barely able to make it through a full day of picking; ruined men struggling to earn the money for the coming night's bottles. *If you cannot surpass men such as these*, he had thought as he glanced at the wretches around him, *then you should rip off your face in shame*. Most of these drunks were white gringos who knew so little about picking produce it was evident they had descended from a better world; they had fallen to the fields, while most of the black and Latino field hands had been in them all their lives.

Today, one of them was working in the row to the left of Julio: a bloody-eyed, green-toothed drunk in ragged dungarees and a filthy plaid shirt. He knew what Julio was trying to do. Every now and then he would look back at Julio at the same moment that Julio looked over at him, and both of them would try to work a little faster. The gringo was only about eight feet ahead of him, and Julio was slowly gaining ground.

You will pass him soon enough, Julio thought, *so do not think about the water, not yet*. He could not recall a time in his life when he had desired water as desperately as he did right now. The thought of water screamed in his head. When the gringo abruptly broke from his row and went to gulp a canful of water from the drum, Julio gained another three feet on him. Now Julio hated the gringo, hated him for having enjoyed the unthinkable pleasure of a cool drink of water while he, Julio, who struggled by rules of honor, was dying of thirst. His imagination smoldered with wrathful fancies. He could see the gringo reel and drop dead, and the vision thrilled him. He saw himself quickly picking his way past the corpse and then dashing to the water drum.

In the row to Julio's right worked Big Momma Patterson, an enormous black woman wearing a wide straw hat, kneepads, and canvas gloves. She had been ten yards ahead of Julio when he'd started to work, and had since increased the distance to more than a dozen yards. She was the best picker in the field today, one of the few who could do this work wearing gloves. The only picker better than Big Momma Patterson was Sammy Bowlegs, who was at the moment in jail for setting fire to a woman's hair and then dousing the flames with whiskey.

Julio's hands and knees were now beginning to achieve their regular rhythm, and as his picking action got smoother, his pace picked up even more. His hands darted in and out of the vines, grasping the tomatoes against his palm, snapping them free with a twist of the wrist and a crook of the index finger, dropping the fruit gently into the basket, pushing the basket forward, moving up behind it on his knees. Over and over. Every time he filled a basket, he stood up, feeling pain grind into his back and bite into his knees, then hoisted the basket up to his hip and carried it to the checker.

He normally carried the basket on his shoulder, but the first time he filled a basket this morning and tried to swing it up on his shoulder, he lost his balance and just barely managed to drop the basket right side up before his feet tangled in the vines and he fell. The gringo picker laughed and said something that Julio didn't understand but took to be an insult, and he nearly strangled on the restraint required to keep from jumping on the son of a whore and beating him to death. Big Momma Patterson looked back at him and smiled and shook her head reprovingly. Gene came stomping down the row, swearing at him for his clumsiness. Julio gathered the few tomatoes that had spilled, then jerked the basket up onto his hip and lugged it off to get his first ticket of the day, ignoring Gene's ranting with the same dedicated indifference he gave to the flies raging around his head. As soon as the checker gave him his ticket, he grabbed up an empty basket and went back to his row.

While working on his third basket, he paused, leaned low over the vines, and vomited quietly. Then, feeling dizzy and gutted, but certainly better, he wiped his mouth with his shirt sleeve. *Well*, he thought, *that's done*.

He passed the gringo while filling his fifth basket, went to the water barrel, and drank like a drowning man.

THE LUNCH WAGON came down the road, its little bell tinkling brightly, and parked in the shade of a stand of cajuput trees. The owner and driver of the wagon was an American named Harold. He sweat constantly and had blue teeth. The wagon's wide serving window was attended by Harold's chubby teenage daughter, a sullen, dead-eyed girl with a face eaten by acne. The wagon came out to the fields every day. Harold accepted cash or tickets for his goods, the entire range of which consisted of hot pork-sausage sandwiches, cold cheese sandwiches, corn chips, unchilled cans of soda pop, Moon Pies, Marlboro cigarettes, and candy bars.

Several of the crew chiefs' whistles shrilled nearly simultaneously. "Eat!" Gene hollered at the pickers in his crew. The workers swarmed out of the fields to form a couple of ragged lines at the lunch-wagon window.

Julio was not at all hungry. He was not even sure his stomach would accept any food. But he knew he had to try to eat something; he would need the energy this afternoon. He was one of the first to reach the serving window. He fought down a swell of nausea as he waded through the heavy smell of hot pork grease, and he paid five tickets — $1.75 — for a cheese sandwich and a can of Dr. Pepper.

He crossed the road and found a comfortable spot of ground to sit on under the shade of a cajuput. He stared morosely at his lunch: five tickets. He had seven left in his pocket. The dozen baskets he had picked so far today were the fewest he had ever picked in one morning, fewer even than on the morning before yesterday, when he had not come out of his drunken sleep until nearly nine o'clock and had not reached the fields till a half-hour later. If he did not do better this afternoon . . . But of course he would do better. Didn't he always? By the afternoon he had food in his belly and the return of his strength, and the afternoons were longer, so he could pick more baskets.

But there was no denying the fact of his great tiredness. He had never in his life felt as weary as he was feeling right now. He did not understand it. This was not the first time he had worked in the sun while his head did penance for the sins of the night before. It was just the worst time. He wondered if he would be able to work all the way through the afternoon without dropping from exhaustion. And then he realized it was the first time in his life he had ever wondered such a thing.

The cheese sandwich lacked substance and tasted mainly of oily mayonnaise on the verge of turning rancid, but he forced himself to chew it, swallow it, keep it down. He saw Esteban de la Madrid standing in one of the lunch-wagon lines, and his stomach tightened in anger. The seven tickets in his pocket, plus about one dollar in change, represented all the money Julio had in the world, and the reason for that sad fact was an unfortunate incident of the Sunday just past; but the cause of that unfortunate incident — and thus the true cause of Julio's present impoverishment — was Esteban de la Madrid.

STOP FOR HIM, he's a *mejicano*!" Esteban shouted as Diego's rusty, smoke-belching Dodge rumbled past a tall hitchhiker dressed in dungarees. "Give the poor fellow a ride, hombre. Don't be a bastard."

It was late in the afternoon. They were on State Road 82, an isolated two-lane blacktop flanked by cattle pastures and stands of slash pine, returning to Immokalee from a wonderful day in Fort Myers. They had been to a movie and shopping in the stores, and had gone to a Burger King and to several bars. Esteban was in the front seat with Diego, both of them wearing straw hats they had purchased in the Edison Mall. Julio was in the back with Francisco, who was not feeling too well. Francisco's eyes were black and red, and both swollen nearly shut; his lips were puffed and purple, and his nose freshly broken. All of these disfigurements had come to him in an alley behind a Fowler Avenue bar, courtesy of a shrimp-boat worker twice his age who had disputed the legality of a shot by which Francisco had sunk the eight ball to win yet another game at the pool table.

"It is a technique much admired in Ensenada," Francisco said in response to the shrimper's protest. "I will accept another Budweiser as my prize." Although he had learned his English in border-town bars from Tijuana to Juárez, Francisco spoke the language better than Diego, who had been born and raised in

Colorado and had learned both English and Spanish in childhood. Unfortunately, as Diego was often reminded by his friends, he had not learned either language very well, and his pronunciations in either were often mysterious, his syntax a tangle of confusion. It was said, perhaps unkindly, that no matter which language Diego used to call his dog, the poor, perplexed animal would simply stare at him from the distance and scratch itself.

"Well, we ain't *in* no fuckin' Ensenada," the old shrimper said. "Illegal shot. You lose. Make mine a goddamn Michelob."

Cockily, Francisco winked at his friends at the bar and said, "Maybe you wish to resolve this disagreement outside, eh, old fellow? Under the eyes of God?"

"Fuckin' A John square," the shrimper said, heading for the back door.

And now Francisco was not feeling so well.

Diego brought the Dodge to a stop, and the hitchhiker came running. "Listen, you," Diego said to Esteban, "maybe I should get rid of this car, eh? Get a bus, maybe? To pick up *all* the damn people you ask always that I pick up?"

"Your kindness has put a smile on Saint Peter's face," Esteban said. "You are ten feet closer to heaven."

The hitchhiker was breathing heavily when he got to the car. Julio moved over, permitting him to have the seat by the door. "Thank you very much, *compañeros*," the man said as Diego put the car back in motion. The hitchhiker was hatless and his hair was cropped short, like a soldier's. His jeans and jacket still held the smell of new denim. He wore low-cut brown shoes, cheap and dusty but new looking, and a beautiful gold watch whose stretch band was slightly too large for his wrist. He said his name was Eduardo. Esteban introduced him all around.

In answer to Esteban's questions, the man said he lived and worked in Fort Myers, that he was a baker's helper, that he was going to Immokalee to visit a girlfriend. He asked about Francisco's savaged face and laughed along with everyone else — except Francisco, who glared at them all through the red slits of his swollen eyes — when Esteban told the story of the fight. The man had a small, dark tattoo between the thumb and first finger of his left hand, a design something like an arrowhead, and when he noticed Julio's silent interest in it, he casually covered his left hand with his right. His Spanish was much like Francisco's, a border-region singsong. He was, he told them, originally from Mexicali.

He asked a lot of questions about their work: Was the pay for pickers good this time of year? Did they get paid every day, as he had heard? They must be doing very well to have a car and go to Fort Myers for a whole Sunday to enjoy the stores and bars. Did they think he could get work in the fields? He missed his girlfriend and wanted to live closer to her.

"Hey, amigo, any poor *chingado* can work in the fields," Esteban said. "All you need is the strength of a burro and the same amount of intelligence. But to give up a job in a nice, clean bakery to work in the fields, *ay, mamá*, that is too stupid to think of, even for love."

"But you do get paid every day?" the man asked.

"There is that to be said for fieldwork, yes. Good pickers like us always have money in our pockets, while guys like Diego here must wait for Fridays to get their money."

"Where you been, hombre," Diego asked, "you don't know these things?"

The man leaned over a little to look around Esteban and through the windshield, then turned and looked through the rear window. Julio looked, too, his curiosity aroused. There was no other traffic in sight in either direction.

"I've been away," the man said. From under his jacket he drew a small, chrome-plated automatic pistol. "Stop the car," he commanded. "Pull off the road."

"What?" Diego said. He looked in the rearview mirror, and the man waggled the gun up where he could see it. "*Ay, Dios,*" he said tiredly. He slowed the car, eased off onto the shoulder of the road, and shut off the engine.

"Who told you to cut the motor, *pendejo*?" the man asked.

"Nobody, I don't know — nobody," Diego said. "I just do it always because the motor gets too hot and sometimes the radiator cap blows off and you see, you see right there on the front of the hood where there is like a bubble on it, there is where the cap always hits when it blows off the radiator and you should see what —"

"Shut up!" the man ordered. He was holding the gun low, out of the sight of anyone who might drive by, but pointed vaguely in Julio's direction. "I don't want to shoot anybody," he said, "but I have done it before. If any of you should be so damn foolish as to try anything brave, well . . ." He shrugged, staring intently at Julio, who abruptly became aware that he was glowering at the bandit and wondering if a bullet from such a small gun would hurt very much.

"Do *you* understand?" the bandit asked Julio, pointing the gun at his face. The black muzzle suddenly seemed enormous.

Julio nodded. His tongue tasted of copper. Beside him, Francisco had gone pale under his bruises.

"Very good," the bandit said. "Now get all your money out of your pockets and drop it on the front seat, between those two." He indicated Diego and Esteban. "All of you, now."

Julio worked his hand around in his pockets and extracted three dollar bills and some coins and dropped it all into the front seat. Francisco, too, leaned over the back of the front seat, groaning with the pain of the effort, and deposited his money between Diego and Esteban, who were shifting about, digging out their own money.

"For the love of God," Esteban said plaintively, "why are you doing this to us? We are not rich. We are *mejicanos*, the same as you. If you must rob somebody, why don't you go and rob the gringos. They have all the money. That is what *I* would do."

The bandit laughed without humor. "Sure, sure. Of course you would. Pancho Villa — that's who you are. Now, all of you, pull your pockets out completely." He leaned forward and looked down into the front seat, checking Diego's and Esteban's pockets. He smiled and nodded. Francisco's pockets, too, now hung limply from his pants like white flags of defeat.

Julio had not turned his pockets out. The bandit stared at him and raised his eyebrows.

"That's all of it," Julio said. "Truly."

"Oh, *truly*," the bandit said, smiling. "Well, please forgive my lack of trust, amigo, but I insist."

Julio stared hard at the bandit, flicked a glance at the pistol, locked eyes with the man again. If he had been asked at that moment what was going through his mind, he could not have said, but the bandit saw something in his eyes that made him lose his smile and draw back the hammer on the little gun. Julio had never heard that sound except in the movies, had never imagined how differently it could strike the ears in the world of flesh and breath. Even on this little pistol, pointed at him from a distance of less than two feet, it was a sound to chill his skin.

"Hey, amigo," the bandit said softly, almost sadly.

A van with dark-tinted windows went whooshing past.

Julio pulled his pockets inside out, and the rest of his money fell to the floor. Seventy-nine dollars. The bandit chuckled delightedly and began scooping up the money with his free hand. "*Truly*," he said, looking at Julio and laughing. "*Truly*." Ten of those dollars had been won the night before at the cockfights; the rest represented what was left of the money he had saved since arriving in Florida. Julio watched the bills disappear into the bandit's jacket.

"Jesus Christ, Julio," Francisco said, his slitted eyes widening the little they could. "You have been robbing banks."

"That was all the money I had," Esteban whined. "Twelve little dollars." He gave Julio a look of reproof. "I was not so rich like some in this car. Listen, hombre, leave us some little bit of money, eh? To buy a beer and a taco tonight?"

"Is this one always so stupid?" the bandit asked nobody in particular, nodding toward Esteban as he gathered money.

"Always," Diego said. "But listen, can you not leave us with *some* money? I have a wife, hombre. I have children, I —"

"Silence!" the bandit commanded, waving the pistol at him sharply before aiming it back at Julio. Diego put up his shaking hands like a movie robbery victim, nodding repeatedly.

"Put those hands *down*, stupid! Sweet Jesus! You fools think you are the only people with troubles? If I told you *my* troubles, we would all drown when your tears filled this car. And speaking of this car, I want you all to get out of it, right now."

"My car?" Diego said. "You are not going to steal my car?" He sounded on the brink of weeping. He had recently paid ninety-six dollars for this car, having saved the money from overtime pay during the past few months at his job as an auto mechanic. Accumulating such an amount had not been easy; almost every penny he earned went toward the support of a formidable wife and seven ravenous children.

"*Steal* your car?" the bandit said. "Are you trying to insult me, *pendejo*? I never steal such trash as this."

Diego appeared both relieved and injured.

"When I steal a car," the bandit said, "I steal something my sainted mother would not die of shame to see me driving. No, hombre, I am not going to *steal* your stinking car. But I *am* going to borrow the flea-ridden thing," the bandit said, "and only because it is now almost dark enough so that no one will recognize me as I drive this rolling disease. Leave the keys where they are and get out — everybody, now. Start walking back in the direction we were coming from. Move!"

They got out and began walking. When they had gone about ten yards, they heard the motor grinding as the bandit tried to start the Dodge. They could hear him swearing at the recalcitrant engine. Finally, the motor clattered to life, and they stopped and turned to look. The transmission gears shrieked. The car lurched onto the highway and began a ponderous acceleration, pouring thick clouds of smoke. Its taillights became dimmer through the trailing haze as it faded down the road.

It was a long walk back to town. None of them mentioned the incident. They hardly spoke at all. Except for Diego, the Chicano, the native citizen of the U.S., they all ducked down in roadside ditches or ran and hid in the pines every time a pair of headlights came flashing down the highway. One never knew when those headlights belonged to *la migra*.

Diego walked with his arm out and his thumb up. But there was hardly any traffic on this Sunday evening, and none of it stopped for him, so he had to walk all the way back with the others — a distance of more than twenty miles. "Why couldn't the son of a bitch have waited until we were closer to town before he robbed us? *Chingado!*" He was enraged by every car or truck that came out of the darkness and flashed by without even slowing down.

"God damn them! Bastards! Sons of bitches! No-good sons of whores! Do they think I am going to rob them, the lousy, no-good bastards? *God damn them!*"

Every time he caught sight of Esteban, he swore at him, too, only more loudly and with greater heat and imagination. Esteban was careful to keep a good distance from Diego. He was careful not to get too close to any of them, for it did not require superior intuition to sense that nobody was feeling very cordial toward him at this time. Much later, he would argue that he was being made a victim of unfairness, but not now. For one of the few times in his life, he was exercising prudence and keeping silent, even in the face of Francisco's accusation: "It's all your fault, you stupid son of a bitch! We ought to hang you from one of these trees." Which were the only words Francisco spoke to him on the entire walk back to town.

The sky was turning gray when they reached town. Esteban quickly faded into the shadows of a side street and vanished. When they turned a corner at Third Avenue, they saw the Dodge parked in front of the public library. Diego whooped with joy and jogged to the car. By the time Julio and Francisco got there, Diego was no longer looking pleased. He was staring in open-mouthed shock at the freshly crumpled left front fender. "Look," he said, pointing a quivering finger at the damage. "Just look at what he did. The bastard stole my car and could not drive it twenty miles without wrecking it. That *hijo de la chingada* should be in prison! Look! Look how he has ruined my car!"

Even as tired and dispirited as he was, Julio could not help thinking that now the left fender more closely resembled the one on the right, but he did not think this a proper moment to mention it to Diego.

Then Diego discovered the parking ticket under the windshield wiper. He howled and snatched it off the car, which, he now saw, was positioned in front of a fire hydrant.

Diego shook like a man suffering from a severe nervous disorder. He stamped his feet. He turned in small circles and whimpered like an overwrought dog. He planted his feet far apart and shook the ticket at the sky, as though demanding an explanation from God himself.

"What kind of son of a bitch —" he began, then started choking on his own spit and bile and outrage, and fell into a harsh fit of coughing for a minute or two before he could regain his breath. Julio, meanwhile, was observing this whole display with a degree of awe and apprehension. He had heard of persons who had died in the grip of such fits of rage. A blood vessel had burst in their heads, or something like that.

"What sort of son of a bitch," Diego resumed, once again brandishing the ticket toward heaven, "would give a man a parking ticket in the middle of the fucking *night*? Oooooh, God! What bastards! What injustice! *Where are you, God?* What *injustice* this stinking world is full of!"

During this time, the town had slowly been coming to life all around them. Several cars and pickups rolled past, their occupants staring at the men with curiosity.

"Hey, Diego, you maybe better go home now," Julio said, "before another policeman comes by and gives you another ticket."

Diego stared at him as though he had grown a second head.

"I mean, your car is still parked there, amigo."

Diego whirled and got into the car and slammed the driver's door shut with such force that the window shattered and showered him with glass. He pounded on the steering wheel with his fist, and his curses rang through the streets. The engine ground and sputtered, and he swore at it until at last it caught and roared and poured black smoke from its exhaust pipes. He raced the engine up so high it seemed about to explode, then yanked the car into gear. The tires screamed as the Dodge shot into the street and veered wildly right and left before at last slowing down and straightening out.

Diego drove away in the rattling Dodge, trailing billows of oily smoke, heading home to a wife and seven children who would shriek and squall the whole while he got ready to go to work, sleepless and empty-pocketed.

"At least the thief left his keys in the car," Francisco observed as he and Julio made their way to the farmers market to wait for the field buses.

"That's right," Julio said. "Diego actually has much to be thankful for, doesn't he?"

Francisco started to laugh, then winced and moaned with the pain the act caused to his battered face.

ESTEBAN HAD BEEN to blame for the robbery because he had been the one to make Diego stop for the hitchhiker. It was as simple as that. Diego had since made it quite clear that he did not want to see Esteban again for the next eighty years. "If I do," he had told Julio, "I am going to drive over him with my car until the tires go flat." Francisco — whose face looked even worse a couple of days later than it had on Sunday — would tremble with fury at the mention of Esteban's name. "He will have no need of pockets for a while," he said, "because every day I am going to take from him every nickel he earns, until I have regained the nine dollars I lost." Julio had not spoken to Esteban since the robbery, either. Esteban de la Madrid had been responsible for the whole thing. It was as simple as that.

Except that now, chewing the last of his awful sandwich and drinking the final swallows of the warm Dr. Pepper, Julio knew it was not at all as simple as that. It was no simpler than giving a name to the vague but implacable anger he had been feeling in the past few days. An anger drawing ever closer to dread — a dread like teeth gnawing around his heart.

He had been brooding about the robbery since it had occurred, and he knew his friends thought his depression was due to having lost so much money to the bandit. That was, of course, part of it. He had, after all, been robbed of his entire savings. But the real reason for his continuing despair over the robbery went much deeper than the loss of his money. The terrible thing about it all, he kept telling himself again and again, was that he had done nothing to try to stop it from happening.

Why hadn't he tried to take the gun away from the bandit? Why? The fellow had been laughing and enjoying himself — especially after Julio had finally emptied his pockets. The bandit had become loose with his attention, scooping up money from the floor, talking to Diego about the damn car. His guard had been very loose. He had been no more than arm's length from him. *I could have grabbed the gun*, Julio thought. *I could have grabbed it and tried to wrestle it away.* The bandit had not appeared to be very strong.

So why hadn't he tried it, eh? Had he perhaps been afraid? Was that it?

Well, of course he had been afraid. Certainly he had. The fellow had held a gun, God damn it. Show him the man who was not afraid of a gun, and he would show you a fool.

And why would such a man be a fool?

Why? The question was beyond foolish. Because a gun can kill you, that's why. Kill you fast and easy.

Julio felt the teeth tearing, ripping through him.

Kill me? Kill *me?* Sweet Mother of God, was that what he had been afraid of? Of being killed? Of being killed fast and easy?

It made him laugh to think about it, and the laughter burned in his eyes.

October 1987

EATING HEAD

Lorenzo W. Milam

Jesús's favorite ice cream is pistachio, but sometimes he'll order the *nuez*, vanilla swirls with bits of nut. He orders for us, and then we sit and eat and watch the people go by on the street. We watch the pale Americans, the drunk Americans, the loud Americans, the unlovely Americans. Intermixed with them are the mahogany Mixtecs and Zapotecs — boys with buckets full of fish, women carrying hazel-eyed babies, children running and laughing, old men dressed in traditional white shirts and huaraches.

"Gringo-watching," I call it. I've been living in Mexico on and off for twenty years, and slowly I'm developing this prejudice, this terrible prejudice, against Americans. *They're so pale and wan — in such a hurry*, I think, trying to forget I'm one of them. Jesús, who brings me down here every evening, isn't prejudiced. He watches the young ladies, watches them all. Those from north of the border are as interesting to him as those from his own hometown. He wants them all.

This morning we had our first rain in almost six months. This is a drought area; the hills have turned sere. I often wonder that the whole region doesn't come down with some great fever from the sky, a grand conflagration that will burn up the fields, take the city of Puerto Ángel, and the tourists, up in one smoky mass.

"What's it like being here, in Mexico, in a wheelchair?" The woman asking this question is thirty, maybe thirty-five years old. Her chin is sharp, her face pale, eyes hidden behind large dark glasses. I am watching the balloon man across the plaza. He has balloons in all sizes and shapes, floating in a bunch above his head. I favor the *monstro*, the monster head, a funny face with tiny squiggly balloons sprouting over its top. I wonder how he makes it look so . . . well, human.

"I have a brother in a wheelchair," she says. "He's a quad. That's why I'm asking." Throaty voice, no accent. She must be from the West.

A brother who's a quad. They always say that. Or: "My dad has diabetes; they took off his leg right here" (slicing motion across the thigh). I hear that a lot. "My uncle's in a wheelchair: Vietnam." The nonspecific description. Or the very specific: "My mom's a stroke victim and her whole right side is useless — she uses a cane and we have to feed her." All these cripples. A world full of cripples. And I get to hear about all of them.

Absolute strangers come up to me to tell me about people I don't know, will never know, can never care about. *I have enough troubles without this*, I think. *They are just trying to reach out, the only way they know how*, I tell myself. "You're not alone," they are saying, sure that I feel alone. "I know someone who's just as badly off as you," they're saying. "Don't you feel better now?"

Maybe I am softening. I no longer pretend I'm a deaf-mute when they start in. I don't — as I did, the last time I was here — pretend that I speak no English. "Lo siento — no entiendo los gabachos," I said. Sorry, I don't understand Americans. There were times, I have to admit, when I just looked at them and said, "Oh?" or, nodding my head, "Mmm."

Thirty-five years have probably softened me, made me more willing to hear: people hurting, everywhere people hurting. This woman hurting, with whatever it is that our families feel: angst, maybe; fear, possibly; guilt, certainly. "Anytime anyone talks to you, it is God talking to you, telling you something you need to hear right now," Stephen Gaskin used to say. Sometimes I wonder if the words are a little garbled, the messenger a little befuddled. What to respond? I know in 1952 or 1966 or 1974 you couldn't get a whisper out of me on this "my brother's a quad" business. Now it's different.

"It's the same," I tell her. "No — that's not exactly right. It's complicated and it can be frustrating. But there are compensations." Like my helpers, Jesús and Diego. They've been working for me for three winters now. They take me about Puerto Ángel, and to the beach to swim. They get me out of the car and down to the water and then back out, rinsed off, in the car, home again. They buy my food at the public market (inaccessible), help me cook it (I'm learning to cook black beans and rice), even help me eat it.

"I have two employees here," I tell her. "They help me a great deal." Two employees. What does that mean? Two young people who tell me about their lives, and I tell them about mine. Jesús with his careful walk, his dark, polished skin, his serene face, his full Mayan nose, the high cheekbones. Diego with his great, slow, wise eyes, the rounded features right out of the Yucatán.

"They're fun to be with," I add. They are filled with that strange mixture of Mexican manhood and sentiment that fascinates visitors. Sometimes they get drunk with me. "They might be the best part of my stay here," I say. "They're very good workers. They're with me sixty, seventy hours a week. I pay them $140 a month. That's more than the minimum wage here." Diego, who makes me laugh when I'm feeling blue; Jesús, who jollies me out of my bad moods. "Qué pasa, gabacho?" What's going on? they ask me, at those times when I don't feel like leaving the house. Together, they get me out the door. I don't protest, not much. At the ocean's edge, I can't be thinking those thoughts anymore, can I?

"They teach me Spanish," I tell her. "I'd have to pay Berlitz a fortune for the lessons they give me." Then I tell her about some of the frustrations. Mexico isn't set up for wheelchairs. Not many ramps, almost no curb cuts. There are very few hotels that have rooms for the disabled. I had to scout out the five restaurants (of twenty) in Puerto Ángel without steep stairs. I have yet to get over my unwillingness to be carried in and out of restaurants, bathrooms, banks.

"It's a concession to being in a Third World country," I tell her. "With my two helpers, it doesn't have to be a big deal." They like showing off their strength, and they can be very funny about it. "Pinche gringo," Jesús will say. "Pesa cuatro cientos kilos." That means I weigh 850 pounds. It is his way of telling me he likes me.

"There's no special parking, or special placards," I tell her. "But when I park my car on the beach here, near the water, the police never bother me." In California, or Florida, or Texas, the police would apologize while giving me a $250 ticket.

"Sometimes I feel like a mountain climber here," I say. Quite suddenly, the sky turns orange-green, then a brilliant green-black. The usual spectacular tropical sunset that happens so quickly, even when you're expecting it. The sudden end of the searing heat and too-bright sunlight.

"How'd he get to be a quad?" I ask her. The street turns dim and dusty. I can no longer see her face, only the outline of her head. She still has on her dark glasses. She tells me about the drinking, the drugs, the general helling around when he was twenty. The accident.

"They didn't know whether he was going to live," she says. She pauses. "I think he's a much better person for it."

I look at the balloon man. He's standing under the one street light of Puerto Ángel. The light has just come on, so the balloons cast a dark cloud over him. He has so many of them straining on their strings, and he is such a tiny fellow, I half expect to see him take off, float away into the starry sky, rise dispassionately into the stratosphere. Our balloon man which art in heaven.

"A much better person," I say, nodding. Then, "That's an awful thing to say."

"I'm sorry?" She's not apologizing — she just doesn't understand.

"I don't like to hear that," I say. She turns her head to watch the balloon man. He creates his monsters right there out of rubber and helium. He fills the balloons from the great, bruised silver tank at his side, twists them into the shape of animal faces, or strange bodies, turning them this way and that. When he fills the balloons with the helium, it makes a raw, screeching sound. She sits down on the curb next to me.

"I don't understand," she says, "but maybe I never will." She tells me she is a nurse in Colorado. Even with her training, it seems the hardest job was learning to deal with her brother's body. "I've done caths — lots of times. And yet, nothing is harder for me," she says. She stops. "He was the one who used to pick on me all the time, never had any time for me, called me 'stupid.' And now there he is, lying flat on his bed. Sometimes I have to give him his catheter in the morning. And he can't even cover himself up."

A little girl, not more than five years old, is trying to push two boys, probably her brothers, up the hill in a scratched red wagon with warped wheels. One of the boys, the older one, is yelling, "Recio, recio!" Faster, faster! The other one is just sitting there, at the front of the wagon, without a stitch on, digging the hell out of his ride. The girl can barely get it to move, what with the hill and all; she's pushing with all her might.

"Maybe he's better for it," I repeat. "But I don't think it's up to us to say that sort of thing." What is it they say? *Only the gods can worship God.* Only the gods can pass judgment on those who are now so different.

Jesús is behind me now. He pushes back and forth on the wheelchair, slowly, rhythmically. There's a fiesta in Tonameca. He wants to get over there to check on a young lady he claims will be his next *novia*. He wants to get me over there because he knows that by ten, when things are starting to happen, I will be drunk and will insist on going home. He doesn't want to miss a thing. I've been to the Tonameca fiesta before. The ground is sandy; people will be pushing in on my wheelchair from all sides. We'll get stuck; people will stare. From two feet away, they'll stop dead and stare. They always do. They've never seen a six-foot gringo in a wheelchair.

I wouldn't even go if it weren't for the fact that, when I'm there, I'm living. *Living,* for Christ's sake. Going out in the world, seeing different people, a different world. The woman wants to know if her brother in his wheelchair would be happy here. *Sure,* I want to tell her. *It'll be all right if he doesn't mind getting stuck, having people staring gaga at him. It's all right, if you don't mind being a freak.* I want to tell her that. But that sounds bitter — and it would be a lie. There aren't many of us around, and we're the object of curiosity; but it's a curiosity coming from some of the kindest, most open people in the world. I think of my times here, the despair that occasionally comes from the inaccessible buildings, places I can't even dream of getting up to, or down from; the times that my bladder is bursting and there's no place to go, and I want to scream, or cry. And Jesús and Diego, with their great wise serious eyes, understanding somehow, always figuring out what is going on, figuring a way for me to make it.

"It's a different world down here," I tell her. "Your brother might like it, or, then again, he might despise it. You can never tell. It depends on him, and how much he likes Mexico, and Mexicans." I think about what happened this morning. Jesús started in to tickling and pinching me, and then ducking away when I tried to grab him. He thought I was turning a bit sour, and he wanted to be sure I got my daily quota of tickles. I wonder if the attendants in the United States — what do they call them? "personal-care attendants"? — are allowed to tickle their charges. Is that written in an attendant's job description? A good, thorough tickling when the patient starts to feel bad?

"To me, they're all gods, so I forgive them everything," I say. I think of Diego, his great round face, that monumental face out of the tombs of Quintana Roo — the lids so heavy, the lips so broad, so compassionate, the eyes so inexpressively expressive.

"What do you mean by that?" she asks, moving her hands vaguely.

I wonder if I can get it across. "To be here, to really enter the country, you have to be willing to leave certain things behind. And not for a week or a month — but for a long time," I tell her. "They have something special here.

"Mexicans learn to love so quickly and easily," I say. "Some of us — if we are open, or lucky — learn to love them in return. We gringos are so much slower, so much more fearful of love. For us it's like taking hostages. Especially for those of us . . ." I want to say, "for those of us like your brother and me," but I don't finish the sentence.

Over where the sun has died on the horizon, there is a bare smudge of rouge, a burning off, so far off. She's standing up, brushing her skirt. "It's so hard for us to learn about it," I say. I want to tell her more, but she turns, waves goodbye, and is gone. I think of that line from Samuel Beckett: "The trouble with her was that she had never really been born."

AN HOUR LATER, Jesús and I are smack-dab in the middle of the Tonameca fair. I get stuck in the sand twice, and at least three hundred people give me a good going-over, not looking, but staring. Jesús has talked me into eating several tacos *de cabeza*. Brain tacos. "It's all right if you don't mind eating all those thoughts," I tell my friends later. "It's supposed to make you smarter."

"Quién fue esa mujer?" Jesús asks me. Who was that woman? He thinks I should make her my *novia* and marry her. He thinks I should marry every woman I meet. He wants me to settle down. He starts to look for his girlfriend in the crowd. I get a glimpse of her. She has long black hair tied in a single thick strand, interwoven with a single thread of bright red wool. Her eyes are as deep and as mysterious as his own, her skin the color of butter chocolate.

"No sé," I say. I don't know. "Otra gringa." Another gringa, I tell him. "Poquita confusada." A little confused. "Como todo." Just like the rest of us.

August 1990

All The Panamas
In The World
And Herb's

T.L. Toma

I'm sitting with Herb in my '77 Datsun pickup. Herb has a top-of-the-line Lincoln. Dark leather and digital controls. A heater like something out of Los Alamos.

The heater in my Datsun is busted. Outside it's a blowing wind. I'm thinking snow.

"We should have taken your car," I say for the fifth time.

Herb doesn't respond. He's hunkered down in the passenger seat, knees brushing up against his chin, peering out the windshield at the door of a small white house with a green roof.

Herb's wife, Carol, is inside. This, at least, is the way Herb's got it figured.

I've known Herb for ten years. We work out of the same

plant. Herb's short, but there's a meatiness to him. He has big hands. Knots of veins in his neck. You come away thinking he's four inches taller than he really is.

Herb's got this way of bringing his right hand, his ball hand, up and over fast, hooking it sharp till his thumb grazes his left ear. His ball always skids a good twenty feet down the lane. By the time it hits the pins, it's like a fist powering right into someone's face. Herb can pick up a seven-ten split heading into the tenth frame like nobody I've ever seen.

If it wasn't for the way Herb looks standing the five and a half steps from the line, toeing the boards and cocking his arm just before he rockets the ball down the alley, I'd say, Herb, give it a rest. Back off already. I'd say, Let's scoot.

Outside, it's going dark fast.

I'm thinking I want to get home to Angie. It's Friday. She'll be making her fish stew.

Angie's not Catholic. Went to church maybe five times in her life, and then it was always Lutheran or Presbyterian. Not a religion you remember for long. It's one of the things I like about her. Me, I grew up going to church summer camp and playing in church softball leagues. Vacation Bible School. Choir. We're talking God, economy size.

Angie has a thing about adopting other traditions. Says she's without enough of her own. Get a sniffle and she's there with a bowl of matzo-ball soup. On Bastille Day she sets off firecrackers.

I'm sitting here in my '77 Datsun with the busted heater, freezing my vitals off, while Herb chews on whether his wife is right at this moment inside, her mouth open and her eyes closed. Her toes maybe aimed at some mountain peak in Asia.

I turn on the radio. An orange glow fills the dash. I flip the dial, but there's news everywhere.

"Must be on the hour," I say into the steering wheel.

Herb isn't talking. He looks out the window and pulls at his lower lip. On the radio there's talk of a U.S. fleet steaming to somewhere or other. Libya, maybe, or Lebanon. It's hard to say because of the artillery punching up against the announcer's voice.

"Turn it off," Herb says. "Will you do that for me please? Just turn the damn thing off."

Herb's worried. He's worried some other guy is in the white house with the green roof, probing Carol's fifteen hundred in dental work with his tongue.

"I can't see a goddamn thing," he says. "Don't you ever wash your windshield? Don't you know how the hell to take care of a car?"

"It's dark out," I tell him. "It's going to snow."

"No wonder your heater doesn't work," he goes on.

"Herb, she's probably at home."

Fish stew puts Angie in the mood, is the thing. We have this big down comforter from Denmark.

HERB FIRST MET Carol at one of those money machines. This was before the machines started showing up three to a block. There were still some kinks in the system.

Carol was trying to pry the twenty from between the metal lips when Herb offered to help.

Herb left Andrew Jackson's right ear and most of the head in the machine.

Herb explained all this to me a long time ago.

About a year ago Carol started to put on the weight. At least that's the way Herb saw it. He'd toe up to the line in the fifth frame. He'd turn around and say, "She's this minute stuffing her face." Then he'd lay the ball on the pine boards and wait for the explosion of white pins.

Me, I never noticed Carol's weight. When I thought of Carol, I always thought of her hair. It's wild, curls on top of curls that seem ready to lift free and clear of her head.

"She's still a hell of a looker," I told Herb.

I was trying to help out here.

Herb wouldn't let up. He started to ride Carol about her weight. He began making cracks in front of other people. In front of Angie and me.

"Him with that gut and all," Angie would say later.

I tried defending Herb a couple of times. I tried hard to

make out like maybe Carol was putting on a few pounds. I had these images that floated in front of my eyes. I gave up in the end.

It was in Herb's head, is what I'm getting at.

CAROL'S AT HOME, I bet," I tell Herb again.

Before he can respond, a little rectangle of light splits wide and somebody steps from the door of the white house with the green roof. It's a woman. I can see that right off. A silhouette with wild hair. Curls up to here.

The rest is blue and gray shadows.

Herb lets out a big puff of air. "OK," he says. "All right." His voice sounds small and far away. Even in the gray light I can see him working the fingers of his right hand.

I've got to be careful here, I tell myself.

The woman moves past the car. She heads up the street. She's wrapped in a thick winter coat. Still, she looks thin.

I don't mention this to Herb.

I'm hoping Herb'll want to blow it off, go home and pour himself three fingers and wake up later like nothing ever happened.

"OK," Herb says again.

"Herb," I say. I shove my hands under my armpits. "Herb."

"You saw," he says. He's looking out the windshield. He isn't looking at the woman, who's halfway down the street, or at the white house with the green roof, or at anything at all. He's just staring into the windshield, like he can't see past it.

"I saw a shadow," I say. "I saw a damn shadow is all."

"This car is filthy," he tells me. He is staring at the windshield. "Disgusting filthy."

"Herb," I say. "We should go. Let's go on now."

He doesn't say anything then, so I crank up the Datsun and put it in gear. The snow has started. It falls in large, lazy flakes.

HERB GOT WORSE about the diet thing. He nagged at Carol. He really chewed on about it. One night, Angie and I were over there late. We were all sitting around, having a few drinks. Watching TV. Nothing special, just nice, like it can sometimes be. We watched a movie about a cheerleader and a football player. This football player was worried that he was gay. This is what the movie was about, I swear. He decided he was going to crawl on top of this cheerleader and prove to the world and himself that he was one straight arrow. This cheerleader was blond and long-legged. Ran around with these big yellow pompoms in her fists. Turned out it was the cheerleader who was gay. This made the football player feel better, and everything ended on a happy note.

The news came on, and we all settled back to watch U.S. troops running across these ditches. They crawled out of one ditch and ran for a while. Then they all tumbled into another ditch. They did this three, four times.

It was Colombia, or Calcutta. Could've been Columbus. I don't remember.

Then Herb got out of his seat and changed the channel. Said it was time for "Carol's show."

We all watched. They had this woman with a microphone in her hand, roaming the audience, like on those morning talk shows. All they talked about was fat, and how to get rid of it. The woman with the microphone moved from row to row, stopping to talk to the women sitting there. It was almost all women. There were only a few men that I could see.

She talked to the women, and each one explained how she'd lost twenty, thirty, forty pounds. In five or six weeks. One woman had this bloated look to her face. All puffy. But the rest of her was all right. The rest of her looked thin. She had long legs.

This woman with the puffy face talked about how she'd lost eighty pounds. Eighty pounds in ten weeks. The audience went wild over this. They oohed. They hooted. They slapped their thighs. I've got to admit, I sat up for a closer look. Eighty pounds is nothing to sneer at. I thought about what it'd be like to bowl with an extra eighty pounds hanging off me. I tried to think about what it'd be like in bed, with Angie, if I had that eighty pounds.

Then I looked over at Carol. I looked over at her and tried to decide if she was fat. Maybe just a little spread in the tummy was all. Not enough to get worked up about. Not like the women on TV.

Everyone on TV applauded. Then there was a commercial, only they didn't break to a commercial, like you'd expect. The commercial was right there in the show, in between the different women who stood up to say they'd lost twenty, forty pounds.

FOR $39.95 THEY'D send you a six-week supply. If it didn't take, if you were still a fatty after six weeks, they'd throw in another six-week supply. Free.

Herb hunched forward in his seat. His elbows rested on his knees. He leaned far into the living room. He brought his hand up and pulled at his lower lip.

"See?" he said when they talked about the special pills. "It's a monster deal."

Herb was talking to Carol. He was talking to Carol, but he was looking at Angie and me. "I don't think they'd do that about the second supply if it didn't work. I don't think they'd do that unless it was for real."

Carol didn't say anything. She stared at her fingers for a while. Then she got up and went into the kitchen for some coffee. When she got up, Herb didn't look at her. He was staring hard at the TV.

I tried to sympathize. I tried to get inside his thinking. If you'd ever seen him working on a two-twenty game heading into the seventh frame, you'd know what I was going through.

I pretended Carol was at that very minute in the kitchen with her head stuck deep in the refrigerator, cramming food into her mouth like you do when you're in a hurry. When you don't care what you eat, and you have to be somewhere. Or like when you want the last of something, just because it's the last of it, and you eat fast, not tasting anything. Just wanting to get it down, because it's there and it's the last of it.

I couldn't figure it. I didn't think Carol was in there doing that.

Later, in bed, Angie turned to me and said, "What a jerk. What an A-1 jerk."

I had my arm around her. I could feel her head on my shoulder, her breath on my chest.

I DRIVE AROUND for an hour, maybe more. Herb isn't talking.

I'm not at all sure what to do. I can't take Herb home. If Carol's there, it might get nasty fast. I could take him to my place. But then there's Angie — Angie and her fish stew. There's the down comforter.

So we drive. I head through town, come out the east side, and then turn right around. The snow's slicing down hard now. Every once in a while I can feel the tires on my Datsun give suddenly, slipping an inch or two over the glazed road.

My left wiper makes this scratching sound, gouging a small scar in the glass.

Herb sits there and stares out the window. Tugs at his lip. Works the fingers of his right hand.

I'm thinking I can't keep this up much longer. It's cold and I'm tired and I can't see.

I dip my head and try to peer through the haze.

We come to an overpass. Street lights hang like flaming balls of cotton. I feel a slight jar, a silent adjustment in the way of the world, as the rear of the car canters to the left. The front grille ticks right. Guardrails dance by. I've got my hands gripped tight on the wheel, but there's nothing to steer. I'm just hanging on. It's quiet. I listen to my own breathing. The guardrails float past a second time, and then we're sitting on the shoulder, my front bumper some eighteen inches from a sign that says, BRIDGE FREEZES BEFORE ROAD.

We sit there and watch the snow kicking up and spiraling in tight eddies.

"Sorry," I say. My voice sounds strange. I glance over at Herb. My knees are jammed far up under the steering column. My fingers ache.

Herb looks up at me. He's got this expression on his face. He's not angry. He's not surprised. He looks the way he looks when he's just powered the ball down the lane and it's no more than halfway through its skid. Herb looks the way he looks when the ball races over the lacquered pine, an ugly bruise set against the wood, and he's already pivoting on his heel and marching back to the molded plastic seats that ring the scoring area. You look at his face and know that he knows that the ball's going to rocket into the one-three pocket and fire pins *whomp-whomp* against the back matting.

That's the kind of expression he has now.

I downshift, ease up on the clutch, and coax my Datsun back onto the road.

CAROL'S ON THE quiet side. Can't say I know her all that well.

I've said a total of maybe twenty words to her without some sort of audience. It was a while after Herb started in about the diet business. He was getting crazy about it.

Carol threw Herb a birthday party. Big three-nine, is the way she put it. There were maybe fifteen people in all, the guys from the plant and their wives, a couple of neighbors from up the street. Not a big party, but nice. Just everyone sitting around drinking a few, munching on chips and licking guacamole from their fingers.

The week before, Herb had accused Carol of sneaking money from his wallet. The way Angie told it, Herb claimed Carol was going through his wallet and his pants, picking up the loose change, pinching a dime here, a quarter there.

I tried to get inside Herb's head on this one. I was willing to understand. I wanted to understand.

Herb figured Carol was running up to the Baskin-Robbins during the day, going for one of those overgrown banana splits with extra whipped cream and fudge topping. He decided she was sneaking out to the takeout chicken place a few miles down the road. He couldn't shake the idea of her slopping up those mashed potatoes heaped in gravy, the kind they serve in little styrofoam containers. It got so that Herb wouldn't let Carol out of the house. He stopped giving her money. Said she could go for a few days without feeding her face.

This is the way Angie explained it to me. Of course, I was only getting one side of it. It wasn't as if I had the whole picture. No way was Herb going to talk about these things with me.

If I were locking my wife in the house, I wouldn't want to talk about it.

At the party, Carol looked tired, but not unhappy. There were a lot of jokes about turning forty and all. Carol had made an angel food cake topped with white icing. "Herb" was spelled out in tiny licorice sticks.

It was nice.

Carol was careful not to eat anything herself.

I noticed, is what I'm saying.

I didn't want it to be this way. I didn't want to think of Herb in this way at all.

It was getting late, and Herb had been generous with the drinks. He'd stand at the dining-room table and mix up a batch of gin and tonics. Then he went with a pitcher of tequila sunrises.

Herb was into it. Herb was having a good time. Thirty-nine didn't seem to give him any trouble at all.

I headed into the kitchen for more ice. Carol was there, washing some dishes.

"KP already?" I said to her.

She looked back over her shoulder and grinned at me. She grinned through those enormous curls. It was the kind of grin you want to stick in your pocket and keep always and pull out whenever you need some cheering up.

In the window behind the sink sat a little portable TV. She had the station turned to one of those twenty-four-hour news channels.

"I hate waking up to warm whiskey in the bottom of a glass" is what she said.

The refrigerator door sucked open. I emptied two ice trays into a plastic bowl. I moved to the sink. I was going to wait until Carol was in between whiskey glasses, and then hit the cold tap and fill up the ice trays.

On TV a helicopter came out of the sky and hovered maybe six feet above the ground. The place looked hot. Dry. Palm trees flapped in the distance.

"Where's that?" I asked, nodding toward the television.

They were jumping from the helicopter. They'd come to the lip of the chopper and pause, then leap out into the open air. When they hit, they flexed their knees and rolled, coming up in a crouch. They looked ready to use their rifles.

Carol shrugged. "I'll be done in a minute," she said then. She smelled of soap.

Carol had on a pink blouse. Her bra straps made these small ridges in the cloth. Every time she bent to reach for another glass, a small crescent of purple poked from beneath the pink. It looked like the edge of a real whopper.

I touched her there. I set both ice trays on the counter and touched the bruise lightly, with my thumb.

She froze. Her back went tense. Everything seemed still and very far away.

She turned to look at me. She did not smile, or say anything. She stared until I looked away.

She turned back to the dishes. "Panama," she said then. "I think maybe Panama."

WE'VE BEEN THROUGH town twice, and Herb hasn't uttered a peep in about thirty minutes. I can feel the hate boiling up in him. Meanwhile, my headlights are useless. High beams, low beams. I even try my parking lights. All I get is blowing snow backed up against construction-paper black.

I figure Angie's eaten the fish stew by now.

I take a left and turn onto the freeway. It's not until we pull into the parking lot that Herb looks up. He looks around for a long time, like he's coming out of a deep sleep.

When I open the door and step out, he doesn't move. I reach back and pull out my ball and shoes.

What the hell, I tell myself.

I'm halfway to the entrance when I hear the truck door slam.

We get lane eleven. From lane eleven you've got a good view of the entire layout.

I throw a sucker strike first frame, the ten pin kicking wildly off the metal brace and topping the seven just enough to send me back to the scoring table. Herb picks up a ten-pin gimme for his spare.

Next ball I hook bad and end up missing the pickup. The six pin wobbles and then stands still as a sentry on duty.

One lane over, a young redhead rolls a pink ball shot through with gray streaks, like thin ribbons of cloud. She can't bowl, but the ball looks pretty as it works its way up the alley.

On the seat behind her sits a folded newspaper. The headline is big, a good inch and a half of bold black. Below it is a grainy picture. From here it looks like a giant elephant. Or maybe just a real-fat guy in sunglasses.

The headline is shouting something about Nagasaki, or Nicaragua. Maybe Nigeria.

Herb stands with his back to me. He's rubbing the ball the way he does, tiptoeing back and forth along the pine boards, lining up. I look at him standing there, and I realize his ass is too big. His ass is too big. Too much sitting. His elbows look like they're two hundred years old, wrinkling into tight little whorls and capped by liver spots.

Next time try a punching bag. Try a wall, fat boy. This is what I want to say.

I watch as Herb sends his ball dead-level into the one-three pocket. The ball rushes there like it's being sucked in. The ball rushes in and is there and it is beautiful and there is no getting around it. There is the play of fluorescent lighting off rubbed pine, the odor of fresh wax cutting the air. There is Herb, already making his way back toward his seat. Herb with that look on his face.

It comes to me then that I'm going to keep my yap shut.

Herb's ball slices into the one-three pocket. The pins seem to pause and gather close for a moment, only to hurl themselves finally *whomp-whomp* against the back matting.

August 1990

Blood For Oil

Chris Bursk

The first few thin cuts
I could forgive, even the gashes at the wrist,
the clawing to get the job over with.
But after the heart refused to give up,
each nerve cell
protesting, this pain
the only way the body knew
to insist on its rights,
how could you have begun to search kitchen drawers
for a sharper blade? And then stab
not once
but three times? Why would anyone
wish to harm the belly,
the old buffoon who wants only to be fed?
What had it done to deserve this butchery?
The act monstrous
not because it's impossible to imagine
but because it *is* possible.
All night on my knees
I crouch over the dark stains, scrub
like someone doing more than cleaning,
determined to rub so hard the blood
is not just gone
but has never been there, this work
clear, specific, a trail
I follow from room to room.
At least your daughters won't have to see
the streaked walls, soiled curtains.
There is no healing Mass for your friends.
All we can do now
is turn the dark clots of your blood
back into pale, watery wine

and flush it. SEIZE THE DAY,
the poster by your front door reads.
HAVE YOU MADE THIS DAY
THE BEGINNING OF A NEW LIFE?
the poster by the back door asks.
What kind of world is this?
you'd cry out, start ticking off items
on a list you faithfully kept
updating: reports of starvation in our own city,
the dismemberment of a young gay man,
a former student of yours killed
while waiting for a light to change,
this war. You overlook evidence
to the contrary, don't want
to be confused any further by living
simultaneously
in a world of incalculable horror
and small comfort. *Blood for oil,*
you said. *Mark my words,*
there will be blood for oil.
Now, on every station
where there's supposed to be music,
there's more news
of the bombings. If I run this faucet long enough
the blood will soak loose
from my rag, the water unravel
in my hands, cold
and clear again. How difficult it is
to forgive this world for being
what it is: accidental
and unfathomable. Apparently
there was no place in your ledger
for my clumsy, sometimes exasperated acts
of concern, no acknowledgment
of your own generous impulses, wry humor,
the body's gallant persistence,

its great courage
that had kept you alive
even with your wrists slashed,
even as the sun fell across your face.

August 1992

A Soccer Hooligan
In America

Carl-Michal Krawczyk

We're on this Greyhound bus heading down to an American football stadium in New Orleans for the England vs. U.S.A. preliminaries of the World Soccer Championships. About ten of us, all told, England supporters every last one. We're feeling beery and belligerent, as always, and we're chanting the usual: "ENG–LAND, ENG–LAND, ENG–LAND."

The other side is supposed to chant back whatever they are. I'll give you a for-instance. When we're in Italy playing the fucking I-ties, they chant, "ITAL–IA, ITAL–IA." In the land of spiked helmets and bratwurst, they give it, "DEUTSCH–LAND, DEUTSCH–LAND." Sometimes you'll get one clown in the German crowd who'll stick his right arm in the air and shout back, "Sieg Heil!"

So we beat the shit out of him because they're not allowed to say that anymore. We hate the Germans. Always have.

But the Americans?

They were chanting for England! We couldn't believe it. Here they are, our adversaries, giving it, "ENG–LAND, ENG–LAND." There was even someone at the back of the bus chanting, "MAR–GARET THAT–CHER, MAR–GARET THAT–CHER."

I'll never, ever forget that day.

Or the expression on Vic Stavski's face. He was our Instigator. He was leaning over his seat, staring back at this mixture of faces, trying to detect a bit of the old El Sarcastico. But there wasn't any. Then the Americans start telling us how much English and Welsh they've got in them. Stavski's got this calculator his dad gave him, and he's going up and down the aisle, laying a bit of the old El Provoko on them. He says, "OK, now, what percentage Scotch did you say your old man had in him when you were created?" And they were going, "Oh, about a third Scotch. A quarter English. Little bit of Navajo . . ."

Then they start asking us questions about *our* royal family. At first, Stavski was laying a bit of the old El Posho on them, telling them that Queen Elizabeth and us were neighbors, on account of the fact that our pub, the Old England, is about half a mile from Buckingham Palace. This one old dear sitting at the front of the bus is lapping it all up, she is. What Stavski was saying, though, was true. It's just that you don't think of the royal family and people like us being neighbors. Whenever we talk about the royals, it's like they're a million miles away. It's THEM and US.

But then Stavski gets annoyed, because people in America eat up stuff on royalty like pizza, but they don't know anything about *us*. So he shouts out, "There are two Englands, you know!" And everyone's staring at him, like they don't know how to respond to that one. So he gets up from his seat, and he shouts in their faces, practically, "Anyone heard of Gary Linekar?" He's England's number-one soccer player. "Paul Gasquione? Barnesy?" He went through the whole England Eleven, and they were chanting, "ENG–LAND, ENG–LAND."

The bus driver was looking at us in that big mirror. He put on his microphone a couple of times, but before he could say anything, we were back in our seats behaving ourselves. We just couldn't get anyone riled up, though. It was like they were disarming us of our ammunition. Until this one wanker comes out of the lavatory at the back and says, "I say, old chap. Jolly hockey sticks. Pip, pip, and all that. . . ." That did have a whiff of the old El Derogatory about it, and we had to struggle to keep Stavski off him or he would've eaten him alive. Victor Stavski's only five-foot-six, but he'll just dive into your face with his head, no messing about. He looks like James Cagney. That was his nickname when we were kids, and he's got a tattoo that says "JC" on his right arm.

We made a pit stop in Memphis, Tennessee, and someone pointed out it's where Graceland is, and the bus got real quiet. It was like we were paying our due respects to the One and Only. Stavski's got every record Elvis Aron ever made, and he got everyone started on "It's Now or Never." He never forgave Elvis, though, for not coming to England, and the Germans never let us forget that Elvis spent a lot of time in Germany when he was a GI. Stavski was telling this to someone standing next to him in the men's room one time, and this bloke said, "Yeah, but Elvis is still alive." Stavi said, "Yeah, well, maybe there's still hope!" And when he turned around to look at who he was talking to, this bloke looked *just like the King*.

We went into this 7-Eleven right next to the bus depot. The bus driver said we couldn't take beer back on the bus, so we got some soft drinks. Stavski pays for his, and he's almost out the door when the shopkeeper shouts, "Come right back, y'hear?"

Stavski's just standing there, mentally searching his pockets for something he might've lifted. (Afterward, when we were all laughing about it, he said the only thing he could think of was that he took a couple of gulps on his drink and filled it back up on account of all the ice they use in America.) Anyway, the shopkeeper keeps looking over at him, and Stavi's just standing there with one foot in the door. (Technically, they can't get you for anything until you actually leave the premises.) Then Stavski shouts, "What's your problem, mate?" You knew just by his tone of voice that something's about to happen. Me and some of the lads are dragging him out, but just as we're leaving the shop, Stavski notices that the shopkeeper is wearing a toupee, and he's shouting, "He's wearing a fucking wig! Look at him! A fucking homo!" I did feel a bit sorry for the shopkeeper, because Stavski was right about the toupee.

Some high-school-basketball kids got on the bus in Memphis. They were also real friendly, which was just as well, because they looked like walking skyscrapers. Stavski was really getting into it with them, going up and down the aisle, hand-jiving all those different ways there are to shake hands in America. Then the bus driver, who's been eyeing Stavski in his big mirror, says, "Sir, I'm gonna have to ask you to sit yourself down."

The bus got quiet, real quiet.

First off, no one that I know of (and I've known Stavski for — what? — twenty-nine years, all told) has ever called him "sir" before. Not unless they're trying to be sarcastic, and then we just clout them around the headgear. There's nothing we hate more than sarcasm. It's not honest. On top of that, we already had a bit of a run-in with this driver fellow when he told us we couldn't spread out our ENGLAND banner. Now he's asking Stavski to sit down, but what makes matters worse is that the whole bus is listening in. It's like the Gunfight at the OK Corral as we travel through the middle of America on a Greyhound bus. Not a murmur coming from anybody.

But Stavi won't sit down.

He's just standing there, staring down at the driver from deep inside his mind, and his mind had been doing some weird things ever since we arrived in America. It was his idea to take the Greyhound bus in the first place.

"What's your problem, mate?" Stavski says.

The driver doesn't say anything for a few minutes. Then he says, "Sir, I'm not gonna ask you again. If you don't sit yourself down, I'm gonna pull over, and there'll be one sorry person standing on the sidewalk looking at the back of a Greyhound bus."

Stavski just looks at me, and I look at him, and then I look back at everyone else. I know exactly what Stavski is thinking, but the driver was not being sarcastic.

He puts on the turn signals, slowing down to pull over. You could hear those turn signals right at the back of the bus: *Tick, tick. Tick, tick. Tick, tick.*

Everybody's waiting for something to happen. But just as the driver begins to pull over, Stavski sits down, and the Gunfight at the Greyhound Corral fades into history. For the first time in his life, Stavski does what he's told and no handcuffs or straitjackets are involved. He continues to stare at the driver, though, real blatantly, like he's going to salvage some pride that way.

After about five minutes, the bus driver calls out, "Hey, Stavski. Come on down front, will you?" Real friendly-like. "I want to talk to you for a minute."

Stavski stares at him hard for a minute or so, but then he goes down front. I follow him.

The driver leans to one side, pulls his wallet from his back pocket, and hands it to Stavski. We couldn't believe it! That was a turn-up for the books if ever there was one — someone giving Victor Stavski their wallet!

The driver says, "Take a look at the name on the driver's license." Stavski pulls the driver's license out, and now everyone's leaning over their seats trying to see what's going on. "See," said the driver, "I'm just an old Polack myself!"

His name was Stavski.

They weren't related, it turned out, but Stavski and him were really getting into it, talking about all kinds of things. It turned out the driver's son was an architect, which is what Stavi always wanted to be before he became a soccer hooligan — at least, that's what he always used to tell our probation officer. Then they start talking about Stavski's old man, and he's never, ever talked about his old man before.

His old man was a Polish immigrant to England. I remember when Vic used to get real embarrassed about introducing him to people. At the Old England, he used to pretend his old man was someone else's dad, because he always spoke with a

Polish accent. Stavski smashed him in the face once and told him never to come down to the pub again. That was after the England vs. Poland match that Poland won. His old man had supported Poland.

Victor Stavski was a real England freak. He believed in England. He followed the team everywhere. The only games he ever missed was when they banned us from Europe after a bunch of fucking I-ties got trampled to death in that stadium in Belgium. He loved England, and if anybody said anything that had a whiff of the old El Derogatory about him being English, he'd deck 'em. One time at the German border, the customs guard's looking at his British passport and says, "So you are not an Englishman. You are a Polish man." That was real snide, but since we can't clout uniforms, Stavi just looks up at him and says, "Yeah, but where I'm from we don't put people in gas chambers!"

But here he is, talking to the bus driver about his old man, like all of a sudden he was proud to be Polish, and his old man wasn't even alive no more, though he would've really liked it in the States. Everywhere we went in America, people were friendly. Stavski said, How can you smash people in the face when they're trying to be so nice? Some college students were telling us that the only reason people were friendly was because of business. Stavski told them, You gotta eat, and if people at the burger joint are nice to you when they serve you, then that's a bit of extra gravy. Nothing wrong with that!

The driver reminded us of John Wayne because he had a deep voice and he talked real slow. 'Course, once we started calling him Duke, he started talking even slower and his voice got deeper. He was telling us all about the cattle trails in America, and where they were driving all those cows on *Rawhide* every Wednesday night on BBC, because none of us knew. He'd point out the window every now and then and say, "Know who they hanged from that there tree?" He talked about Billy the Kid and Al Capone like he knew them.

Stavski asks him how come outlaws in America got to be heroes. Duke couldn't explain it, except that nobody was all bad, that there was a little bit of good and bad in everyone. Stavski really liked that, because, like I was saying, in England there's only THEM and US. THEM is good, and US is bad, and in between you have all these pathetic white-collar people who just bow and curtsy to THEM and run a mile when they see US coming. Then, just to confuse things even more, one of those basketball kids was telling me and Stavski that in America "bad" was good. They kept saying, "You bad, man," but they really liked him.

For a while there, it was like Trivial Pursuit without the board, because we were asking each other questions back and forth. Stavski got the bus driver, and he got him good. He says, "Who said this: 'Look at it, Adam. Feast thine eyes on a sight that approaches heaven itself'?"

"Gotta be something outta the Bible, right?" fast-talking Duke says after about ten minutes.

Stavski came right out and told him, "It was Ben Cartwright talking to Adam back in 1959, in that very first episode of *Bonanza*!"

Duke says, "Are you shittin' me?" and everyone on the bus is going, "ENG–LAND, ENG–LAND, ENG–LAND."

WE WON THE match.

It was England 4, U.S.A. 0. We avenged the 1950 defeat when the United States knocked England out of the World Cup. By the time we got to New Orleans, though, it was like Stavski had forgotten why we were there, like he had forgotten all about England. He didn't even see the match. He was shooting the breeze with the locals on Jackson Square, drinking coffee refills, and making free local calls from our hotel room. If you want to know the God's honest truth, Stavski was really getting hooked on America. He told me that people in America talked to the real *him*, which was kind of strange because before, he never, ever knew who the real *him* was. We used to talk about that a lot with our probation officer.

Back on the bus, going back up to Chicago, Stavski was the quietest I've ever known him to be. He was just staring out at America, watching the wheat fields, and the riverboats moving southward in slow motion. It felt like someone had just died. I should've known something was going to happen, because usually by this time he's tearing up the seats and tossing a few of the locals out the window.

We had a different driver. He was all right, but nothing like Duke. Duke gave Stavski his phone number and told him to look him up when he was in Chicago again.

About a hundred miles shy of Chicago, everyone on the bus is asleep except me and Stavski. It was dark and all you could see in the night sky was McDonald's golden arches. Stavski turns to me and says, "If Gothic cathedrals could talk, what accent would they have?" I say, "Posh," and get a smile out of him. It's something we used to say when we were kids sitting on the steps, watching people go into Westminster Abbey. I was surprised that he even remembered. Then, for the rest of the journey, he's comparing Gothic cathedrals back home with these golden arches. He was going on and on about how easy it was to understand things in America, and how Americans speak our language.

The next day, at O'Hare Airport, as they're calling the passengers for our flight back to London, Stavski's nowhere to be seen. We're searching everywhere for him — the lavatory, the pinball room — and then I see him, over by a circular postcard rack near a phone booth, looking at a huge telephone directory.

"Stavi!" I scream. "The plane's about to take off!"

He doesn't even look up, and I'm getting a bit embarrassed, because for a split second I'm thinking it isn't him, and everyone's looking at me. I walk up to him. "Stav," I say, realizing something's wrong, "what's the matter?"

He won't even look at me. He's shaking. "Stav," I say, "the plane's about to leave."

All of a sudden he turns to me and shouts, "I'm defecting!"

"You're what? What do you think you are, Russian?"

I tried to put my arm around him, and that was when he smashed into my face with his forehead. All I remember is falling over, blood pouring out of my nose, picture postcards flying through the air, and my mate Victor Stavski running back into America.

April 1993

WHY SCHOOLS
DON'T EDUCATE

John Taylor Gatto

L aments about our schools are nothing new; everyone is an expert, it seems, when it comes to education. While most critics
point to the lack of funding or the shortage of teachers, John
Taylor Gatto insists the problem goes deeper; we've turned our schools,
he says, into "torture chambers."

If that sounds abrasively radical, consider this: Gatto, with almost thirty years' experience as a public-school teacher, has just been
named New York City's Teacher of the Year for 1989.

Gatto teaches seventh grade at Junior High School 54 on
Manhattan's Upper West Side. Something of a local legend, he's a
chess player and a songwriter — and he grows garlic. He was once
named Citizen of the Week for coming to the aid of a woman who
had been robbed. He has lectured on James Joyce's Ulysses at Cornell

University and has taught philosophy at California State College. Perhaps it's not surprising that he's been approached by a film company interested in making a movie of his life.

Gatto once ran for the New York State Senate on the Conservative Party ticket, and some of his ideas are quite traditional: he stresses "family values" and questions increased funds for education. But he's too much of a maverick to be easily labeled. At a recent hearing in New York, he castigated the school system for "the murder of 1 million black and Latino children," and was met with a standing ovation.

What follows is the text of the speech he gave upon being named Teacher of the Year.

— Ed.

I ACCEPT THIS award on behalf of all the fine teachers I've known over the years who've struggled to make their transactions with children honorable ones: men and women who are never complacent, always questioning, always wrestling to define and redefine endlessly what the word *education* should mean. A "Teacher of the Year" is not the best teacher around — those people are too quiet to be easily uncovered — but a standard-bearer, symbolic of these private people who spend their lives gladly in the service of children. This is their award as well as mine.

We live in a time of great social crisis. Our children rank at the bottom of nineteen industrial nations in reading, writing, and arithmetic. The world's narcotic economy is based upon our own consumption of this commodity. If we didn't buy so many powdered dreams, the business would collapse — and schools are an important sales outlet. Our teenage-suicide rate is the highest in the world — and suicidal kids are rich kids for the most part, not poor. In Manhattan, 70 percent of all new marriages last less than five years.

Our school crisis is a reflection of this greater social crisis. We seem to have lost our identity. Children and old people are penned up and locked away from the business of the world to an unprecedented degree; nobody talks to them anymore. Without children and old people mixing in daily life, a community has no future and no past, only a continuous present. In fact, the term "community" hardly applies to the way we interact with each other. We live in networks, not communities, and everyone I know is lonely because of that. In some strange way, school is a major actor in this tragedy, just as it is a major actor in the widening gulfs among social classes. Using school as a sorting mechanism, we appear to be on the way to creating a caste system, complete with untouchables who wander through subway trains begging and sleep on the streets.

I've noticed a fascinating phenomenon in my twenty-nine years of teaching — that schools and schooling are increasingly irrelevant to the great enterprises of the planet. No one believes anymore that scientists are trained in science classes, or politicians in civics classes, or poets in English classes. The truth is that schools don't really teach anything except how to obey orders. This is a great mystery to me, because thousands of humane, caring people work in schools as teachers and aides and administrators, but the abstract logic of the institution overwhelms their individual contributions. Although teachers do care and do work very, very hard, the institution is psychopathic; it has no conscience. It rings a bell, and the young man in the middle of writing a poem must close his notebook and move to a different cell, where he learns that humans and monkeys derive from a common ancestor.

OUR FORM OF compulsory schooling is an invention of the state of Massachusetts, from around 1850. It was resisted — sometimes with guns — by an estimated 80 percent of the Massachusetts population, with the last outpost, in Barnstable on Cape Cod, not surrendering its children until the 1880s, when the area was seized by the militia and the children marched to school under guard.

Now, here is a curious idea to ponder: Senator Ted Kennedy's office released a paper not too long ago claiming that *prior* to compulsory education the state literacy rate was 98 percent, and after it the figure never again climbed above 91 percent, where it stands in 1990. I hope that interests you.

Here is another curiosity to think about: The home-schooling movement has quietly grown to a size where 1.5 million young people are being educated entirely by their own parents. Last month the education press reported the amazing news that children schooled at home seem to be five, or even ten years ahead of their formally trained peers in their ability to think.

I don't think we'll get rid of schools any time soon, certainly not in my lifetime, but if we're going to change what's rapidly becoming a disaster of ignorance, we need to realize that the institution "schools" very well, but it does not "educate"; that's inherent in the design of the thing. It's not the fault of bad teachers or too little money spent. It's just impossible for education and schooling ever to be the same thing.

Schools were designed by Horace Mann and Barnas Sears and W.R. Harper of the University of Chicago and Edward Thorndike of Columbia Teachers College and others to be instruments for the scientific management of a mass population. Schools are intended to produce, through the application of formulas, formulaic human beings whose behavior can be predicted and controlled.

To a very great extent, schools succeed in doing this. But our society is disintegrating, and in such a society, the only successful people are self-reliant, confident, and individualistic — because the community life that protects the dependent and the weak is dead. The products of schooling are, as I've said, irrelevant. Well-schooled people are irrelevant. They can sell film and razor blades, push paper and talk on telephones, or sit mindlessly before a flickering computer terminal, but as human beings they are useless — useless to others and useless to themselves.

The daily misery around us is, I think, in large measure caused by the fact that — as social critic Paul Goodman put it thirty years ago — we force children to grow up absurd. Any reform in schooling has to deal with school's absurdities.

It is absurd and anti-life to be part of a system that compels you to sit in confinement with only people of exactly the same age and social class. That system effectively cuts you off from the immense diversity of life and the synergy of variety. It cuts you off from your own past and future, sealing you in a continuous present, much the same way television does.

It is absurd and anti-life to be part of a system that compels you to listen to a stranger reading poetry when you want to learn to construct buildings, or to sit with a stranger discussing the construction of buildings when you want to read poetry.

It is absurd and anti-life to move from cell to cell at the sound of a gong for every day of your youth, in an institution that allows you no privacy and even follows you into the sanctuary of your home, demanding that you do its "homework."

"How will they learn to read?" you say, and my answer is: "Remember the lessons of Massachusetts." When children are given whole lives instead of age-graded ones in cellblocks, they learn to read, write, and do arithmetic with ease, if those things make sense in the life that unfolds around them.

But keep in mind that in the United States almost nobody who reads, writes, or does arithmetic gets much respect. We are a land of talkers; we pay talkers the most and admire talkers the most, and so our children talk constantly, following the public models of television and schoolteachers. It is very difficult to teach "the basics" anymore, because they really aren't basic to the society we've made.

Two INSTITUTIONS AT present control our children's lives: television and schooling, in that order. Both reduce the real world of wisdom, fortitude, temperance, and justice to a never-ending, nonstop abstraction. In centuries past, the time of a child or adolescent would be occupied in real work, real charity, real adventure, and the real search for mentors who might teach what he or she really wanted to learn. A great deal of time was spent in community pursuits, practicing affection, meeting and studying every level of the community, learning how to make a home, and dozens of other tasks necessary to becoming a whole man or woman.

But here is the calculus of time the children I teach must deal with:

Out of the one hundred sixty-eight hours in each week, my children sleep fifty-six. That leaves them one hundred twelve hours a week out of which to fashion a self.

My children watch fifty-five hours of television a week, according to recent reports. That leaves them fifty-seven hours a week in which to grow up.

My children attend school thirty hours a week, use about eight hours getting ready, going, and coming home, and spend an average of seven hours a week on homework — a total of forty-five hours. During that time they are under constant surveillance, have no private time or private space, and are disciplined if they try to assert individuality in the use of time or space. That leaves twelve hours a week out of which to create a unique consciousness. Of course my kids eat, too, and that takes some time — not much, though, because we've lost the tradition of family dining. If we allot three hours a week to evening meals, we arrive at a net amount of private time for each child of nine hours.

It's not enough. It's not enough, is it? The richer the kid, of course, the less television he or she watches, but the rich kid's time is just as narrowly circumscribed by a broader catalog of commercial entertainments and his or her inevitable assignment to a series of private lessons in areas seldom of the child's own choice.

And these things are, oddly enough, just a more cosmetic way to create dependent human beings, unable to fill their own hours, unable to give substance and pleasure to their existence. It's a national disease, this dependency and aimlessness, and I think schooling and television and lessons have a lot to do with it.

Think of the things that are killing us as a nation: drugs, brainless competition, recreational sex, gambling, alcohol, the pornography of violence, and the worst pornography of all: lives devoted to buying things, accumulation as a philosophy. All are addictions of dependent personalities, and that is what our brand of schooling must inevitably produce.

I WANT TO tell you what the effect is on children of taking all their time — time they need to grow up — and forcing them to spend it on abstractions. No reform that doesn't attack these specific pathologies will be anything more than a facade.

1. The children I teach are indifferent to the adult world. This defies the experience of thousands of years. A close study of what big people were up to was always the most exciting occupation of youth, but nobody wants to grow up these days, and who can blame them? Toys are us.
2. The children I teach have almost no curiosity, and what little they do have is transitory; they cannot concentrate for very long, even on things they choose to do. Can you see a connection between the bells ringing again and again to change classes and this phenomenon of evanescent attention?
3. The children I teach have a poor sense of the future, of how tomorrow is inextricably linked to today. They live in a continuous present; the exact moment they are in is the boundary of their consciousness.
4. The children I teach are ahistorical; they have no sense of how the past has predestined their own present, limiting their choices, shaping their values and lives.
5. The children I teach are cruel to each other; they lack compassion for misfortune, they laugh at weakness, and they have contempt for people whose need for help shows too plainly.
6. The children I teach are uneasy with intimacy or candor. They cannot deal with genuine intimacy because of a lifelong habit of preserving a secret self inside an outer personality made up of artificial bits and pieces of behavior borrowed from television or acquired to manipulate teachers. Because they are not who they represent themselves to be, the disguise wears thin in the presence of intimacy, so intimate relationships have to be avoided.
7. The children I teach are materialistic, following the lead of schoolteachers who materialistically "grade" everything — and television mentors who offer everything in the world for sale.
8. The children I teach are dependent, passive, and timid in the presence of new challenges. This timidity is frequently masked by surface bravado, or by anger or aggressiveness, but underneath is a vacuum without fortitude.

I could name a few other conditions that school reform will

have to tackle if our national decline is to be arrested, but by now you will have grasped my thesis, whether you agree with it or not. Either schools, television, or both have caused these pathologies. It's a simple matter of arithmetic: between schooling and television, all the time children have is eaten up. That's what has destroyed the American family; it is no longer a factor in the education of its own children.

WHAT CAN BE done?

First, we need a ferocious national debate that doesn't quit, day after day, year after year, the kind of continuous emphasis that journalism finds boring. We need to scream and argue about this school thing until it is fixed or broken beyond repair — one or the other. If we can fix it, fine; if we cannot, then the success of home schooling shows a different road that has great promise. Pouring the money back into family education might kill two birds with one stone, repairing families as it repairs children.

Genuine reform is possible, but it shouldn't cost anything. We need to rethink the fundamental premises of schooling and decide *what* it is we want all children to learn, and *why*. For 140 years this nation has tried to impose objectives from a lofty command center made up of "experts," a central elite of social engineers. It hasn't worked. It won't work. It is a gross betrayal of the democratic promise that once made this nation a noble experiment. The Russian attempt to control Eastern Europe has exploded before our eyes. Our own attempt to impose the same sort of central orthodoxy, using the schools as an instrument, is also coming apart at the seams, albeit more slowly and painfully. It doesn't work because its fundamental premises are mechanical, antihuman, and hostile to family life. Lives can be controlled by machine education, but they will always fight back with weapons of social pathology — drugs, violence, self-destruction, indifference, and the symptoms I see in the children I teach.

It's high time we looked backward to regain an educational philosophy that works. One I like particularly well has been a favorite of the ruling classes of Europe for thousands of years. I think it works just as well for poor children as for rich ones. I use as much of it as I can manage in my own teaching — as much, that is, as I can get away with, given the present institution of compulsory schooling.

At the core of this elite system of education is the belief that self-knowledge is the only basis of true knowledge. Everywhere in this system, at every age, you will find arrangements that place the child *alone* in an unguided setting with a problem to solve. Sometimes the problem is fraught with great risks, such as the problem of getting a horse to gallop or making it jump. But that, of course, is a problem successfully solved by thousands of elite children before the age of ten. Can you imagine anyone who has mastered such a challenge ever lacking confidence in his or her ability to do anything? Sometimes the problem is that of mastering solitude, as Thoreau did at Walden Pond, or Einstein did in the Swiss customshouse.

One of my former students, Roland Legiardi-Laura, though both his parents were dead and he had no inheritance, rode a bicycle across the U.S. alone when he was hardly out of boyhood. Is it any wonder that in manhood he made a film about Nicaragua, although he had no money and no prior experience with filmmaking, and that it was an international award winner — even though his regular work was as a carpenter?

Right now we are taking from our children the time they need to develop self-knowledge. That has to stop. We have to invent school experiences that give a lot of that time back. We need to trust children from a very early age with independent study, perhaps arranged in school, but which takes place *away* from the institutional setting. We need to invent a curriculum where each kid has a chance to develop uniqueness and self-reliance.

A short time ago, I paid seventy dollars and sent a twelve-year-old girl with her non-English-speaking mother on a bus down the New Jersey coast. She took the police chief of Sea Bright to lunch and apologized for polluting his beach with a discarded Gatorade bottle. In exchange for this public apology, I had arranged for the girl to have a one-day apprenticeship in small-town police procedures. A few days later, two more of my twelve-year-old kids traveled alone from Harlem to West Thirty-first Street, where they began apprenticeships with a newspaper editor. Next week, three of my kids will find themselves in the middle of the Jersey swamps at six in the morning studying the mind of a trucking-company president as he dispatches eighteen-wheelers to Dallas, Chicago, and Los Angeles.

Are these "special" children in a "special" program? No, they're just nice kids from central Harlem, bright and alert, but so badly schooled when they came to me that most of them couldn't add or subtract with any fluency. And not a single one knew the population of New York City, or how far it is from New York to California.

Does that worry me? Of course. But I am confident that as they gain self-knowledge, they'll also become self-teachers — and only self-teaching has any lasting value.

We've got to give kids independent time right away, because that is the key to self-knowledge, and we must reinvolve them with the real world as fast as possible so that the independent time can be spent on something other than mere abstractions. This is an emergency. It requires drastic action to correct. Our children are dying like flies in our schools. Good schooling or bad schooling, it's all the same: irrelevant.

WHAT ELSE DOES a restructured school system need? It needs to stop being a parasite on the working community. I think we need to make community service a required part of schooling. It is the quickest way to give young children real responsibility.

For five years I ran a guerrilla school program where I had every kid, rich and poor, smart and dipsy, give 320 hours a year of hard community service. Dozens of those kids came back to me years later and told me that this one experience had changed their lives, had taught them to see in new ways, to rethink goals and values. It happened when they were thirteen, in my Lab School program — made possible only because my rich school district was in chaos. When "stability" returned, the lab closed. It was too successful, at too small a cost, to be allowed to continue; we made the expensive, elite programs look bad.

There is no shortage of real problems in this city. Kids can be asked to help solve them in exchange for the respect and attention of the adult world. Good for kids, good for the rest of us.

Independent study, community service, adventures in experience, large doses of privacy and solitude, a thousand different apprenticeships — these are all powerful, cheap, and effective ways to start a real reform of schooling. But no large-scale reform is ever going to repair our damaged children and our damaged society until we force the idea of "school" open — to include *family* as the main engine of education. The Swedes realized this in 1976, when they effectively abandoned state adoption of unwanted children and instead spent national time and treasure on reinforcing the family so that children born to Swedes *were* wanted. They reduced the number of unwanted Swedish children from six thousand in 1976 to fifteen in 1986. So it can be done. The Swedes just got tired of paying for the social wreckage caused by children not being raised by their natural parents, so they did something about it. We can, too.

Family is the main engine of education. If we use schooling to break children away from parents — and make no mistake, that has been the central function of schools since John Cotton announced it as the purpose of the Bay Colony schools in 1650 and Horace Mann announced it as the purpose of Massachusetts schools in 1850 — we're going to continue to have the horror show we have right now.

The curriculum of family is at the heart of any good life. We've gotten away from that curriculum — it's time to return to it. The way to sanity in education is for our schools to take the lead in releasing the stranglehold of institutions on family life, to promote during school time confluences of parent and child that will strengthen family bonds. That was my real purpose in sending the girl and her mother down the Jersey coast to meet the police chief.

I have many ideas on how to make a family curriculum, and my guess is that a lot of you will have many ideas, too, once you begin to think about it. Our greatest problem in getting the kind of grass-roots thinking going that could reform schooling is that we have large, vested interests profiting from schooling just exactly as it is, despite rhetoric to the contrary.

We have to demand that new voices and new ideas get a hearing, my ideas and yours. We've all had a bellyful of authorized voices on television and in the press. A decade-long, free-for-all debate is called for now, not more "expert" opinions. Experts in education have never been right; their "solutions" are expensive, self-serving, and always involve further centralization. Enough.

Time for a return to democracy, individuality, and family.

I've said my piece. Thank you.

June 1990

Giving Away Gardens

Dan Barker

A wise man fills bellies.
— Tao Te Ching

Idiot wind, it's a wonder we can even feed ourselves.
— Bob Dylan

A Crip gang member approached the woman for whom I was building a vegetable garden — an old woman on welfare, an ex-prostitute, ex-waitress, ex-chicken-butchering-plant worker. He said he was tired; pimping was hard work. I kept to my hammer and shovel, hearing the woman's tubercular laugh, and repressed a moral urge to bash his brains in, instead muttering something nonsensical about individuated karma and samsara, knowing and glad that the garden will persist longer than he will.

THE MAN WHO helps me would have been a great pioneer; he hates lawns, and jobs. He is nearly fifty and has never been able to tolerate employment for longer than three months at a stretch. To him, lawns and civilization are the same phenomenon, ass-backwards and fit for no good use. When he and I are working together, we are a team; we get it done. We frequently use the language of work learned in Vietnam.

Work starts in March. There is limited time — spring — in which to fulfill the contracts among the Home Gardening Project, the funding sources, the city government, and the new garden-ers. The purpose of the work is to foster self-reliance and im-proved health and well-being.

If the weather is looking like the rains will break, I order the two truckloads of two-by-eights and two-by-twos we'll need to build the gardens (125 in 1990, as many as we could afford to build), order the mushroom compost to augment the three loads of weed-free organic soil that are mixed and delivered to us each week, beg for a front-loading tractor, then spend a day cutting the lumber to size. Warm up for the next two and a half months of constant building, and move on. Three ten-cubic-yard dumps a week are loaded three cubic yards at a time into the back of our pickup, along with soil frames, trellis, tools — and a wheel-barrow tossed on top and tied down. At Ms. Wittingham's, the house looks well kept. She comes out, a cataract clouding her left eye, and says her husband of forty years used to love to do the garden, but now he's disabled with a stroke. Sorry to hear that; that's why we're here. In the back, over the old garden site, she says. We shoulder the lumber back to the plot beside the garage, lay out the two-by-eight frames, bang 'em up; the sound of the hammer drums the neighborhood awake. My helper starts at one end, and I at the other; he builds two, I build one, strong-box fashion, the way my old carpenter friend taught me. He builds the trellis while I line out the frames square to the world and knock down lumps of weeds. I leave a three-foot path between frames so the garden can be comfortably tended, then we set the frames by nailing twelve-inch stakes at each corner. My helper does one side, I the other. Work output is divided equally, down to the erg.

We start wheeling in the soil, two wheelbarrows a turn, six to each soil frame: back left, back right, middle back, middle front, front left, front right — times three; the man who started goes second next time. Place and nail the trellis; one man strings while the other rakes the new seedbeds smooth — the Zen of that garden, the rest and quiet. While I talk how-to with Ms. Wittingham, my helper cleans up, stows the tools, and plugs in a fresh chaw of Redman, maybe tunes in Paul Harvey. Talk ain't work as far as he is concerned.

I teach Ms. Wittingham block and succession planting, seed conservation, composting, watering, and fertilizing, and tell her I'll be by again in early May, when the weather warms up, to deliver starts for tomatoes (four varieties), eggplant, peppers, basil, and flowers. What's a garden without flowers? She says thank you. She's shy about the new gardening techniques; she's never availed herself of a social service before, but now she's older, and the money is gone. Careful to preserve her dignity, I accept her thanks for making it so easy, one phone call. This is truly a Christian thing you're doing, she says, and I say, Glad you like it; I think, but don't say, that it's also Buddhist, Taoist, Pagan, Dyak. Men putting in gardens is a phenomenon fifty thousand years deep.

On to the next one. Birdi Johnson has arthritis; she's been into the wine already this morning, doesn't come out of the house much. Neither does her fifteen-year-old daughter; too much danger lurks out here. We build the second garden of the morning, slipping on the mud and dog shit in the backyard. All she wants to grow is tomatoes, beans, collards, and corn, and I say, Great, sure, grow what you like to eat; it's your garden. If your hands are too painful to do the planting, I'll send someone by to help you. On to a burger palace that will let us wear rain gear dripping with mud while we eat. At the Burger King, the children are proud of their wounds. They take delight in having a close friend who took a bullet in the neck or had the flesh of his shoulder carved off by a glamorous commando knife. Vietnam brain gauze gone crack crazy.

Then on to Roger Kerns: he's a paraplegic, an ex-athlete whose back was shattered in a drunken car crash. He's into SEVA and lecturing high-school kids on the net results of cars and fun. He tries not to show bitterness at being the dupe of some universal force that took him off at the seventh vertebra. This garden is a double high — one frame on top of the other — so he can reach it from his electric wheelchair. He's supervising the placement, wants the overhanging maple branches pruned, the yard debris cleared away; we do what he wants, a little extra work because he can't and we can and we're here. I do the training talk; he says now he can further advance into vegetarian living, *namaste*.

And on again to Thelma Cason's; she's eighty-eight, still going strong, has done for herself all her life. I'm lucky to know her and her example, but the kids from the crack house down the block busted through her door, knocked her down, broke her hip, stole her thirteen-inch TV (black-and-white to boot) and her food stamps. She heard I was a "good man," and that is what I'm trying to be. At least the little punks can't steal the garden. I'll drop by later this summer to see how you're keeping, I tell her, and I'll call Senior Services for you, have them come and fix that back door like they should have last winter; your life and my life is our life, even this brief meeting, but you get to keep the garden, yes, it is yours, a gift. A gift should be well-made and leave nothing wanting. Art.

Those last two went fast, on the flat ground, time for one more if you've got the oomph. Sure, why not, long way to go and a short time to get there. Let's do Kris S., who's got three kids by three fathers, no men around now, Aid to Families with Dependent Children and food stamps, run out of her last two houses by crank monsters. She wants to teach her children vegetable gardening to ensure that they will never starve. We build it in the sunny spot next to the kitchen door. She wants to be employed as a part-time planter, and I give her Birdi Johnson's address.

The day is done; we'll do it again tomorrow and tomorrow and tomorrow, all week the same people, the same stories, until two and a half months pass in a kind of hazy dream exertion that manifests 125 new gardens, 125 renewed lives. On the way back to the operations site, where the soil, tractor, and lumber are stored, a BMW misses a stop sign and nearly crunches the truck, our most essential tool; the brothers drinking wine and waiting for customers down on the corner of crack alley stare at us big dumb white devils.

I INVENTED THIS work by consolidating my experience working in nurseries, construction, and writing poetry, while trying to recover from a divorce and being robbed at gunpoint in a wino grocery store, knocked unconscious by a blow from a .38 butt, the muzzle pressed against my occipital bone for ten minutes, the hammer cocked; the gunman's accomplices couldn't figure how to open the computerized cash register. Three months later, I see the robber buying primo vegetables at the local gourmet-produce store, gut-wrenching fear like being back in the war, in a firefight, my only weapon a quarter to call the cops. But they don't want to go to the trouble of busting him and his girlfriend accomplice; he wore a mask, impossible to visually identify him in court, they contend. But I knew, and vowed never again to put myself in the victim's position. Meanwhile the schools are failing, the jails are full. The Dalai Lama tells how to quiet the demons grown from hate and fear and act out of compassion. I and thou are one, like it or not.

OUT IN THE suburban slums, where the ground has been in turf for a hundred years, or the housing was built on the gravel of an ancient flood plain, it is not enough just to go in and pass out seeds. To be effective, it is necessary to bring in the whole garden. Most of the impoverished, elderly, and disabled in need of additional sustenance are no longer able to till that depleted soil. They can plant, weed, water, and harvest, though, if they've got a garden to work in. In my city alone, there are thousands who need and would use the proceeds from a garden. Any of us could become one of them.

You never know how the cards will play: riding high in April, down to the Social Security office in May. If after twenty years of little but bad news, some bozo says, Hey, I've got a free vegetable garden for you — that's right, soil frames, weed-free organic soil, seeds, starts, fertilizer, instructions, cooking tips, yes, I'll come to your house and build it for you, no strings — you might be glad to see me coming. Another spark of human joy alive in the soul-stream.

Most of us have little connection to the manipulations that constitute business/government. Many of my clients' more immediate concerns are whether the Meals on Wheels girl is going to show up and if she'll have time to smile or, better yet, chat. They worry that the medical department will cut off their gangrenous diabetic legs. Many say the government just wants them to die. And in their defiance, they go out to tend their new gardens. All summer, that is where we find them: outside, working, making sure that the gardens are just right, harvesting their evening meals. For some, so alone, it is all they have. The life-spirit is winning.

The work is religious, and there are three sides to it: one involves arduous physical labor — building soil frames, wheelbarrowing soil, four gardens a day, four days a week, until the goal is reached or surpassed. A second side absorbs the incessant tales of suffering, sees the misery and despair, the rotted teeth, the heart problems, the amputations, the disfigurements of body and soul. The third side demonstrates the capacity to run a business dedicated to the alleviation of that suffering. The books and the reports must be straight so I can create a garden where there wasn't one before, and with a bit of work, a new occupation of time and spirit is opened.

Many observers consider giving away complete vegetable gardens — to the poor, the disabled, the single mothers, the aged — to be the best idea they've ever heard. It is real, it works. They tell me so; they bestow their blessings on me. For some, it seems a miracle that another human being would do this work at no cost to them. For others, it is only their due for life having fouled them. How did getting a vegetable garden become possible? I decided to make the gardens available, and they decided to take them. But no matter the motives or reasons, excuses or philoso-phies, the result is now 525 new vegetable gardens — real change in the real world. The soul, the well-being of several thousand people, has been substantially enhanced. They are lifted by their practice of self-provision into the greater miracle of life.

SOME NATIONALLY SYNDICATED columnist recently opined that we live in a world that no longer rewards virtue. No reward, indeed. The first lie is that we live *in* the world. We *are* the world, the planet spinning, the galaxy, the universe. We are each other. The aged are our mothers and fathers, the young our children, their mothers our sisters. The reward for doing virtuous work is the same as it has always been: you grow out of your own abyss, dissolving the existential distance betwixt I and thou. Nirvana is not a reward, it is a condition.

Vegetable gardening is one of the ways that Americans prac-tice virtue. To grow a vegetable garden is to proclaim your hon-esty and self-reliance, your dedication to the fundamentals of living a decent life. It is a connection to the seed of life growing and nourishing the body soul. By my own efforts, in harmony with nature, I am able to sustain myself and my neighbor. I and we are worthy beings.

The charitable trusts and foundations want to know if the gardens work. Do they dissolve the current anguish ripping the dignity from the impoverished? I can't say with certainty that one gang kid has been deflected from his run toward a violent end or prison, or that I've passed out sandwiches to people who have no reason to vote, or given shelter to homeless families. But I've saved thousands of people considerable money, time, and trouble, trips to the doctor, despair, sessions with their thera-pists, and longing for death. I tell the gardeners that this is the store you don't have to go to. You get hungry, come on out and pick yourself a meal. When you plant, use three seeds — one for you, one for your neighbor, one for God. They always laugh when I mention God, or silently let the word slide on by. I go home knowing that I've planted the possibility of self-caring. But the donors want a figure; I tell them each garden is capable of producing at least five hundred dollars' worth of food a summer, if you don't count gas, time, etcetera, and that 95 percent of the gardens are productive the first year, 85 percent the second — I don't keep track after that, though often I run across a garden still producing after five or six years. Some people even load their gardens onto trucks when they move.

What is more difficult to convey is the health and joy evident in a seventy-year-old woman showing me her beans and tomatoes, or the pride of accomplishment beaming from the face of the twelve-year-old son of an ex-prostitute whose mother put him in charge of the garden. Or the envy of neighbors — I put down a garden, and the next year two or three neighbors will call for theirs. We're strictly word of mouth. I wouldn't know how well it was working otherwise. There's never been a shortage of recipients, only a shortage of money, time, and energy.

THE ORIGINAL IDEA was the diaspora of the perpetual garden, a way to reverse what is so celebrated now, the deprivation of the many for the gain of the few. Too ambitious a thought. The free-market/welfare system victimizes those unprepared for its complexities; it's too large, too pervasive to be countered by something so small as a garden, extended metaphor or no. Still, the notion contains the whole cycle of life, incorporating use of local materials (dairy and racetrack manures, construction subsoil, compost, surplus seed), reducing use of fossil fuels, reconnecting people with life — thus serving all. Everything necessary is already in place: parks departments have tractors, trucks, working space, and greenhouses, much of the time underused; thousands of people desire to be of service to their neighbors; workers could be recruited from extension agents and agricultural programs. All we have to do is put it together and get it paid for. One announcement on TV and there would be no end to the requests for gardens. People in need need all the help they can get. They will be the ones, and are the ones, who quiet the neighborhood. They will endure and will invite peace from others.

It's taken me seven years to get the project into the black, and it couldn't have happened without the goodwill and generous hearts of my wife and friends. We lift ourselves. Accolades go to the foundations and trusts that have sponsored and believed in the work. They call it charity, but it is simply service, a providence that can even be employed by the recipients, as shown by several older women who wanted — and got — double or triple gardens so they could provide vegetables for the entire neighborhood.

They ask me why I do this, and I say it needs to be done. Don't you need a vegetable garden, one you can get to, one you can use and maintain without too much physical effort? There, now you've got one, good luck, happy to do it. Or, once, when I was tired and being interviewed for the *Statesman Journal*, the young reporter asked why, and I said I'm out to change the world. And when she asked, What do you do in real life? my tact left me, and I replied, Don't you think giving away gardens is real life? Don't you think trying to lift the weight of suffering by one micron is real? To affirm the good in you, in life, the Tao speaks of neighbors who do not tread on each other, but live their lives in quiet wonder, grow old, and die. And the way to affirm the good life is to deliver it. If such an act challenges the men on the corner, good; shovels are easy to come by.

What is bothersome is not that giving away gardens is so wonderful, but that it is so rare.

December 1990

SCAVENGER'S RUN

David Grant

In Guangzhou, China, I once saw two men row through the muddy waters of the Pearl River to pick up floating leaves of cabbage. Now, a few years later, that's what I do: make the scavenger's run.

Every weekday morning, one of us from the Family Kitchen visits a half dozen groceries and restaurants, picking up what others have left behind. By 4:30 P.M. we are serving one to two hundred people a free meal in our downtown soup line. We ask no questions and preach no sermons. We don't even practice, to my mild dismay, a moment of silence before the meal.

I make the scavenger's run three days a week, usually with my toddling daughter, Amara. We often refer to it as "gleaning." But that's a mite genteel. It is the most ordinary work, going through garbage — a humble job, simplicity itself.

I've learned a lot, though. I've learned that the smell of rotten onions and potatoes can nearly make you vomit. Cucumbers grow the most disgusting molds, citrus the prettiest, cabbages the slimiest, and sweet potatoes the fastest moving. I savor the opportunity to rescue the merely ugly, the broken and the dented, the slightly bruised, and the firm but aging. I'm especially happy when I find some cut or potted flowers. Bread and begonias! Ah! It is a fine feeling, saving the deserving from oblivion.

Daily I am reminded of our society's wastefulness, but also of the generosity of this land and its people. As I drive the rounds, I nibble cracked-wheat baguettes donated by the deli at the "natural" grocery store. Amara likes the cinnamon rolls from the politically correct bakery. We live largely off the food that we collect; our lifestyle is voluntarily simple and sometimes elegant. We have managed to avoid the involuntary and graceless poverty of those whom we serve and with whom we share the discarded bounty.

Though in theory I adhere to the injunction to carry neither gold nor silver and to travel the open road lighthearted, unpocketed, and free — and though, at various times, I have experienced unstinting generosity from destitute people — I find it easier to be openhanded when I possess abundance. I shudder to think how dependent our generosity is — the generosity of the nation, as well as my own — upon our position as global aristocrats.

I often select a few of the choicest items for our community house. I like to delight everyone with honey-sweetened eggnog that didn't sell, overstocked but perfect organic broccoli, unsulfured molasses that spilled out of the self-dispenser, sweet carrots whose only sin is crookedness, live clams whose time has come but not quite gone, a few pounds of pecans donated for unknown reasons, full-tasting mushrooms barely edged with dark brown ripeness, a bag of succulent mixed citrus found out of place and too much trouble to sort.

We deserve it, of course. We're volunteers. Educated. Six of the seven who collectively run the kitchen are white, and the other one — me — is what my black compatriots refer to as "high yellow." We could be making millions, for God's sake.

We're an ornery bunch. We actively support boycotts of General Electric, Morton-Thiokol (table salt), grapes from California and Chile (even those donated), and the top fifty military contractors (when we have a not-too-painful choice). Unlike most Catholic Worker–inspired programs (all totally autonomous), we accept government commodities — grateful for the butter, cheese, hamburger, canned meat, flours, grains, cereals, and beans. Not surprisingly, we squabble over details and haggle about what's right. That's another reason we call it the *Family* Kitchen. We do the work that needs to be done, but pragmatic ambiguity prevails.

Since the Kitchen was begun spontaneously, without institutional support, we are beholden to no one but ourselves. Only two founding members remain. In thirteen years, the Kitchen has grown considerably. At the end of the month, when our clients' checks from pensions and welfare and Social Security have run out, we near our capacity of 250. The whole thing runs on the honor system, and it is an honor to be part of it.

We pride ourselves on our fresh produce and make special efforts toward nutritional balance and eye appeal. We call the Kitchen, perhaps unwisely, "the best free meal in town." Actually, it was an anonymous woman who stood up one day in the welfare office and proclaimed: "The Family Kitchen is the best free meal in town!" We've silk-screened the slogan on T-shirts and pullovers — morale boosters that we sell at cost. My ambivalence stems not from the boast, but from the bruiting of "the best."

The Catholic Worker position, from which our Family Kitchen derives, calls for "a harsh and dreadful love." It calls for blurring the distinction between the server and the served. It calls for a total restructuring of, in Dorothy Day's famous dictum, "this filthy, rotten system."

Mine is, at times, a contrary spirit. I enjoy eating off the bottom sometimes. But often, as I said, I bring the best stuff home. Some also gets picked out by volunteers during preparation, before it goes through the windows to our "guests." (Of the thirty volunteers who work once a week, some are poor, some not.) All this self-rewarding pilferage still does not amount, amid the cornucopia, to a hill of beans.

THE CORE OF all ethics, it seems to me, emanates from the charge to take personal responsibility for whatever problem is at hand. Our economic system runs on the premise "Seek ye out, at least cost, the best stuff!" A bargain hunter is what every consumer conspires to be. "Hey, Jack," jives the man in the sharkskin suit, "I've got a good deal!" He is showing me something bright and shiny, though nondescript. "Something better, man! And cheaper, to boot!"

The adman expounds our culture's formula for progress. No matter that he is selling oil at the price of our soil. No matter that the man's bananas hardly come from a "republic." No matter that his cheap flight is a horribly noisy plane thundering over billions of living beings who have ears. No matter that *all* his flights — and all his massive uses of energy — inexorably, under the law of entropy, exact terrible, unmitigable tolls over great distances. No matter that his automobile entices us away from enjoying and improving our own neighborhoods, that it spreads pathogenic fumes, and that governments are planning wars over the price of the fuel it demands. No matter that his herds are destroying rain forests so we can have cheaper hamburgers, since we've already permanently eliminated vast portions of the North American temperate forests for their grainfields. In purely materialist terms, the free-market man's code — "the best for less" — is a cover for thieves.

In *The Gospel of Swadeshi*, Mahatma Gandhi asks for an economic principle of neighborliness. He asks that transactions between producer and consumer be as local and direct as possible — in order that all the consequences of manufacture and delivery be clearly known. He asks that one's neighbor be patronized *even if* the goods are inferior and *even if* they come at a higher price.

I have yet to accept this fully, much less to act upon it consistently. I am highly sensitized to the dangers of cronyism and to the shoddy depredations of the make-a-buck crook. I admire and support craftsmanship. My wool U.S. Navy pants, for example, are durable and well-made. They were a near gift, bought at a local surplus store for two dollars. At the other extreme, my wife, Barbara, and I spent more than two thousand dollars on our bicycles and accouterments (trailer for cargo and daughter, halogen night lights, panniers, and clothing). Figured over a ten-year period, that comes to about 3 percent of a Mercedes or 10 percent of a junker — and does not account for the diametrical differences in environmental impact or personal health.

Common sense and common decency are overwhelmed by the ubiquitous, voluminous propaganda of advertising and diplomacy. We can't see the neighborhood forest for the multinational trees. In the pavilions of banks and markets, sybarites play shell games. Little men huddle, hoarding money — bits of paper, shreds of trees. They chortle that they own the forests.

But we scream, "Soon there will be no forests! There will be only specimen trees, props for the urbanites!"

The little men's noise machine laughs on cue. They keep playing in their pavilions, shouting at each other: "Trade balance!" "Debtor nation!" "Budget deficit!" "Deterrent defense!" These men in tailored suits get to shout these slogans because they consider themselves the progenitors of a *developed* nation.

But what does that mean in light of *swadeshi*? At one grocery store (a cooperative) where we scavenge, the produce that *we* discard is set aside for a city woman who uses it for compost and to feed her ducks and geese. This, I would say, is a sign of true *development*.

The best free meal in town" skirts an abyss. On the one hand, it takes the adman's pitch to its absurd extreme. For this, we are to be congratulated. (Catholic Workers have been accused, by the ill informed, of not laughing enough.) On the other hand, the slogan trumpets superiority — back to that shout: "We're number one!" Unfortunately, "America, first and best" means "The rest of the world, last and worst."

Why not give up that race? Why not give away more, lots more, and lower ourselves further? What would it be like to be "the worst"? Would we attain the saintliness of Simone Weil or Layman P'ang or Francis of Assisi? Or would it be just another ego trip? Ah, well, seek the middle path. It would do us, and the world, a world of good.

Anyway, what is "the best"? The composer John Cage contends that *every* seat in the concert hall is "the best." Two hundred years ago, only the poor of London were deemed socially low enough to eat then-abundant oysters. In the lowland tropics today, white potatoes are a sign of prestige. "The best" often means "the fashionable."

At the Family Kitchen we have introduced rice cakes, whole-wheat bread, raspberry kefir, peach yogurt, crème fraîche, raw goat's milk, and chocolate-covered bars of gourmet ice cream. At the end of every meal, when leftover perishables are distributed, there is occasionally a mild crush for these things, which are now considered "the best."

We are choosy beggars. We must be. The white powder on the pears I picked could have been mold, dust, or pesticide residue. We ask questions and wash everything. Our final criterion is that we use only what we would eat in our own homes.

For our benefactors — the grocers, bakers, and restaurateurs — the Family Kitchen is probably as much a helpful blessing as it is a noble cause. We are reducing their garbage bills. One store even gets paid for recycling the cardboard boxes we toss in their compactor. But we're also a bother. Lucky for them — and us — that the state protects their charity with a good-Samaritan law: giving with an open hand and a good heart is exempt from lawsuit.

For some, we are a convenient dumping ground. One woman gave us her quarter of a cow and half a pig because she decided that she didn't like frozen meat. I do not denigrate her motives; my own are similarly suspect. But for the guy who donates locker space in his freezer (to keep the cow and the pig), for the newly-weds who direct all of their nuptial endowment to us, for the local ice-cream company that buys us a nearly new van, for the hundreds of others who give their money and their time — for these, I can think of no payback other than the joyful knowledge that they labor for the love of humankind.

So what's in this for me? Room and board — the equivalent of about four dollars an hour, including Barbara's work as coordinator of volunteers. And time. Time to write, time to make family masks for Mardi Gras and Halloween, and time to sit

still. Time to practice a mix of Gandhian bread labor, Zen atten-
tiveness, and Christian beatitudes. Time for composting, knit-
ting, basket weaving, gardening, making music, and being
peaceful. Time to walk and to bicycle.

As humbling — and fearful — as it is for me to say it, there
is, however, no merit in what we are doing. No failure in this life,
no success. At the fundament, we could be making bombs and
still redeem ourselves. The work's deepest joy is simply being
able to say, in the name of so many others, "Thank you," and,
"You're welcome."

What we do is bandage the system's walking wounded. We
treat symptoms, not causes. There are some who criticize this as
"charity," who say that in the long run this extends suffering.
"Bread lines should have ended in the thirties," these adequately
fed critics say. "Let it bleed."

Bread lines are, no doubt, a disgrace. Hunger, it is widely
agreed, can be abolished. All it takes is the political will. But in
the meantime, with our stew and bread and salad, I don't think
we have anything for which to apologize. To someone who is
hungry, critical distinctions are moot, if not obscene.

I don't know how much longer my family will be a part of
the Family Kitchen. It is a luxury, I know, to spend working
hours accompanied by my daughter. I can take her on a side trip
to see the toy store's pirate ship. Holding her hand as she "helps"
me carry a bag of surplus cookies down basement steps, I can go
at her speed and work at my own — delighting in the moment,
meeting whatever challenge presents itself, mindfully, slowly.

But despite the best intentions, despite the good work, I
sometimes act the drudge. Once, to pull myself out of the dol-
drums, I embroidered the Family Kitchen's child-drawn logo onto
an apron that Barbara had sewn for me. Later, I added "Scavenger Dave" below. Still, I'm nagged by the feeling that depend-
ence on discards is parasitic — even if parasites are indispensable
to life, even if our "recycling" is righteous, even if we do it for the
benefit of others. Maybe I'm tired of cleaning up what seems like
somebody else's mess.

Being enmeshed in the metropolis is part of our problem. It

is a problem of spiritual connections, of land reform, of controls. We'd rather be teaching Amara how to make a shelter, how to grow food, how to survive in harmony. Is it hopelessly romantic to yearn, on a cloudless night, to see the stars? After three years at the Family Kitchen, and four years amid concrete, we want to return to rural land, to dig in, to take root.

I am privileged to be able to walk away. When I'm old and decrepit, who knows but that I might be there again, serving myself.

I hope never to walk away from it altogether. I intend always to have one foot standing at the back door, to be the lowly beggar next to the stinking dumpster, empty cardboard box in hand. I intend always to have the other foot walking away in disdain from the can't-beat-it deal, to be the do-gooder nobleman supporting a neighborly, perhaps more expensive, trade.

We're all scavenging parasites. We're all glorious nobles. Let us be.

September 1989

AMAZING
CONVERSATIONS

John Rosenthal

Sometimes people ask me a question to which, over the years, I've given a lot of thought. They ask me: How come, if you're a photographer, you don't carry a camera around with you? Of course, what they mean is: How come you don't carry a camera around when you go to see the Bulls play baseball in Durham, or when you walk around Chapel Hill on a Sunday afternoon, or when you go to a party and lots of your friends are there? How come you don't bring your camera to concerts where people are already dressed up as if begging to have their picture taken? They'll ask me: What happens if you're walking down a street without a camera and something incredible happens, say, a beautiful woman in a blue dress starts singing in the middle of traffic, or a tiger escapes from someone's backyard and finds himself trapped by the fire department on top of an abandoned gas station? Don't you wish you had a camera with you at these times? Well, the answer is: Probably. It would also be nice if I had color film in my camera so I could at least catch the blue of the singing woman's dress, or if I didn't have color film at least have a yellow or orange filter, so the tiger pacing up and down on the roof of the gas station could be seen against the illusion of a blue sky, since if you use black-and-white film in a camera without an orange or yellow filter, a blue sky turns out pure white. I'd have to have a few lenses, too, to cover emergencies. What if the fireman wouldn't let me get close to the tiger, and all I had was a wide-angle lens? Later it would look as if I'd taken a picture of a gas station, period. It gets complicated after a while, carrying a camera around, constantly making decisions about what kind of photographs you're going to take of the world.

Well, I don't carry a camera around much, and here's the reason, which is a personal one. Thirteen years ago, when my child was born, I had decided to be present at the birth, to help my wife in her labor as well as I could. There were techniques for proper breathing which I could help with, encouragement, conversation to pass the time in between contractions. I also had my camera by my side, primed with Tri-X film and a suitable 50mm lens. Now, as far as births go, it went pretty well; but needless to say, when the time came for the actual birth, I felt pretty tense. At such a moment, even in the world of modern medicine, questions of life and death are not beside the point. Since my role in the birth had been taken over by the delivery-room crew, it was time to reach for my camera. To be truthful, I had almost forgotten it in my anxiety.

I found myself a proper angle of vision, checked out all my readings, and started to compose the photograph that I would take at the moment of birth. Behind the lens, I noticed something interesting: my anxiety was miraculously dropping away. I was thinking about how I was going to get a good picture. In other words, I had left the world in which I was an emotionally involved participant and had become instead a spectator, one who was making art out of a moment of travail. I found myself at such a distance from everything in front of me that you could say I wasn't even there. And when the doctor finally held a gasping infant in the air, I took a picture. Was it a good picture? Well, all newborns look pretty much alike, don't they? The question that remains, however, is this: Is the photograph we get always worth the distance we put between ourselves and the world? I answer: Sometimes; sometimes not.

■

A FEW YEARS ago I found myself walking down a street on my way to visit a friend who was staying with her grandmother. I had no reason for feeling jubilant that day, but the minute I got out of my car, that's precisely how I felt. It suddenly seemed as if my eyes couldn't get enough of the day. I was on a narrow street of tall two-story houses where only a small driveway separated the parlor of one house from the bedroom of the next. This was in a part of New York City called Forest Hills, and Queens Boulevard, with its loud commercial traffic, was two blocks away, but here there were only the sweet sounds of an ordinary day. Birds were calling back and forth to each other; the shrill cry of children on their bikes could be heard, along with the sound of a car driving by. For some reason, it seemed like paradise to me. I felt momentarily as if my eyes had never looked upon a more beautiful world; they were cataclysmic with joy at what they beheld. And what did they behold? Well, nothing beyond the ordinary: leaves falling in the small yards of narrow-gabled old homes, uneven sidewalks broken in places by the roots of the oak trees that lined the simple street. And a few hours later, when I told my mother where I'd been, she told me that I had spent the first four years of my life less than three blocks away from the street on which my friend was staying.

I never understood this experience, nor even realized that I had *had* an experience, until last week when I looked at Gregory Conniff's new book of photographs, *Common Ground*, and read his amazingly eloquent prefatory essay, "Why a Camera." In this essay, Conniff describes how, as very young children, we undergo an immense experience with our eyes, which is nothing more or less than seeing the world before language has dictated to us the meaning of that world. Now, if that doesn't sound like much, that's only because ever since childhood we've allowed words, our own words and other people's words, to define what we're feeling and thinking and hearing and smelling and tasting — words without which we can hardly imagine a human culture, but which paradoxically, by their very dominance over our senses, have greatly inhibited the possibilities of human experience. What do I mean by that? Simply that we grant words permission to represent things they only stand for, as when we see a landscape pulsating with color and energy, and we say, "Isn't it beautiful," even though at that moment we are not experiencing beauty, which is a kind of wild harmony of the senses.

Words bind us to other men and women by permitting us to agree about the identity of things we have in common, which later we will argue or sing about, but in the very act of creating shareable meanings, we commit a kind of crime against ourselves by letting words stand for such things. The *world* may benefit by such an exchange, but we, actual human creatures, pay a heavy price. The specificity of our life on earth is exchanged for abstraction — color, line, shape, stink, and blossom. Everything becomes words, and as we grow up, we put away our laughter and our silliness and our childish noises, the great sensory hilariousness of our young lives. We pick up a few notions about proper behavior, like what books to read and how to go about getting married and buying a home and being polite and having cocktail parties, even knowing how to act at them, and the next thing you know, the little child — who was also an enormously alive sensory apparatus — is just another boring adult going to work in a seersucker suit with a briefcase. Now, this may be the way the world is able to continue, but I have my own suspicion that every time a grown-up picks up a paintbrush, or sits down at a keyboard, or feels the desire to take a photograph of something, that it's really a protest against this conversion of real experience into mere words, a protest not against growing up, but against growing up without our senses.

Now do you see why I felt jubilant that day I walked down an ordinary street in Forest Hills? My eyes, usually dulled by habit, were stunned by seeing for the first time in years the world on which they first looked. They were surprised to find themselves looking at the one spot of earth where they first became aware that they were eyes — aware of color, of movement, of the space between things, and the things themselves. We all have our spot on this earth where our senses came into their own. It could be the city, the country, the suburbs, but wherever it is, we've probably lost track of it. And, I suppose, most of us never find it again, spending our lives in the world of words and the notions that words create. But if someday you feel yourself unaccountably happy, then open your eyes and look at what's in front of you, because what it probably is, is yourself.

■

When I was a little kid, I had one advantage over all other little kids: my father was Joe DiMaggio's lawyer. I can't think of any fact in my life that was so important, as basic to my existence as the air I breathed. Around the age of eight or nine, when you would belligerently exchange facts with your friends about your father — how much weight he could lift, or how fast he could drive — the fact that my father was Joe DiMaggio's lawyer was so big a fact in 罗 that my friends wouldn't even believe it. Whenever these friends would come over to my house to play, they would ask my father, "Mr. Rosenthal, are you really Joe DiMaggio's lawyer?" And when my father would answer that he was, they would look at him for an extra second or two, actually see him for a moment through that vague blur that all adults are to children.

Of course, to us kids coming into baseball awareness right at the end of DiMaggio's career, even Joe was a kind of blur, a legendary name like Ty Cobb and Babe Ruth; it was only as we got older and realized how difficult it is to be really good at anything that we got to know something about DiMaggio. He was not only good at something, but great and wonderful. His 1 9 4 1 hitting streak of fifty-six games was ended in front of a crowd of fifty-seven thousand in Cleveland when Ken Keltner made two spectacular catches of DiMaggio line drives. But the hitting streak ended only for the record books; Joe went on to hit safely in the next twenty games — a feat beyond talent, for it describes character, rock-hard character. And it is just this sort of character that makes his record in today's game, a game played by high-rolling multimillionaires, the one record in baseball that will never be toppled.

When I was ten and my brother was fifteen, a year after DiMaggio retired, my father took us to the stadium for a game, and just before it started he whisked us into a room, and there was the man himself, the Yankee Clipper. Then I shook his hand, and let me tell you something, *awesome* is the only word to describe it, not the awesome kids use today to describe the way a car looks, but the awesome that was once reserved for things like Mount Everest. To this day I remember that immense hand of DiMaggio's, the size of a baseball glove.

And then about five or six years ago, when I was visiting my family down in Atlanta after my father had mostly retired, he turned to my brother and me and said, "By the way, DiMaggio's staying down at the Hilton for the Old-timers' Game, and I've got to go down there this morning and talk some business with him. Would you like to come down there in a couple of hours and meet Joe? Would you still get a kick out of that?" My brother and I looked at each other as if to say, *Can you believe this man?* So a couple of hours later, there we were, outside DiMaggio's room at the Hilton, and an imperious voice inside asked, "Who is it?" We announced ourselves awkwardly, and then the door opened and there was Joe DiMaggio again, the Bronx Bomber, a tall, graying man, tense and unwelcoming, whose handshake now seemed like any other handshake and whose hand now seemed no larger than any other hand. Well, it must have been a bad day, for Joe was unhappy; he was physically hurting, and underpaid for the Old-timers' Game, and maybe he wouldn't do it anymore — he was too old for these things — and then there was some dissatisfaction with the Mr. Coffee contract. Finally I said, "Hey, Joe, who was the best pitcher you ever came up against?" and he didn't take a second to say, "Carl Hubbel," and I said, "What did you think of Mantle when he came up?" and he said, "Mickey Mantle, well, he was a very nice, strong kid," and I realized I was going to have to be content with that. Then Joe went back to grousing about this or that, and a little while later we all excused ourselves and left Joe looking anxiously out the window and down into the parking lot.

Now, you might be saying to yourself, I know what he's going to say now: how sad it is that our childhood heroes grow old and lose their heroic qualities — "Joltin' Joe has left and gone away," that sort of thing. But then you'd have to think there was something fair about my brother and me sitting there in that room with the Yankee Clipper with all of our childhood expectations still intact. There was nothing fair about it. What's fair about having to conclude one of the greatest careers in baseball history after only thirteen years because of injuries? Do you think Mr. Coffee compensates for *that*? For that matter, what can compensate any of us for lost love, but particularly DiMaggio, who lost it forever? No, you won't find me saying anything about the Bronx Bomber but this: when you play ball like he did as a young man, you aren't playing it just for yourself, you're playing it for everybody. And why for everybody? Well, for the simple reason that the day arrives for all of us when we have to have something to remember, something to call us back to ourselves, something that is not likely to be better than the young DiMaggio, Number 5, moving gracefully through the solitary spaces of center field, or standing at the plate, a bat in his hand, opposed forever in his youthful perfection to all the powers of mediocrity.

■

THERE WAS SOME indication recently that I had reason to fear for my health. I didn't know at the time how serious the threat to my health actually was, but there's no question that it seemed serious. There were tests to be run, which took days. Once, I requested some information about preliminary findings, which turned out not to be a good idea, since they were ambiguously negative, whereas the final judgment was positive, or should I say, happy. Data was sent off to another state, another institution, for an answer to a problem I might or might not have, and all that data was misplaced for ten days, and all the while I waited and wondered and grew somber and assumed that the confusion ultimately had its foundation in bad news. When you enter the land of the ill, you do what you can do. Perhaps everybody knows that. What you don't know, and what you're likely to forget once you know it, is how the world looks when your hold on it has become a little fragile, a little tenuous.

With the thought that I might have a bout with ill health that could last for quite a while, I didn't think to myself, *Oh, no, I won't be able to make photographs*, which is supposedly the thing I love to do most, at any rate the thing I'm best at. I thought instead, *No, I'm going to lose contact, for a while anyway, with the* world, *which is what I take photographs of.* I didn't think how sad it was that I wasn't going to get to take pictures of winter land-scapes, whether they be in North Carolina or New York City; what I thought was *I'm not going to have the winter, the winter itself; forget the photographs.* I wondered, *Am I going to be in a room somewhere, perhaps feeling badly, unaware of cold, bright, piercing afternoons when the sky is so blue that it hurts even to think about it, as blue over North Carolina this one time of year as it is over Texas?* Boy, do we take the world for granted, and isn't it almost impossible to know that we're doing it?

No, it wasn't photographs or the world of art that I was go-ing to miss; to tell you the truth, from the perspective of ill health, art seemed to me just a pleasant, middle-class activity, and all my sensitivity to it just something that came with the territory of having been a lawyer's son. As artists, we convert the world into a spectacle from which we are denied participation, by virtue of being outside of that world, observing it, structuring it. To me, it sometimes seems almost impossible to live authentically in the world while trying to convert it into art. Am I perhaps not doing that right now? All those hours spent in the darkroom perfecting a certain gray tone, all the discipline of timing, rhythm, and spontaneous calibration, and all of the hours used up in figuring out where one is going to point the camera: well, I guess this is the way I live my life and this is what I do. But wouldn't it be wonderful if I could sometimes have the world more on its terms and less on mine? I've spent so many hours thinking about pho-tography that I can't remember taking my wife dancing even once in the last year. With the thought that I might be ill for a while, that was what bothered me the most — that I hadn't been dancing in a year. I mean, how did I ever get it in my head that being an artist was more important than dancing with my wife? Most of my life I've heard that art is the intensification of life itself, but I've had the suspicion (and I have it now more than ever) that the people who say that sort of thing have long ago stopped paying attention to the shape of an ordinary afternoon, or to the gift of sunlight on an unmade bed.

Well, the word is in now that I'm OK and will probably enjoy good health over this Christmas holiday and into the new year. And of course, being human, I'll probably let my insight into the primary beauty of life itself grow vague and disappear from my consciousness. Maybe it's not even a very good insight. Maybe we don't get to have the world the way I want to have it right now unless we leave it, go into seclusion without all the impingements of our confusing responsibilities, our families, our car payments . . . and simply celebrate the obvious things in front of us, like rain or winter or the smell of earth. Maybe we have to become ill before we can do justice to the beneficence of good health. I don't really know the answer to this question.

■

ALL MY LIFE I've heard the expression "A photograph doesn't lie." But the real truth is that photographs do in fact lie about some things, and not about others. Is this what Diane Arbus meant when she wrote, "A photograph is a secret about a secret"? I loved a woman once who left me with very little warning. Had love blinded me, or was I merely self-centered? For five weeks I felt a lover's anguish. And then it was gone. I questioned myself closely: were feelings being hidden, disguised, outgrown?

Five years later, I looked over the contact sheets of photographs taken during the years of our romance. Her image appeared everywhere — from the beginning, when everything was a sweet tempest, to the end, when long silences occurred regularly. I thought to myself: These pictures will tell me something, at least something about myself. I have had five more years of sense and experience; surely I stand a good chance of finding out *something* about her that I didn't know at the time.

With all those small images spread out before me, I began to look. There she was, as I photographed her, large eyed and dramatic. I remembered when I took every picture — moments of mutual travail were recorded, and moments of high-on-the-hog. In these photographs she was most presently there, a good image for any photographer to come upon.

But I began to notice something I had never seen before — and it is fair to say that it was the secret of these photographs, the one they concealed from me until I least needed to learn it. To put it simply, her image was there, but she wasn't. What I beheld was a version of self which continually changed, and always for the benefit of the camera. For just a moment I felt a huge abyss had opened up at my feet.

Like many women who have spent a great deal of time in front of mirrors, she had become precisely aware of herself from the outside; in fact, she had come to dwell in that outside image of self. The version of self that I recorded in my photographs was only whichever one she deemed appropriate for that occasion. It was, in other words, what she allowed me. When it was time to be sad, she was sad like an actress is sad; when I dared the camera to find an image equal to our joy, there she was, as if looking at an angel for approval.

Am I suggesting that she wasn't as alive as I was? There is plenty of proof to the contrary. Nor would it be fair to say that, wearing so many masks, she was a creature of artifice, less real, say, than myself. Her tears, after all, tasted of salt and left a trail down the sweet dust of her face.

But if I may be permitted a metaphor to say something that is otherwise almost inexpressible, I would say this: her inner landscape was determined by conditions which didn't really exist for me. Because she knew how to react to almost everything, her skies had a finished look to them, a blue that promised not to cloud over just as the guests were arriving. She was not averse, it seems, to being perfect. This was the secret of my photographs, which came to me five years too late. But too late for what? The fact is that in my inner landscape it rains a great deal — and often at the wrong time: on days when a trip to the beach is planned, when the picnic is packed, when love has made a bed for itself in the garden. How often it rains at the wrong time. But, oh, how I love it when rain pours out of the sky, ruining everything, all plans and projects and schemes and expectations, leaving me sitting in a dark house having amazing conversations with myself.

■

I THOUGHT I saw Beth the other day. I could have sworn it was her — a tall young woman in a full skirt, wearing leather boots, her hair long and rippled like the sand after high tide, her long sleeves rolled up to her elbows. She even walked the same way, taking big, long strides, as if the only thing she ever wanted to do was walk — not those huge strides which suggest defiance, but more a merry walk that borders on self-admiration.

What reminded me of Beth also had something to do with where I saw her walking. It was one of those places where you might see Beth, but where you would never expect your other friends to be. Years ago, you might suddenly see Beth crossing the interstate miles from town, coming out of some nondescript patch of woods, some stray dog rambling in front of her. Or you might be driving down a small road in Chatham County, and there she'd be, bending down to pick up a piece of quartz or mica that had caught her eye. The woman I saw the other day was walking by herself, with no other pedestrian in sight, beside a detour around a highway-construction site near Duke Hospital.

Similarly, you would never be able to anticipate when Beth might show up in your life. Once, when I was living in the country outside of Chapel Hill, renting a huge house by a pond for seventy-five dollars a month, Beth showed up during a snow-storm asking for a place to stay, announcing with tears in her eyes that she'd just had her heart broken. One summer, I was staying with a friend in Boston in a little apartment he was renting on Beacon Hill, spending the summer there in the hopes of writing one or two poems that might be deemed acceptable according to the ludicrous standards I was then setting for myself. I received a phone call from Beth, who told me that she was waitressing out on the Cape in Hyannisport and that I had to take her camping that very weekend on the dunes by North Truro.

That weekend, high on the dunes above the unfamiliar sound of the ocean, on the night of a full moon, I woke around three o'clock and saw Beth hundreds of feet below me, walking along the edge of the bright water.

Looking back now, it seems that while the rest of us were protesting the war and going after degrees and developing those abilities that in later years would enable society to define us, Beth was insisting on her right not to take herself too seriously; to make, as it were, a small art out of wandering around, house-sitting, and putting together odd collections of found objects. She should have been an artist; in fact, she should have been a photographer, what with all that perambulation and unortho-dox appreciation for the everyday. But as you already may have gathered, Beth believed that truth — at least when she would think about it — was to be found only in the heart of spontane-ity, and that all the learning of technique which any art requires would eventually put a cramp in her soul. And furthermore, she liked to sleep very late into the morning, and after she woke up, there just wasn't that much time to get into the intricacies of self and self-expression.

No, it wasn't Beth I saw the other day, striding along the road, her hair flying, ruffled in all directions, her sleeves rolled up to the elbow. It was a young woman who reminded me of her, that's all. Beth's forty years old or thereabouts, married to some fellow in microchips, and living outside of San Francisco. Some-one who saw her a couple of years ago told me that her hair was about two inches long. And as for her blessed spontaneity — I don't know. She hasn't dropped in lately, though of course she knows my door is always open.

December 1986

Born Too Young: Diary Of A Pilgrimage

Sparrow

Sparrow, we're flying," she says, which is a good sentence.

"Look out the window — it looks like the end of the world," I tell her, "an eerie barren coastline."

"That's the wing of the plane," she says.

A year from today we will straggle into Calcutta and see Baba — that's the plan. Until then, there's this.

WHO AM I? An aspiring writer traveling to his guru in the Age of Madonna. I'm thirty-two — which is a little old for this — and I'm traveling with the woman I love, and I'm less afraid of flying because she is the one I want to die with.

But first I want to see Baba and offer myself to the Lord. I'm not sure he's the Lord — part of this journey is to find out — but whether he is the Lord or no, or whether *anyone* is the Lord, or whether there *is* a Lord, I want to present myself to the Lord, and the place to do it is where Baba is. Why? Because, as a member of the Ananda Marga Society, I've been dancing around his picture for eleven years, and he's come to represent the Mystery.

The Trip to Bountiful just started. A woman with blue hair is talking to a man without making a sound. For three dollars you get the sound.

"It's good, but it's salty," Jeanne says of the quiche.

I'M REALLY TRAVELING, after two years of planning. It's Jeanne who made me go. She said, "Set a date," and this is it, July 20. Three months ago, she called me on the telephone, cried, and asked me to sleep with her — and I told her no, stay with your husband.

And now we're married, and I'm wearing the turquoise ring she and her old boyfriend Nelson bought in Mexico; I've wrapped wire around it so it won't fall off.

"We're flying into the dawn," Jeanne says, and Geraldine Page is crying, in a silly hat.

I go to sleep in Jeanne's lap, then worry myself awake.

"Look, there's land," she says now. "Between those clouds — can you see? It's green. I wonder if we're over Ireland."

But I can't see. I'd intended to buy glasses on Dyckman Street before leaving, but there wasn't time, so I won't see too much. I'm not sure how much I *want* to see.

"See anything?" I ask.

"Yeah. Fields. Look."

Just like America, I think: pieces of green and brown stitched together, white roads.

"Fluffy white clouds are really lots of water that will soon fall down," Jeanne tells me.

"What's that, something you remember musical scales with?"

"It's a haiku I wrote in fourth grade."

A way out is what I'm looking for. A way out of *what,* I'm not sure.

■

IN LONDON, JEANNE changes our money: "It's $1.51 to a pound!"

she says. "That's terrible!" I like the money; it looks Chinese. Our finances are already deeply interwoven: I bought her quiche; she paid for my luggage. It's my secret hope our finances get so interwoven that we never part.

Lately, I secretly feel we will *not* be together forever. Staring at her introverted WASP good looks, I think, *No, not her.* She's told me she feels the same. She, leaving a marriage of seven years, wants security but not finality; I, leaving a lifetime of aloneness, want finality but not . . . proximity.

The dharma brought us together and the dharma will bring us apart, at the proper time.

A sign on the bathroom door says, TOILET ENGAGED.

"This is fun," I say.

"It's like they made this whole place just to amuse us."

When I try to define the dharma, I fall into abstractions I've read in dictionaries.

"There are no panhandlers here," I tell her in Victoria Station. Then a fourteen-year-old comes up to me and says, "Kin you lend me sixpence for the tube?" I give him ten pence, and he gives me back two.

People in London write the same things on benches:

Dave

+

Claire

"Sparrow, there's a clock with a sign over it: 'This clock is three minutes fast.' "

What is dharma, the hero of this poem?

∎

JEANNE AND I are bored with one another, bored with London. She quit her job hoping to relax, and now this traveling without direction is worse than working nine to five. Jonathan, her former professor who writes for the *Observer*, made us feel like fools. How can we expect to get a job when there're three and a half million unemployed? Beats me. Jeanne seems to want to be in India, or at least Africa.

Of course there's the dharma, watching over all this like a lion, the dharma that moves only to stretch its jaws.

My new plan is to unload all my spiritual advice immediately, to get to the real part, whatever that is. (Spiritual advice to myself: Go the Royal Wedding; find a pot to make rice and beans in; write this book. Try to serve.)

The dharma is what you do that doesn't result in pain later. The inverse is the definition of karma.

I REFUSE TO cut my hair; I'm not sure why. It's similar to the way I refuse to spend Kennedy half dollars — not because I respect him, but because I respect the way I once respected him. I respect the John F. Kennedy of my childhood, who died in my fifth-grade class when Mr. Hassett came in the door looking worried, and later David Wolfthal cried, because, he said, "Today is my birthday."

Garry Wills points out that Kennedy, the youngest President, and Ronald Reagan, the oldest, were born only eight years apart — and though one is a saint and the other a fascist, their policies were almost identical: Nicaragua, Vietnam; the invasion of Grenada, the Bay of Pigs.

But at twelve, Jessica Gorton and Debbie Ross and I made chocolate-chip cookies (this was before you could make them from a box) to raise money for something — a library? And we giggled a lot and were united by the Cause. As I look back, it was like working for Stalin — noble in every way except its purpose.

THERE'S NOT MUCH hostility toward Americans, though the couple we met at the airport said, "For a few weeks, when there was a lot of terrorism, you didn't see them," as if they were talking about an infestation of mice.

(And when I said, "As soon as we decided to go to Europe, everything got radioactive and the dollar fell," they smiled proudly. The English are pleased to host calamity.)

■

IN HER JOURNAL, Jeanne mentions returning to New York; it activates my fear of being left. Joan left me for Paul Rosensweig, Marianne left me for Anurag, Anat left me for Israel — and though they all came back and then I left *them*, their leaving is what I remember. Will Jeanne leave me? Will I force her to? That's why I'm going to Baba, to get off this turntable — and that's why I married Jeanne, so she couldn't leave me, though you can't stop the leavers from leaving once they decide to leave.

Things are about as bad as they can get:

1. Tomorrow we lose our room with little prospect of finding another.
2. We have no contacts here, and we're meeting fewer and fewer people all the time.
3. Jeanne is losing faith.
4. We're both dully mad at each other.
5. I want to go to the Royal Wedding even though we have lots more important things to do.

[*Sparrow and Jeanne leave London headed for the U.S. naval base in Faslane to join the Peace Camp, the site of a permanent, ongoing protest against American military presence.*]

NOW WE'RE IN the village of Milton Keynes. We rode a coach two hours from London, discouraged by the countryside; it looked like Pennsylvania. An eight-hundred-year-old hedgerow doesn't *look* any different from an eighty-year-old one.

MY WORRY TODAY is that I'm falling out of love with her. She's a good listener — a *thoughtful* listener — so it's taken me this long to run out of things to say and start listening to *her*. And she doesn't have much to say. She thinks bumblebees and flowers are beautiful (I suppose they are); she didn't like *Hannah and Her Sisters*; she's confused a lot, and confused about why she's confused; she didn't like London, and she doesn't like Milton Keynes because she doesn't "belong here." ("That's the idea of traveling," I explain. "You go places you don't belong. If you belonged there, you'd already be there. Personally I know where I belong, and that's why I'm not there.")

■

WE JUST HAD an enormous fight, and Jeanne shouted, "Shut up!" and ran away. I assumed we were finished, that she'd gone back home — but she was on a ridge, crying, all curled up. I told her, "I'm afraid that without sex and without romance there isn't enough to hold us together," and she said, "Don't you see, Sparrow? *This* is what holds us together."

WE'RE BOTH A little sick and vaguely despondent. We've reached the part of traveling where you no longer think everything's cute, but you can't tear yourself away from looking.

■

MORNING IN GLASGOW — according to our clock, which is wrong. Porridge simmering on the stove. Jeanne making the laundry clean downstairs.

I was desperate to get to Faslane for the protests on Hiroshima Day; then I discovered that yesterday was Hiroshima Day. (I'd thought it was the eighth. I make that mistake every year.)

Luckily there's Nagasaki Day, in two days. But we don't know if Faslane even exists. It's not on a map. Some people we met here, a couple from Madison, passed a Polaris submarine on a loch on their bicycle trip. They stopped and took each other's picture in front of it. But they didn't see a Peace Camp.

I'M IN LINE at a discount store in one of those fascinating areas of the modern world: a brand-new slum. Everything is plastic and fluorescent, and the buildings are high-rise, but there is about the place an unmistakable air of poverty and restlessness.

I'm feeling the unity of the colonized. (I think I've been subconsciously angry at the English for once keeping my country in bondage.) And everyone's dark and looks Jewish.

"Jewish?" asked the man at the bus station.

I said yes, and he half closed his eyes. "Good, good," he said.

■

NOW WE'RE IN Helensburgh. Jeanne ate meat this morning! That upsets me.

"You said if I wanted to eat the local food, I could."

"What about the poor animals?"

"Oh, shut up."

Now I'm shutting up, and she's walking along the curb. She's breaking my rules one by one. One day she doesn't meditate. The next day she won't read *The Secrets of Chinese Meditation*. The next day she eats meat. The sun is setting on my British Empire; she is India, and I must let her go.

[*Sparrow and Jeanne find the Peace Camp and are arrested during a demonstration.*]

THEY DRAGGED ME awhile, cursed me, and my hat fell off. "You've made your point," they said. "Now we'll give you a chance to walk."

I felt sorry for these guys, puffing away. They stood me up, and I started to half walk. We had a good quarter mile to go, and by the end I was practically supporting them, one on each arm. They were so out of shape, just walking up the hill bushed them.

"You did the right thing, cooperating," my officer told me.

"My quarrel isn't with you, anyway."

"Yeah. Everybody's got to make a living," said my arresting officer.

They took me out and sat me under a pylon at the base of the hill, where everyone else was. People were divided into Those Who Walk and Those Who Are Dragged. Those Who Are Dragged were higher caste — they got cheers from the supporters at the gate and looked more cheerfully absurd — but I'd already become a walker, and it was too late to go back.

The officer filled out my description without my cooperation. Two of them worked together to guess my height (five-foot-ten) and my weight (eleven stone). They described my hair as "long and tousy."

The captain started telling me that if I didn't have a passport, I was violating the Aliens Act and would be deported — at my expense — like the five people they deported last week for demonstrating. "Just send 'em back," the captain said.

"But if you don't know what country I'm from, how do you know where to deport me?"

"Maybe to Russia," he suggested.

I became friends with my second arresting officer, who started telling jokes: "Where did the first Englishman come from? When a monkey screwed a pig. And the monkey's suing!" He guessed I was Hispanic and called me Jesús Rodriguez, though he later revised my ethnicity to "Hispanic with Jewish tendencies."

"I'm gonna arrest you for pot," he told me, gesturing at the cooking pot in my bag. Jeanne liked that.

"I think if I had twenty-four hours with you I could get you around to my way of thinking," he said, but the opposite seemed to be happening.

"He's a Zen Buddhist type of guy," he said to another cop.

"No, I'm into yoga. I'm with the Ananda Marga Society."

"I'm with the Mafia," he shot back. "Tell me, this meditation that you do — is it Transcendental Meditation, or is it beyond transcendental?" he asked.

"It's beyond."

"Well, how high is it, would you say? Around thirty thousand feet?"

"That's about right," I said.

He read my book and criticized my handwriting, my spelling, and my punctuation. The best part was when I told him I agreed with those who didn't walk.

"Why?" he asked.

"They obstruct the wheels of justice."

"There *is* no justice," he told me low. "There's justice for them that have, but them that have not have no justice." He smiled at me. "You know, if I hadn't been an officer, I would've been a freedom fighter."

"It's not too late," I said.

"What do you think I've been doing all these years? How do you think I know so many languages?"

"You've been a freedom fighter?"

He nodded.

"In Angola?"

"No. South Africa."

"For the blacks?"

"For the honkies."

AFTER DINNER I walked by the loch; it was the first time I'd seen it in a green-silver dusk light. It was like when your girlfriend dresses in an evening gown and comes down the stairs in candle-light and looks just like Janet Leigh, and you think, *Do I deserve this?*

Jeanne is like that, so pretty sometimes I bite my lip. (And that I love her is in there somewhere, that in some way I *do* deserve her. And somehow I deserve the loch now — because I got arrested for her?)

I went up to the water — the sky still light blue over the hills, but the water already dark — and in my mind I let go of Jeanne, scattered her to the universe. *I've loved her; she's loved me,* I thought. *That's the best we can do. Eventually we'll part. Why try so hard to ensnare her?*

I HAD A turtle in fourth grade. It was small and green, about the size of a silver dollar. Its shell was not even hard. I bought it at Woolworth's. It lived in a clear plastic basin, smaller than an LP, with an artificial palm tree perched atop an elevated area meant to resemble an island. I kept the turtle at school for some reason. I don't remember its name. I used to take it out and pet it a lot. When it died, Mrs. Dunne told me I'd petted it to death.

And that's been with me ever since: the fear I'll pet everyone to death.

SEE, I KIND of stole Jeanne from her husband. That is, our eyes met — hers and mine — after one of Alice's poetry classes, and we spoke on Ninth Street on the way to the Yaffa Cafe with the gang. I told her to meditate.

After that, we took to walking through Tompkins Square Park. Spring was coming out of the earth. It became a kind of joke that we were in love. We never said it to each other, but my friend Sheila saw it in my eyes.

One day, Jeanne threw her arms around me just before she rode off on her bicycle. I was elated.

I knew she was married — she wrote a poem thanking her husband for doing the dishes — and I met him at Alice's reading at the Sixty-third Street Y: a lanky, older-looking artist type. She loved him, I could tell.

So I was surprised when she called me at my parents' house

one Sunday. I'd never given her my number.

"This is Jeanne," she said. "Scott and I have been having trouble lately."

She asked me to sleep with her and started to cry. I said, "Stay with your husband; you'll have the same problems with any man. Maybe we could just go to the movies?" We wrangled affectionately. Finally, she hung up.

When she moved out, she asked me again. I walked around for a week, wondering, *Is it adultery to sleep with Jeanne?* and asking my Jewish friends for advice.

Jeffrey, my oldest friend, said no, absolutely don't do it.

Sheila said it sounded very exciting.

I asked John if I could use his apartment, and he said sure.

Then I did it and went around asking the rest of my Jewish friends if it was adultery.

Eli, the Jewish yogi, said that it *wasn't* adultery, but that flirting with her when I knew she was married was morally questionable.

I had started referring to her as "the adulteress," but after that I stopped. I trusted Eli.

Then one day, at Jeanne's suggestion, I called her husband, Scott. He sounded just like her; he kept starting sentences and not finishing them. Finally he said, "I guess what I want to do is just cry" — and he cried. I didn't know what to say. Should I apologize for stealing his wife? Tell him I'd been in his position? That it wasn't really my fault? I remember being very calm.

He told me about being at the beach by himself, out on a jetty, and a bunch of sparrows flew by and he thought, *Even here, the sparrows.* And I thought, *Jesus Christ, for the rest of his life this guy's gonna hate sparrows.*

JEANNE WAS IN the Group. She joined within a month or two of her move to New York, in 1978, and had two "mates" before Scott. One was a young guy named Mark — cute but unreliable. The other was Louis, who played the piano. They got along miserably, but she drew a loving cartoon about him that she still has.

When she and Louis broke up, Louis and Scott became a couple — but they were even more miserable.

Jeanne wanted Scott, so they got together. Then a week later she didn't want him anymore, but she stayed with him.

Everyone had a mate in the Group, mostly heterosexual but occasionally not. There were thirteen in the Group. Stanley, who founded the Group, had two mates.

Stanley was a Jew in his fifties who was once a renowned Beat poet. His idea was that, since everyone gets bored with their mate eventually, you should have sex with other people. I think they had some kind of rotation within the Group. (I'm afraid to ask about a lot of this.) They also had sex with people they met on the street, people they were trying to recruit into the Group.

Jeanne once said, "We'd meet people on the street, bring them back, talk for a few hours, then flip coins to see who would sleep with whom, and go off and have sex."

The idea was total honesty. (Stanley's booklet was called *Stop Lying.*) If you were completely honest, you could overcome all the conflicts that arose. When Stanley was there, everything worked. He had a way of balancing out conflicts. Jeanne once implied that he was her guru.

Stanley believed in Reich's theory of the primacy of the orgasm; he believed that people exist to give each other pleasure. But in the period Jeanne knew him — and, of course, had sex with him — Stanley was essentially impotent. Later he became an insomniac, developed a mysterious, undiagnosed disease, spent a year in Saint Vincent's Hospital, and died.

After Stanley died, the Group lost direction. They were all supposed to take care of the person with the greatest need at any given moment, but it was too hard. Everybody was in love with somebody else but could never quite *have* them. And jealousy never gets easier, Jeanne says.

She wrote once that the most satisfying moments were crying on someone's shoulder.

Stanley died because they were too dependent on him, she believes. If Stanley were alive, she would still be in the Group.

My best friend, Miami Steve, met the Group. His brother Larry got involved with them in 1976 because he liked the idea of getting laid a lot. He took Steve to meet them. "They were all these droopy, depressed people who thought they were so much better than everyone else," Steve told me. "And the guy, the leader, was just into power. He looked me in the eyes and said, 'I can help you. Will you join us?' It was sick. It was really sick."

JEANNE ACTUALLY SAYS horrible things like "I need to feel my body more," things I vowed I would never have a girlfriend who said.

She has had sex with hundreds (thousands?) of people. She said she once tried to count them and failed. And though she emphasizes that the sex was often mechanical, for years she got to sleep with anyone she was attracted to. You can tell she thinks it unfair she can't now.

ANANDA MARGA HAS a belief that one should marry a fallen woman and raise her up, and I feel like I'm doing that — *teaching* Jeanne monogamy — but I wish she'd renounce her past, or at least admit that Stanley was a bastard as well as a saint.

SEPTEMBER 10
Dear Sparrow,

I started to make you a card for our anniversary, and I suddenly realized I've been frozen with regard to you. Emotionally frozen. Since when? Maybe since we got to Faslane. Why? Fear of losing you? Remember how much I loved you? I feel like a shadow of that now. I can't remember what it felt like; I remember I loved you more than anyone, and now I'm half turned away. How do I turn back? It would hurt so much to lose someone I loved as much as you. Somewhere underneath, I still love you, although I've somehow convinced myself I don't. I have to dig it out.

Jeanne

I'VE BEEN THINKING we won't stay together forever," Jeanne just said.

"Why?"

"Sometimes you feel . . . insubstantial to me. Although a lot of the time I feel insubstantial to myself."

[*Sparrow and Jeanne take jobs picking fruit. John, a fellow worker, offers them rides each day to the orchard.*]

PERHAPS TEN TIMES a day I have these auguries, like *If this plum falls, our relationship is over.*

I had her and John sit up front coming home today. They acted like nervous teenagers on a date. She ate one of his sweets (a bad augury), and I sat in the back. I felt like a pervert who hires guys to fuck his wife.

JEANNE TOLD ME more about the Group today, in a big uncharacteristic torrent while leaning against the apple bin. When Stanley was alive, they'd all go over to his house almost every night. They'd talk for hours about their feelings, then choose whom to sleep with. "We were encouraged to express a preference, but if two people wanted the same person, they would flip a coin."

"Did you ever have, say, eleven people together?"

"No, the most we ever had was five. We tried six once."

"How was it?"

"Great for about five minutes. But we rarely did more than three. Four is hard to coordinate; if two are doing something, the other two have to do something." She looked at me. "So what do you think of my checkered past?"

"It seems almost innocent when you talk about it," I said. "And kind of foolish. Just a bunch of people chasing their tails — or chasing each *other's* tails."

But now, writing this, it seems horrible, almost satanic. I feel how Jerry Falwell does — that as Rome falls, people run around screwing as much as possible and call that enlightenment.

I FOUND THIS book titled *Tell Me Why* in our room. It explains everything: the first newspaper was published by Julius Caesar in 60 B.C. (*Acta Diurna*); every year, 2 million people die of malaria; one rarely stammers before the age of five; during sleep, the only muscles that don't relax are the ones that keep your eyes closed; and the honeymoon comes from the practice of stealing the bride by force. ("That's kind of what we're doing," I told Jeanne.)

WE'VE BEEN FIGHTING a lot. I'm angered by her attraction to John. Everywhere we go, she gets obsessed with some guy. Then I panic, cling to her, and she treats me with contempt. I explode, and we leave.

She wants this other man, though she knows rationally she'll never get him. *I'm* the other man; she got me, now she wants the other other man.

WE WERE HIGH up in some damson-plum trees — thirty-five feet. On the ladder, I shook with fright.

If I were a true devotee, I wouldn't be afraid to die, I thought. *Good thing I'm not.* And I climbed back down the ladder.

[*Jeanne departs for Wales, leaving Sparrow in Glasgow.*]

OCTOBER 7
Dearest Satyavani,

The Peace Camp, right now, is nine people in the Communal Caravan who've just finished playing cards and eating "pieces" — sandwiches. I like it here; you can retreat to your room and read two hours, then enter the parlor and hear people tell jokes like "How do you know if there's an elephant in the refrigerator? Elephant tracks in the margarine."

Jeanne and I are fighting to see who's in charge of the relationship; neither of us likes to lose. I love her, and I'm happy she's off meditating in Wales. I just finished Dante's *Inferno* — great book, if a bit ghastly. I look forward to starting Purgatory tomorrow.

Love,
Sparrow

I CALLED MY parents and learned they're coming to England — or France, or wherever I am — on Christmas.

The separation from Jeanne is easier than last time. Though the thought that she'll run off with a virile Buddhist *has* crossed the threshold of my mind, I have refused to give it a home.

My biggest fear is she'll want to be alone.

SHE'S IN A youth hostel in Chepstow, just over the border in Wales. She told me last night, "It struck me that I really do love you and have nothing better to do than spend the rest of my life with you."

LAST NIGHT I read the *Notebooks* of Camus: "The moral is that we never realize for which crime we are punished."

OCTOBER 21, THE Ashburne Guest House. A nondescript man with a mustache answers the door.

"I believe my wife reserved a room for me."

"Yes," he says vaguely, looks at a list, and takes the keys for number 6 off the board. In the room, I see no one and four beds.

"I thought she'd already arrived," I say.

"Oh, perhaps. I'll go check."

I sit for four minutes. The door opens. "I'm sorry. She's in number 1."

She opens the door with a laughing smile — tall, in a too-big sweater, her eyes keener than I've ever seen them. "You're here!"

I put down my bags and we hug, awkwardly.

"We should carry pictures of each other so we can make sure it's the same person," she says.

BUDDHIST PSYCHOLOGY IS so fascinating!" Jeanne explained. "The idea is you're constantly getting these sensations — pleasurable sensations that you want to continue and painful sensations you want to avoid. They begin inside of you, but eventually they reach the surface. They normally kind of dissolve away, because all experience is transitory, but if you hold on to them, you push them back inside you, into your unconscious, and they become *sanskaras*, and you have to experience them again. The trick is to catch them when you first recognize them, and realize they're

transitory, and not hold on to them. It's hard, but I'm just starting to get it."

I cried at her devotion to the practice.

Now for the bad news: as meditators, we are utterly incompatible. We sat together for ten minutes, and she got up and said, "I
feel like this bright light is shining on me. It reminds me of when
my father used to take home movies."

November 4, in London: I feel very close to America. I feel that
if I'm not careful, if I slip on the floor, I'll be back in America.
And I don't want to be in America. For one, I'll be embarrassed
that I never got to India.

And I like being an expatriate. It's glamorous and funny.
You walk into a neighborhood with your pack on your back, and
within two days you've found the post office and the library and
you're lamenting gentrification with the locals.

I'm afraid to go into therapy, afraid I'll shout at her, "You've
been a millstone round my neck since the moment our eyes connected on East Tenth Street!"

Counselor: Please explain the difficulty, Mr. . . . . Sparrow.

Mr. Sparrow: We've been traveling through Britain for four
months now. Everywhere we go, she tells me she's infatuated with
someone and wants to sleep with him. Finally, it culminated in
Cheltenham, where she was in love with our friend who drove us
to work every day and found us new places to live, and who was
consistently sarcastic and abusive to me. We decided to have a
talk with him, to get things "out in the open," which resulted in
his inviting her to Glastonbury the weekend after I'd gone.

Then she went off to meditate for two weeks at a Polish Boy
Scout camp. After that, we got back together. She says she's not
in love with me. I told her romantic love isn't everything; she's
supposed to be a Buddhist now. We spent a lovely day together,
singing in the rain, hitchhiking. At the end of the day, in a loft
bed in Bristol, she told me she's still not in love with me. I got
angry, we made up. We had sex, great sex. The next day she

looked at me like I was Tyrone Power. The *next* day she told me she's thinking of going on to India alone.

If she doesn't want to be in a relationship, fine. I'm sick of convincing her to be with me. I'm sick of being her minister/healer/dog!

Counselor: And your side, Mrs. Sparrow?

Mrs. Sparrow: We decided when we came together to be frank with one another — he *wanted* that, or so he said. I've never committed adultery, never even touched another man. All I did was tell him my thoughts, and he told me his. He told me hurtful things, too. He doesn't like to talk about it. Just as I was starting my meditation retreat, he told me about how many women he was flirting with and how big their breasts were.

Mr. Sparrow: But you can understand —

Counselor: Excuse me, Mr. Sparrow.

Mr. Sparrow: Sorry.

Mrs. Sparrow gives Mr. Sparrow a piercing look.

Mrs. Sparrow: We always worked everything out. How could I *not* tell him what I'm feeling? And I felt that we were getting closer and closer — taking risks with each other — and suddenly he shut down.

Mr. Sparrow: I got tired of her crying on me all the time.

Mrs. Sparrow: I wanted *him* to cry on me. I liked him confiding in me.

Mr. Sparrow: That's what you *said*, but you were too fucked up most of the time even to take care of yourself, let alone me.

Counselor: If you're so angry at her, why don't you leave her?

Mr. Sparrow: (*Breaking down.*) I'm too proud. I want revenge. I want to try again. I don't have the courage. I'm confused.

Counselor: I see.

Mrs. Sparrow: I knew he wouldn't be able to handle it.

Mr. Sparrow glares at Mrs. Sparrow.

Is it fair to endlessly retreat from her and tell her, "Wait for me"? This is how I lost Marianne. But when I see Jeanne, it's worse.

The choices seem to be periodic fights that eat at my guts or a fight that never ends.

■

I'M LOSING SOME basic respect for Jeanne. Ever since she told me she wanted to leave for India, her body seems like a coat. It doesn't draw me, except when we're making love, which has been "hot," as she says. And she thinks my dick's gotten bigger.

NOVEMBER 25
Fred,

Yes, traveling is its own career. Suddenly no one asks, "What are you doing sweeping the floor in a gymnasium at your age?" but instead people say, "Don't you realize you're doing what everyone dreams of?" Certainly if one is going to be a failure one might as well be a failure in motion.

I'm in London, trying to get rich in the First World so I can spend it in the Third World on things made in the Second World, but it's all going wrong — the money, that is. Lost it in a phone booth. (Sixty pounds!) But the life's pretty OK: frost on the window (the *inside* of the window), but plenty of time to write, and for some reason the novelty of double-decker buses doesn't wear off. I find myself hoping for the death of Reagan, but then I think, *Gee, you don't even believe in killing gnats, and Reagan's life is worth as much as a gnat's.* And besides, then I'd have to wish for the death of George Bush and the Speaker of the House and on down to people like Alexander Pike (the principal of my high school). You asked for my political analysis: the Third World will rise up and save us with inefficiency and devotion, the only cures for the twin evils of civilization — money and sex. Now can I be in the movies?

Love,
Sparrow

WE BROKE UP for twenty minutes. "Which idea are you trying to give me?" I shouted into her face, and she walked away.

"I don't know what to say. You know, I've been thinking about all this a lot, and I've had a lot of stray thoughts 'cause I have a lot of conflicting . . . It's really . . ." I stopped writing her words down.

Today she's 90 percent sure she's going to India alone. She'll meet me there, but she *may* sleep with someone else. Then she changed that; she won't. When we'd broken up, I thought, *I'm happy for her.* Fancy that. *She does have courage.* Fancy that.

I HOPE HER lovely hair falls out.

I hope she reaches nirvana — fast.

I'll be OK. I just need a drink. Waiter, the usual.

All the dicks to her are singing: "You are lonely and we are, too."

YOGA IS NOT prison, but it will, all the same, help you pay for your crimes.

"Pain . . . well, I *remember* it. Nowadays I just feel bliss. Pain, as I recall, felt like hunger — only more *localized.*"

I hate women. They smell bad. Unless they're smelling good. That's the problem with women: they're always smelling bad or good — as opposed to men, who simply *smell.*

So why do I like her? I like that frozen look of pain when I stick my cock up her butt.

I'm a terrible bowler, but I don't keep trying to bowl. Why do I keep trying to have girlfriends?

And sure enough, I petted her to death, like all the others.

WHEN I WAS a boy:

I was bit by a monkey.

I was bit by a duck.

I got my head stuck between the bars of a fence.

I touched the bottom of a swimming pool and got a fish-hook in my foot.

I went through a period of kneeling next to my bed at night to pray for members of my family. I'd seen Dennis the Menace do it on TV. I got my sister Anna to do it, too. One time, Grandma saw us and told us it wasn't Jewish.

GOD, BLESS JEANNE. Keep her from harm — particularly from me. May she find what she seeks, everywhere. Watch over her;

give her a hint from time to time. Teach her how service works. Let lots of people see her sweet heart. May she find her work, and work hard.

WE JUST TRAVEL in a circle. There *is* a sort of perpetual motion to it. Maybe we could patent it in Rube Goldberg form: He puts penis A into vagina B. She begins to cry into handkerchief C, which pulls open trapdoor D. He falls through into pit E. He swears revenge, throws spear F up through trapdoor into target G, which causes noose H to loop her around neck I. She cries, "Help! I'm dying!" He ascends ladder J to save her. They kiss. He puts penis A into vagina B. . . .

ODD HOW TWO people can start out in *love* and after a time they *despise* each other.

If I tell her, "Be true to me in India; do not sleep with Other Men," she'll resent me. If I say, "Sleep with anyone," I'll hate her — and smash in her face when I reach Dharamsala.

THE BEAUTY OF the ego is it can weep as much over a fried egg as over the death of its mother. It's always saying, "*This* will save me — if I can just get this egg to turn over."

IN THIS AGE, one must pray for prayer — for the ability to pray.

■

A LOT HAS happened since I last wrote: I peed on Jeanne, and my parents came.

It was Christmas and we were alone and I turned to Jeanne and said, "I've been thinking; maybe I should pee on you," and she said, "That's funny, I've been thinking that, too," and I said, "I guess in the bathtub," and she said, "Yeah," and I said, "We could bring up the heater," and she said, "OK," and I brought up the heater but there was no outlet in there — you know, the British don't *worship* the bathroom like we do — so she said, "We could run some hot water in the tub," and I said, "It's hard to have sex in water," and she said, "We'll keep it low," and I said, "OK," and we climbed in and I stood over her and she started to fondle my thighs. She was kneeling and the water was washing my feet. It was hard to do. If she touched my penis, it'd begin to stiffen, and then I couldn't pee; and it was cold, and it's hard to pee in the cold; and I'm not used to peeing standing up, because Ananda Marga says not to. Besides it's hard to pee in *front* of someone, let alone *on* someone.

Later, my parents called. They'd arrived that morning, slept till three. They sounded drunk.

DECEMBER 28
Dear Macs, Estelle, Bob et al.,

Mom and Dad and I have been keeping up a killing pace these last four days. We saw *Breaking the Code* and *Ivan the Terrible*; I fell asleep in a room full of Raphael's cartoons in the Victoria and Albert Museum. I don't know which I enjoyed more, taking a hot bath or eating pizza. (OK, I miss the U.S.A. a *little*.) They're pretty charming and hip, those old progenitors of mine, but between finding them restaurants every few hours and keeping them from being transformed into Cubist still lifes by passing buses, I've got my hands full.

The great moment was when Mom and Dad came to the squat where I live, helped break up crates for the fire, and drank with us. It was a touching bohemianism-across-the-generations gesture. (I wanted to take them to Marx's grave, but Dad got cold feet.)

Happy 1987,
Sparrow.

HAPPY NEW YEAR! I kissed Jeanne at midnight. She was wooden and shrunken-looking, the way she is when we fight.

Dad told me a joke the other day, near Piccadilly Circus: "This guy Moishe from Brooklyn goes to visit this Texas rancher, and the rancher tells him, 'My land is so big I can't even get across it in my car.' And Moishe says, 'You know, I used to have a car like that!'"

And Dad stops and laughs on the sidewalk, delighted, wagging his head from side to side, letting his cane dangle. Then he repeats the punch line, just to make himself laugh some more.

Yesterday at Kim Leuh, Mom tried to tell the Chinese waiter how the last time she ate in a Chinese restaurant she broke out in red splotches. But Dad shushed her in time.

Dad told a joke he saw on TV the night before: " 'God loves atheists,' the comedian says. 'No, really. God loves atheists. Because atheists never *ask* him for anything.' "

We both stopped and laughed, near Piccadilly Circus, both our canes dangling.

NEXT TO THE Thames, Dad asked me in a confidential tone, "Is there a tax on breasts here?"

"Not as far as I know."

"Because I've never seen so many flat-chested women in my life."

"Must be the recession."

We stood on a bridge over the Thames, watching the water go by. The Thames is a beautiful river. It was just past dusk. I felt this love coming from my parents.

Maybe they're gurus, I thought; *gurus in drag*. Mom appears scattered, and Dad seems absent and deaf, but maybe it's just a front.

MY MOTHER SAID:

"Promise me you'll buy glasses."

"Don't sleep in any farmers' fields. You know how farmers are. And you were telling me about this guy who was killed in a field."

"I'm through shopping."

"Don't you want to take a bath?"

Mom wears this floppy blue fake-wool hat; she's fat and talks all the time — she's a definite "talk first, think later" type. "Don't I look younger than I am?" she asks. (She's sixty-three.)

"You don't *look* younger," I tell her, "you *act* younger," and she laughs.

Being around five-year-olds much of her adult life has had, let us say, a certain influence.

Dad: My personal feeling is that you're not making the best use of your year, just floating around Europe. If you really think Baba has something to tell you, if you think he's more than an ordinary person, you would go there now, and *then* write your book. Baba's giving personal contact now; that's what gives you a personal relationship with the guru. He may not still be doing that in the summer. Your desire to see Baba doesn't seem that strong.

Me: See, I read this book, *The Divine Comedy*, and in it Dante goes on this voyage and finally comes to God. And that's what I'm doing: I'm traveling to God.

Dad: Yes, but Dante had some kind of vision before he wrote the book. That's why I'm saying: go there first, then write the book. People read books because they feel the author has some deep understanding of the world. That's exactly what you'll get from seeing Baba.

Me: Well, I don't think everybody reads books to get the Big Answer. I know I read books to find out what's going to happen. That's why I'm reading *Huckleberry Finn*. I want to find out if Jim is going to stay free or not. And this is a situation of suspense, too: I'm heading to my guru. What will happen?

Dad: All you'll put in the book is your confusion, and people don't want to read about someone else's confusion. They have their own confusion.

[Jeanne takes a plane to Bombay, and Sparrow moves on to France.]

Not only was this house standing during the French Revolution, but a hundred years before. I wonder what the people in this room thought of it: "Good-idea, that French Revolution"? Did it topple the local duke? Did it take two weeks for them to hear about it? Did they think, "That's something far-off in Paris"?

It strikes me that I *don't* believe Baba's the Lord of the Universe. First of all, it's unprovable. If you say, "If you're Lord of the Universe, then end it," and he raises his hand and ends it — well, it *could* be a coincidence. But how can a *person* be Lord of the Universe and also have hemorrhoids (as he does)? You mean he controls every blade of grass that grows on every planet? One guy? Or is all that on automatic and he just has to do miracles and earthquakes and the like? What exactly does the Lord of the Universe *do*?

I'm lost here in Purgatory. My friends keep telling me: "Go to India now!" but I'm dawdling, looking for something to do. This book seems lost, too. In my meditation, I rush up to Jeanne's jungle habitat in India, and she embraces me. "I love you, Sparrow," she says, "but it's not going to work out." I turn right around then, coldly: "Saddle the llamas, Sabu. We're moving on."

Or I ask her, "How many men did you sleep with?"

"Only six," she says with her teenage giggle.

Should I cheat on her? But what if she's faithful to me? Baba says give each error its opposite. Even if she sins, I should be chaste. But aren't I sick of that philosophy?

FEBRUARY 4
Barbara,

I'm in France, living in a house that predates Voltaire. In the rain, the wood and stones are beautiful, and in the sun, the sky is pale. My "work" here is to bag up raw sugar and spiced almonds.

Now that we're adults, I can say, "I've been meditating twelve years." I feel like I haven't made any progress. (I mean, it helps my concentration, my eyes are brighter, I *think* it improves my reflexes, I'm neater — but you can get most of that from doing the exercises in *Reader's Digest*.)

My latest girlfriend took off for India three weeks ago, and I'm thinking of how I was with her: gallant at first, then possessive, always the Hero — till she got sick of it and left.

That's why I'm going to see Baba.

Krishnamurti says you can't open the door. The best you can do is clean up your room and wait for the door to open, by luck. I like that.

Love,
Sparrow

IT JUST HIT me: I am in exile — political exile. I've been think-ing, *One year doesn't make you an expatriate*, but I *did* leave (partly) because I was sick of the States, sick of seeing the number of homeless grow each year as more money is spent on Impression-ist art, sick of watching the cops invade whomever they choose next out of the Rolodex — Grenada, Tripoli, Nicaragua — and, worst of all, sick of my friends sitting in cafes talking about whether to take cocaine, whether to have sex, and not one of them ready to say with their whole body, "This shit don't bounce!"

LATEST JUNGLE SCENARIOS:
1. I walk into her mud hut and say, "Where is he?" as a joke. It ruins our relationship.
2. I walk in and say, "Where is he?"

   She, laughing, says, "Where is who?" Finally, she ges-tures to a rattan wardrobe, and he walks out sheepishly.

   I say to him: "I'm sorry, but I'll have to ritually ex-ecute you." And I do.

LATELY, I GET the feeling Baba's watching everything I do, think-ing, *That Sparrow, now he thinks there's no God. Ahhh!* or, *That Sparrow, now he's crouched next to a road in Brittany mourning his lost love. Ahhh!*

But when I look at Stephan, with whom I'm staying, he seems to have it *made*: he's got his three-hundred-year-old house, his adoring wife, his slightly flaky but faithful kid, his cats, his busi-ness, which is also his obsession. He meditates about fifteen min-utes in the morning and fifteen minutes at night. There's no hassle. But you can't help feeling he's being *duped*. Like some-body said, five meals stand between the judge and the criminal. He's one day away from disaster.

I guess I'd rather have the disaster and get it over with.

FEBRUARY 15
Dharmaviira,

As soon as she got to Bombay, Jeanne sent me a postcard that sounded like she was writing to her fucking *aunt*: "The hydrangeas are blooming across the road. . . . When in Paris, I recommend the Rodin Museum. . . . I think of you fondly." (After I'd written her these wrenching letters — *three* of them: "I'm desperate. . . . I'm determined. . . . I watch for you in the night.") I almost cried right there on the Rue de Louvre. Then, in a postcard store, I looked through a Man Ray book and felt better. Suddenly I had an intuition: *She's sleeping with another guy. I can feel it.* And I went through six Buddhist hells, tourist class. (Night of the full moon.) Next day, I could sense it was an illusion: *She's waiting for me — I know it.* The only thing saving me from this bullshit is the thought that every single person on earth has also gone through it, and the ill-defined belief — born of reading too many books about reincarnation — that if you go through it enough it stops.

Love,
Sparrow

JANUARY 15
Dear Sparrow,

It's late at night and I can't sleep, thinking about everything under the sun. No, not really, just about myself, my life, my future, you. Bombay is difficult. Manic, crowded, hustling. Indians at their worst. As soon as you step out on the street in Colaba, it's "Yes, madam?" "Change money, hashish, heroin?" "You want room? I know cheap hotel." I didn't know it would be so hard to adjust. And the guilt when I see the beggars and the worries about money, the constant sense that I'm being cheated and everything seems more expensive than before. And the smells and sounds bring back memories of the last time, when I was so lonely and doubtful. They strike fear in me, and I have to keep reminding myself why I've chosen to do this.

All I've done is walk the street and run errands — bought a shirt, a Hindi book, envelopes, kohl for my eyes; checked out the train to Igatpuri. Sat on a bench near a huge well — Bhikka Behram Well — and watched people come up to it, drink water, and pray. The old caretaker came over and told me it was a sacred Parsee place. He told me I must go to the zoo, the aquarium, and the hanging gardens. But I looked in someone's guidebook and decided to try Krishnagiri National Park, where there are some ancient Buddhist caves. Anything to get out of Bombay. That's my plan for tomorrow.

I'm lonely, but now that I've moved into the Salvation Army hostel, a little less so.

My farts smell different here.

This morning there were sparrows in the bathroom.

Later: I'm on the train to Borevili (a town one and a half kilometers from Krishnagiri). Beautiful dark girls walk through the ladies' coach selling hair clips and little stick-ons, I guess for the forehead. These Brahman women have gotten lazy and use little fuzzy things instead of colored powder.

Sometimes I wish you were here with me, sharing the sights; I'd be less lonely, braver. But then I think it wouldn't be all that different — I'd be self-conscious in front of you, focused on your judgment instead of my own; you'd be averted when I was depressed, and I'd be resentful of you instead of myself.

I'm trying to remember that I'm making two big adjustments at once — to being in India and to being alone. It helps a lot to know you love me and that I'll see you in a few months. This is making me cry, and it feels good.

In Borevili: I suppose it was too much to expect a park this close to the city to be nice. Schoolchildren in checkered uniforms on picnic, unsavory characters trying to lure me into the bushes, dust and trash. Hotel Ellora in town is one hundred rupees per night. Oh, well, back to Bombay. At least the town was nice: stores in shacks, bullock carts, etc. And it's refreshing to be the only white person in town, although it makes me feel I'm in the wrong place.

Boy on roof of railway platform, trying to fly a kite.

I love you very much.

Jeanne

FEBRUARY 16

Jeanne,

Thanks, that was just the letter I needed (not that I realized it).

Yes, I love you, and I'm certainly willing to suffer a good deal more for it.

(Incidentally, don't tell me about any alliances unless they directly affect our relations — I mean *directly*. I've been much happier not knowing about your attractions.)

I write from my new home, two hundred feet from the Cathedral of Notre Dame (I see it out my window) — the second floor of a bookstore, Shakespeare & Co. It seems like my *place*, with books instead of walls — though it's obvious the owner, Henry, is an emotional fascist.

Write Poste Restante, Barcelona — I expect to quit this joint fairly soon (unless some easy money appears). At the risk of repetition, I love you.

Sparrow

I'VE JOINED A writers group — quite a writers group, too. There's this fellow, Michael. You know the type: bad complexion, preppie, says "epistemological" when he means "epistolary." Then there's Emily, a lady poet of fifty who writes about the Seine; and Annie, who defended literally every poem read.

It ended with this twenty-year-old Irish kid reading for about an hour; he kept *trying* to stop but we wouldn't let him. He's not a good poet, but once or twice he'd break through, for three lines, to the Other World. Why is it I like everyone from Ireland?

But it's exciting; none of them are any good, but they really are *starving poets in Paris*, at the *very bookstore that first published James Joyce*. It's so pleasant to say things like "Chekhov's so good I can't read him!" and have twenty-three-year-old Australian girls open their eyes wide in surprise — particularly after seventeen days of trying to figure out how to ask, "When does the postman come?" in French.

EVER SINCE THAT elf-pretty Australian told me there are bedbugs here, I've been scratching. Face it, I'm a hypochondriac. Also, I may have bedbugs.

MORE SNOW ON the *rue* today, and the second morning looking out at Notre Dame, I thought, *They ought to put a* clock *on that thing.*

I'VE JUST DISCOVERED I'm living with three shelves of the *New Yorker*, the most recent dated 1960! (This is my new house, near the Rhino Cafe. Angry Henry seems to have limitless real estate holdings. I'm building shelves for him, in repayment.
    "Have you done carpentry before?" he asked.
    "Sure . . . a little," I said.
    "When?"
    "Well, shop class.")

EITHER I'VE GOT bedbugs, the worst case of psychologically induced itching in the annals of medicine, or, hopefully, fleas. There *is* a cat, and the itch is mostly on my wrists and ankles.

■

IN MARK TWAIN'S time, America still had a species of English culture: ruder, plainer, more religious, and with more fishing poles, but *English*. The houses and town squares, the books they read, their plays, the very lack of ornament. Men didn't style their hair then, or eat yogurt, or go on vacations to get suntans, or go to discothèques, or wear twelve-hundred-dollar suits. See what I mean? All this New York City jet-set culture that is insidiously infecting American life (and which was indirectly launched by — of all people — the hippies) is French, to the bone. Sophisticated, indolent, and nasty, the French virtues are spreading over the U.S. like so many unisex hair salons. Now, when you go to England, you think, *This is how America used to be*: guys in gray pants standing in bars, drinking. And when you're in France, you think, *This is how America will be*: boutiques full of designer sunglasses on every block.

■

I CAN'T SEEM to work up a passion for females since deciding to have affairs. I'm a six-month-on, six-month-off man. After fighting and loving with Jeanne since spring, it's such a relief to sit on a loft bed and look at my *feet.*

■

TODAY I'M THINKING, *This is the relationship I've always* wanted. *I can put the make on any young Canadian I encounter and still have my real tootsy waiting in the Indus Valley!*

I figured out why Buddhist monks have shaved heads — damn *bedbugs!*

I THINK OF Anne Frank here a lot, because I'm always saying, "Un franc."

Figured out why the French don't have children. The women don't want to get *fat*.

It's amazing to think feminism was born here.

■

I'VE BEEN BUILDING shelves this week: sawing up miserable dumpster wood, soaking off the paper and plaster (*outdoors*, while it snowed), hammering nails, bending the nails, pulling them out. It's all been remarkably boring — though it was the first shelf I'd actually built. I *invented* this shelf, and now you can put a grapefruit on it, just like on a *real* shelf! It was three pieces of wood against the wall; now it's like the immutable shelves of my childhood. It's the opposite sensation of having your house washed away in a flood.

MY MAIN PROBLEM has moved from bedbugs to loneliness, which in some ways is more difficult. For one thing, you can't buy a canister of powder that purports to remove it.

LOOKING IN A *New Yorker* from 1958, I see a brunette in a kimono, her rear turned to me. *I want her*, I think, and then, *Oh, yes, she's sixty now, or dead.*

DO YOU *NEED* a guru or do you just like having one?" an Australian advertising man asked me at a tea party today.

"Well, I suppose I need one."

"Why?"

"I've always wanted to be a mystic."

He laughed. "I thought you had to be born one."

"Well, perhaps, but I've always wanted to be one."

"What do you mean by 'mystic'?"

"Someone who's . . . mystical."

"I can see you've given this a lot of thought."

"You ever have that feeling that everything's fine, that there's nothing to worry about?"

"Four or five times."

"It's like that."

■

LAST NIGHT IN the writers group: Joseph got into an argument about "What is reality?"; Annie's short story was too sentimental about marijuana for my taste; Daniel burst in, apologizing profusely for being two hours late, and then was miffed that we'd started without him. Then this preppie blonde in the corner suddenly became animated. It turned out she was in the wrong place; she thought this was a poetry reading, and began to knock out these poems of her own with these great lines like "I turned five times." She's a Texas academic — something to ponder — and her name is Doña. Her poems don't *mean* anything, which is so refreshing. She sounds drunk when she recites.

Dominique and Stephen and this Texas Ranger and I bought three pizzas and some wine and lugged it all to my seventeen-story walk-up, where I showed them my *New Yorker* collection from 1954 to 1960 — all three hundred of them — and they assumed I carry them around Europe with me. That's the danger of wearing horn-rims.

I *HOPE* THAT preppie blonde arrives tonight as she promised. My fantasy is saying to her, "Want to come back to my house and read old *New Yorker*s?" and then actually doing it.

I *WANT* THIS Doña, am half in love with her, and Jeanne seems suddenly like a memory as faint as *The Four Musketeers* (with Raquel Welch).

MARCH 1

Jeanne,

I just finished *American Notes,* by Charles Dickens. He ends by suggesting the violence in America is due to slavery (after quoting a whole series of newspaper ads for runaway slaves that say right out, "Branded on back," or, "Shot through foot").

I got your Groundhog Day letter and was disappointed. For one thing, it was rather peevish: I was "guilt-provoking," not serious about my meditation. The sense I got was you'd rather be alone and figure out yourself than have me to worry about. Walking three miles to my mail pickup, I'd hoped you'd say, "I'm desperate without you; come right away," and I could say to my pals in the Expatriate Corral, "Well, I gotta saddle up tomorrow; there's a lady in Jamalpur who *needs* me." Instead I see my own uncertainties in you, which seems to make them larger.

At this point, my conviction is that we're not through with each other. Perhaps if you don't see it as a red-light/green-light choice, it'll be simpler — I'd say our condition is decidedly yellow.

(It struck me I haven't talked much about dharma lately — let me. I still say, *I will be with you as long as it serves the dharma.*)

We interrupt this letter to bring you an announcement: I just called my parents from the Café du Notre Dame, and they say you're desperately in love with me. So be it.

Yes, it's cold in Europe (Mom read me your card), and, in fact, I *don't* have heat, but this sleeping bag continues to surprise me. (If I can just get the bugs out of it.) Anyway, thanks to a good diet and tough ancestors, I haven't had even the sniffles.

As for the rest, I've become a slow-moving carpenter in return for a room, and I'm starting to wonder if I'd enjoy *every* occupation as long as it's two hours a day. Oh, yes, I'm putting out *Big Fish* and founding a poetry movement: the Optional Dreams School. In short, I'm again the hickory-haired gadabout you fell in love with in the cafes of St. Marks Place. Care to split a bowl of minestrone?

All my love,
Sparrow

■

IN THE SPACE of twenty-four hours, I was in five countries — more than twice as many as I've visited in my life — and now I'm on the ferry from Nyborg to Korsor, getting a free lunch (with wine!) because it's the third birthday of the ship. I can see blue water with a white halo in the distance — icebergs. Here's what I thought of the five countries. France is great, though it's hard to imagine anyone there being your friend. Belgium exists in my notebook as three words: Namen, Luik, Antwerpen. Nederland — shyness is their weakness. (The saintly safety inspector closed his eyes when he'd turn to me.) Germany is chiefly amazing because the border cops are just what you'd expect — leafing through your French-English dictionary to see if you've hollowed it out to hide drugs. I honestly believe they asked me, "Are you carrying any explosives?" in innocence of the fact that a terrorist would lie to the police. The rest of the country looks like everyplace else — that is to say, Pennsylvania.

This clean, open city of Copenhagen is the least-cynical collection of a million people I've ever seen. We're too far north to be frightened of anything — they call the United States "United Bluff."

It was midnight and I was on the highway, surrounded by snow, in the cold rain, cars and trucks doing ninety and no one stopping. I realized the post with a yellow light next to me was an emergency phone. I lifted the lid and heard a voice say something like "Rijlvod."

"Hello?" I said. "I'm hitchhiking, and I wonder if I can sleep at the police station."

"No. There's no room."

"What do you suggest?"

"In Calding there's a place you can stay."

"For free?"

"Yes, free."

"Where is it?"

"You must follow the road back five kilometers."

I started walking, smelled wood smoke, and went to examine. A big white house. I knocked at the door; no answer. Behind it, stables — couldn't open them. Behind that, a long house with many doors, like a hotel. I found the door the smoke was coming from behind. I knocked. A dog barked. That was all; a dog was living there.

I walked on.

Finally I got to the exit sign for Calding — and another police phone. *Things are going quite well*, I said to myself. "I'm at the exit for Calding," I told the cops.

"You still have five, six, seven kilometers," the cop said.

"How do I find it once I'm there?"

Pause.

"Go to Tryblorn."

"Is it on the main street?"

"There are many main streets in Calding. It's a big place."

"Well, where do I go from the main road?"

"I can't explain on the phone."

I reached another house. Knocked on the door; no answer. *Perhaps all the houses in Denmark are vacant.* Beside it, a garage, big door up. It was surprisingly warm in there. And dry. *This'll do.*

I pulled out my sleeping bag — soaked! *What if the owner drives in at night and crushes me like a Ritz cracker box?* I put the pack in front of me, changed my socks, dried my feet. The new socks were wet, too. I pulled the bag over my head. *What if I get pneumonia?* It's hard to sleep when you're thinking that.

But I must have, because there was a gap, then light under the door. It had snowed — and was colder than the night before. My shoes were completely wet. My teeth chattered.

So Jeanne is either with someone and not writing, or writing to Barcelona Poste Restante, as I directed her. I think she *has* slept with someone by now and probably still *is* in love with me — that's my guess. (*I'm lucky with women*, I tell myself.)

I've been reading Trotsky's *Journal d'Exile*. With Trotsky by my side, loneliness is actually an honor. (Having lost touch with Rakovsky, his last companion of the struggle, Trotsky writes, "There is no one to argue with but the newspapers.")

■

A Sardinian gal agreed to hitchhike with me Monday.

She's a small, pale jurisprudence student with thin cheeks and mountains of curly hair that fall in her face and an electric style of arguing — and a deep, tuneless voice capable of propelling a brokenhearted Sardinian melody across the room.

Reaching for bread at dinner, she brushed against me with her small breasts, and I would've taken her then — and will, most likely, if she wants, and if I can overcome my sexual distaste for the new. I want revenge on Jeanne, though I worry one shouldn't screw on a pilgrimage.

Now, FIRST RIDE gone, with a Balkan dancing enthusiast, the Sardinian asks me in French, "How old are you?"

"Thirty-three," I reply, also in French.

"Great! The same age as Christ! It's a good time for you to be crucified!"

"Yes! Every day I wait! And you? Are you the same age as Mary?"

"Mary? Which Mary?"

"Mary Magdalene."

"Ah, perhaps. Ha ha!"

Now, AFTER A ride with an Iraqi, we're rubbing against each other. Mary Magdalene is certainly entertaining, but one worries one is only "multiplying one's mental protoplasm," as Baba so memorably, if uncomfortably, puts it.

AN ARGUMENT ABOUT AIDS: Fewer people pick up hitchhikers because of it, she thinks. "I live in the city with the most AIDS cases in the world, and people don't think about it much," I brag. "What about those parents who didn't want that kid with AIDS in their daughter's class?" she asks. Christ, news travels fast to Sardinia.

She's a communist (though not a Marxist), she says, and I read her some Trotsky.

IT WAS LATE, and we were cold. We found an unlocked car and decided to sleep there.

A car is in some ways the best place to sleep in winter. The seats are like beds and require no ground cloth; it's the smallest habitable structure, therefore the warmest, and the windows frost up, protecting one from being seen.

She settled into the front seat, I into the back. Besides being scared, I thought it seemed the wrong time for any *move*. We wriggled into our bags, then that silence of two people in a car.

Suddenly she said some hesitant things, mostly in Italian, but ending up in French with "Je ne suis pas content." I leaned into her seat, kissed her on the face, tiny kisses, and held her hand through the bag. She gave me her intelligent look of assessment; also of Italian appreciation for manhood?

"C'est bien?" I asked.

"C'est bien."

I kissed her thick hair. She gave me her look again.

I hugged her, through the seat, but felt no answering contraction. In Sardinia does the man do everything?

She leaned back. I withdrew, toward sleep.

■

GOD, THIS IS a mess. "We should've taken the train last night," she keeps saying. "We'd be out of Denmark, and it'd be easier to get a ride. There's no place for the cars to stop."

All because I kissed her last night — and violated the dharma.

She complains about this morning's Danish bread and predicts a bout of vomiting on her part: "For me, it's like eating plastic!"

THE SUNSET WAS fine, a kind of Danish-modern sunset: a red ball pressing down into the line of the land.

IN THE DANISH fog, among thorny trees — lovely as a Kurosawa film — she's "cold," and the cars will "never stop."

At the ferry she's *still* miserable. But her little snaggletoothed smile is so disarming. ("I wonder if we'd have this much trouble if we spoke the same language," I think aloud. She's silent, but presently says, "From now on, I'm speaking only German.")

I get a kind of Emma Goldman feeling of greatness from her, but I'm always getting that from undeserving dames.

LAST NIGHT WE slept on the floor, in sleeping bags. She was doing yoga when I lay down. *Best to give up on seducing her*, I reasoned. *If she wants me, let her wake me.* I put my head under the pillow.

"Sparrow!"

I woke.

"I think we don't have to go through Frankfurt," she said.

"OK," I said. "For that you woke me?"

She burrowed into her pillow with a girlish smile. "I was too happy to lie alone."

"Oh."

I turned away, and she took my hand in her small, warm one. I put my arm around her and rubbed very slowly a piece of her back. She also moved her hand on my neck, little motions.

"I've been thinking of love," she said.

"What kind of love? Physical, mental, spiritual? Of *all* love?"

"Of love."

"Do you want physical love?"

"I want love masochistic, nonmasochistic, masochistic," she said, stretching.

"You want love masochistic, nonmasochistic, masochistic?" I asked.

She smiled, then propelled herself onto me, moving in the way petite dark women have of kissing and contracting.

I held her at arm's length. We looked at each other.

"No, I've changed my mind," she said, lighting a cigarette. "Call me a schizophrenic."

"I am afraid of making love," I said. "Afraid that I want it only as a conquest, and as *revanche* on my girlfriend. But you are a profound woman, and I respect you and love your company. It would not be just for sex."

"There is someone I love," she said faintly into her pillow, "more than anyone in the world."

"Who is that?"

"He doesn't exist."

WE'VE STOPPED SPEAKING. I walked off to hitch separately. She refuses to speak French now, only Italian, and won't continue if I understand her.

Something living in my belly dreams up these women, orders them out of a catalog.

APRIL 5

Jeanne,

I'm in a phone booth in Genoa, Italy. Mom just read me your letter. Forgive me, I'm not going to Barcelona. It's disappointing that you're not in one place concentrating on writing, but this is already overlapping with my major worry: that you're not sitting around waiting for me to save you. Mostly I just *miss* you — despite your thousand defects, I've never met anyone so committed to relationship. (Keep writing my parents; I expect to arrive in early May — unless the boat from Saudi Arabia takes a month, as some know-nothing in Paris said.) I want to be a little mushy at this point, but I've said all these words too much. Please insert your own favorite line of adoration here:

I still feel like an arrow pointing at you. I mean, we have — how can I put it? — more packages to open.

The moon is like a crooked smile here.

I spent a lot of time in Genoa looking at the water, thinking, *Christopher Columbus sat here, dreaming of going to India — just like me — and out of that America was born*, and wondering what to make of that. (I don't blame him — the first time *I* looked, I thought I saw something out there, too.)

I've been sleeping in a large drainage pipe on the back of a truck — but the cops are prowling tonight, so that may not be doable.

I hope I'm not too in love with you for your taste. Looks like rain.

Solidarity,

Sparrow

BEING HERE IS making me appreciate the American personality: more fluid than the northern Europeans, more efficient than the southerners. America literally united the tendencies of Europe, with just enough of Africa to make Bill Haley possible. And if our ability to change has become an addiction, it's better than having to say, "Mamma mia!" for the next twenty-five centuries.

(The Italians really *do* say, "Mamma mia," the French really say, "Ooh la la," and the English really say, "Cheers, mate.")

Suddenly I don't want to see Jeanne. I don't care what she's been doing — except whether she's cheated on me. She writes lousy prose. (You've read her letters.) Most of the time she floats from place to place, thinking, *I'll be happier at the seashore; I'll be happier in the mountains.* I don't want to see her. I am just keeping an appointment — an appointment I had magnified in my mind into love.

April 15
Dear Julia, Sean, and Sharon,

These Italians! They're like your family — you love 'em and you hate 'em. You know what I mean? Like yesterday: There's this crazy little hotel, Hotel Castello, about 150 feet from where I've been stuck hitching for two days, in Marina de San Salvo. I went in: marble floors, no guests, just workmen listening to the B-52s' new song (it's great). I sneaked into the toilet.

Coming out from the patio, I saw a cardboard box full of cigarette butts and soggy pretzels. *Great,* I think, *I can get rid of this garbage.* (I've been collecting peanut shells and wax from cheese.) I started to decant my stuff into this box, and a guy I hadn't seen before came up and said, "What're you doing?" in Italian. I tried to explain (my *Italiano*'s a little weak): "Io faire auto-stop . . . mangere [gesture for 'eat'] . . . papiere [gesture to my trash]." He said, "No," and walked off.

"Christ, you can't even throw out your garbage in this benighted nation," I mumbled, heading back to my hitchhiking station. Then I heard this whistle behind me — a kid with a plastic bag, coming out of the hotel. (*Oh, they brought me a bag for my garbage — that's nice.*) Inside the bag were nine pieces of fruit!

"Grazie, grazie," I said.

He figured I wanted to *eat* his garbage.

Love,
Sparrow

■

THIS IS HOW I'll always remember coming into Greece: the elephantine mountains rising out of the blue-black sea.

JUST MET THIS little Greek policeman who learned, somewhere, opposite English: "I'm sorry," he says, "the banks are open."
  "Closed?"
  "Yes, closed."
  "But I heard there was a place open till ten."
  "Yes," pointing right, "just go to the left."

APRIL 17
Sulaymaan,
  On the boat to Israel, the problem is where to sleep at night. The seats aren't comfortable, on the floor people step on you, and on deck it's cold and the door slams all night. I'm trying to accept that Jeanne and I are finished, but can't.
  Love,
  Sparrow

APRIL 18
Jeanne,
  I thought you'd like this card. [A cat walking into the green sea.] Since I've spoken to you, I've meditated at the tomb of Dante and the prison of Socrates — one forgets, discussing gurus with Peace Corps veterans of Togo, that there are so *many* gurus. Almost all the great works that survive are Scripture.
  Last night I couldn't sleep — thirst from that feta cheese. I sat up and looked at the waves. After an hour, I realized it was the wake of the ship and felt cheated. Then I went to sleep and saw you. You had branches coming out of your head and were singing through a long tube. I followed, but we couldn't speak — there was glass between us. You were happy and with a truck driver.
  Love,
  Sparrow

APRIL 19

Leila,

Landing in Haifa: half-moon and one sea gull. It's more industrial than I thought: cranes and refineries. The flag, so clean and symmetrical. *Will it last?* I wonder, watching it ripple. It doesn't look like a flag that'll last.

"So this is your homeland," said the Englishman who was watching the casino with me.

"Well, my ambiguous homeland. My mother converted to Reform Judaism, so I'm not really a Jew, according to the Orthodox. Also, I think it's an imperialist state."

Love,

Sparrow

I DON'T FEEL a thrill of nationalism here, like Dad does. He thinks, *Wow, a country full of Jews.* I think, *Oh, no, a country full of Israelis* — another language I don't understand.

The army kids with machine guns are more reckless than the French or Italian cops. The guns aren't part of a *uniform*, but slung behind their backs like canteens. It's scary; they're going to be *used*. (Though it's nice to see two soldiers hugging.)

I DON'T FEEL in danger here, but there are signs of tension. Just now, biting into a big falafel, I tasted meat, spit it into my hand, and dumped it in a box of trash in front of an Arab store. A man came running out: "What did you put in there?"

APRIL 22

Jeanne,

Strange for it to end like this. I can't *bear* to go to India and see you. I think of your flat, jowled face, crucified by your own boredom; of our sex, where my only pleasure was humiliating you. I'm sure you're screwing someone, and you're better off with him, whoever he is.

I *like* Israel, and I'm not in a hurry to roast in India, over a grill, with you. Half a year with a woman, half without — that's the way, perhaps, to go to one's guru. I'm interested in trying to respect the rules of pilgrimage.

I tried in my last postcard to silently say goodbye to you, but I'm afraid I just sounded jealous.

"I'll stay with you as long as the dharma wants." Well, the dharma wants *servants*, not lust. Lust isn't *evil*; it's just a *lie*. Because we're *not* bodies — if we were, sex would make us happy. Well, goodbye. I hate you in my belly, but I also wish you — even there — peace.

Love,
Sparrow

[Sparrow learns that Jeanne is in love with a camel driver in India.]

APRIL 26
Jeanne,

It's time I spoke frankly. I've been writing two series of letters and only sending you one. Now I'll try to write both at once.

I traveled from Copenhagen to Milano with a Sardinian woman named Anna. She was small, with curled hair, and such a profuse personality it's impossible to say if she was pretty or not. We slept in the same room three times: in a white car in Denmark, in a living room in Lübeck, and in a rail car. The second time we *might've* made love, had it not been for my misgivings. I liked her cursing at the cars that passed and that she'd read Nietzsche. She loved to argue and, at first, to brush against my elbow, and we shared bread and a cold.

What I'm putting off saying is that, in four days, she and I degenerated into a failure so similar to ours that . . . perhaps the therapist was right; I need to be with *me* for a while.

But it gave me the accomplished feeling of having had an affair, to keep up with you.

I don't think I love you, and I don't think you love me. Mostly I feel regretful that I wronged you and want to set it right.

Of course, I'm angry, and I'd *hoped* you wouldn't fall for

someone, but I certainly felt it coming. You prepared me for it quite well, don't worry.

Anyway, good luck with your new adventure. Write my parents. Please forgive me, but my mom read me your letter.

Yes, this is hard. I'm not going to reassure you that you're doing the right thing. I got you some presents — for the moment I'm not throwing them away.

Love,
Sparrow

I THINK OF her giggling and showering with him, in love again. "I must pursue this to learn that desire leads to suffering," she wrote in her smug, self-pitying tone.

But wouldn't I choose the same, if I could?

I'VE MADE AN extraordinary discovery. Bic pens in Jerusalem are sold only by Arabs. *Every* Jew sells pens of this Israeli make — Sharit, or something. All morning I've trooped down Jaffa Road without luck — because the Arab Bic I bought the other day started leaking. (That's a good pun: "Arab Bic.")

Soon I'll be at the Y'Shuv, this armed yeshiva camp on the West Bank where all the Torah scholars walk around with machine guns, so I need a new pen.

CHOGYAM TRUNGPA DIED sometime in the last two months, the guy who drove me out said. (I'm in the shul, a prefab, off-white structure with three soldiers outside. This whole community is made of identical, single-story, solar-energy-equipped structures, like a California suburb — and on the road, unreadable signs and women wrapped up like salamis.)

So Trungpa's dead — poor guy, drank himself down. I went to his lectures at Naropa in 1976. He was solid like a furnace, swilling sake and illuminating Buddhism. I had several short conversations with him, only one of which I remember:

"What do you think of Judaism?" I asked.

"The food's good."

"Could you be more specific?"

"I'd have to see a menu."

He'd pee out in the street after the lectures, in front of everyone.

I never really knew him. I remember the sad woman who hung around our apartment; she'd had sex with him on the floor of a bathroom and believed he was going to marry her.

SUNDAY MORNING AND the sprinklers are on. It's ludicrous, the collection of spiny extrusions these guys consider their lawn. Suburbia is inherited, I think.

This place is dominated by kids. One is crying at any moment, and sometimes they set each other off, like dogs, and one imagines how it would be if they all cried at once — like a helicopter landing on your roof.

Orthodox women are supposed to be unattractive, and they're doing a good job: bald, fat, rags on their heads, publicly loudmouthed. The rabbi, who's a kind of guru here, has a wife who could break glass; it seems to be a greater mitzvah to have a more miserable mate.

A lot of people here have dabbled in Eastern religion. In Ananda Marga, there's a collective pride in doing spiritual practice while the other religions do empty rituals; here there's the inverse pride: "Anybody can get high. That's easy. The beauty of serving God is doing things you *don't* want to do." Marriage — particularly with eight kids — is part of that. There's gotta be a logical flaw in there somewhere, but a practice where none of it is enjoyable does have its allure. (People would point to me and say, "He gave up great ecstasies for this." And I do have ecstasies. I haven't told you about them because I don't know how.)

I'M TRYING TO decide if the yeshiva I'm staying at is a cult. They definitely think their rabbi is no ordinary man — you get the feeling if he levitated they wouldn't be surprised — but they don't have his picture on their walls. They have this settlement on the West Bank that's in a strategically delicate position, and they have a little more respect for the mitzvah of dying for God than I am comfortable with. A yeshiva, in theory, is a school, but no one ever *leaves* this one. Is the definition of a cult "a college from which you never graduate"? No, I guess an ashram is that, too. None of them has listened to a new rock-and-roll album for ten years. I give up. Being the last rationalist in the New Age is sometimes a burden.

MAY 19
Jeanne,

Ysraeli Salanti said, "A true rabbi, his congregation should want to throw him out. If a congregation doesn't want to throw him out, he's not a rabbi, and if they *do* throw him out, he's not a man!" Tell *that* to your guru!

Sparrow

I'M AT THE Wailing Wall; arrived with a bunch of preteen chanters in orange hats banging on cheap drums. First, an out-of-work brass band played onstage, and now some interminable speech.

It's the twentieth anniversary of Jerusalem's reuniting. I wonder if the bank's open.

MAY 18
Dearest Sparrow,

I've been here and horribly sick for two weeks and just began to recover three days ago, the same day I found out it's hepatitis and the same day I sent you letters in various cities of India, begging you to take the first flight up because I need you desperately. Then I spoke to your mother (about an hour ago), who said you're still in Israel, and I can't quite write you the same letter there. For one thing, I feel like I'm writing into the void. That's not meant as a reproach; it's just that none of your letters has reached me since March, and I have no idea how many of mine have reached you. Actually, I'm wondering if you have an Israeli girlfriend. And fearing if I say, "Please, Sparrow, rush up to Katmandu and take care of me," you'll say, "After all you've done to me? Who do you think you are?"

Katmandu is awful; it's full of tourists. Nepal is a tantalizing place just out of reach, and I'm not well enough yet to get there.

It's been hard being sick and alone and always having to ask people for things, and you have been the person I've wished were here to take care of me. A French girl, Christiana, has just moved in with me, and my strength is beginning to return, along with my peace of mind, so technically I'm not worried. The doctor gave me syrups full of wonderful herbs, and my main problem is getting the right food without expending too much energy. I think I'll be stuck here for the next few weeks.

Later: I've been out with Dawn to buy a little kerosene cooker that should solve my food problems. Don't know what the hotel owners will feel, but I have a little balcony, so there shouldn't be a problem. I'm really thrilled — I've been wanting my own stove for so long.

So I'm being taken care of. But it would still be nice to see you again soon. I'd like to see some of India with you, too. Please write soon and tell me your plans, your schedule, your frame of mind; tell me whether I still exist.

At night I dream about swimming pools. Once, diving with you in a big, cool, clear swimming pool. Then I dream about Nepalese people standing in long lines on the mountainside and me just standing there with them, waiting. I really hope I can get out in the countryside. The country people I've seen are *so* beautiful. Maybe I'll rent a room in Patan, the town across the river — no tourists.

Everyone I meet who's been to Israel says it's wonderful. How are you liking it?

Please write soon.

Love,

Jeanne

THIS LETTER FROM Jeanne has shaken me. I'd just accepted solitude with some grace, even begun to forgive her — as well as taken her pictures out of my wallet to send home — and now this lead-lined letter, into which no one can see. She doesn't *apologize*, but simply begs — remembering, perhaps, my taste. Then, realizing I won't come soon (for her need is not a boyfriend but a *nurse*), she backpedals and offers to "travel" with me — not a particularly enticing offer, given our history.

"Write me quickly," she says, so she knows whether to look for *another* savior. At the bottom of the letter there is more and more space between the lines, as she leaves out more and more.

So now I'll have to worry she'll leave again. I don't want to have an *affair* with her. But I'm determined to try again — without hope, hopefully. I like the sound of "My girlfriend has hepatitis in Katmandu."

I WAS TALKING to David the Deadhead the other day about whether or not my guru is God. I explained, "See, yoga believes everything is God — not a creation of God, but *is* God itself." I spread my hands, indicating the rubble and little plants we were passing — rubble still left from the Six-Day War! "All this is God! But there are different levels of awareness; this cat is more aware than this Volkswagen. And it works up to the pure consciousness that directs the universe. When a person is an embodiment of that awareness, he's God. Or she." David nodded. It made perfect sense to him. I wished he could explain it to *me*!

The idea of a person being God seems quite dangerous. Suppose one *is* God, but *stops* being God for five minutes, and in that time one instructs one's devotees to invest in the wrong stocks or start a thermonuclear war. The results could be disastrous. They could lose thousands of dollars! My advice to anyone who has the choice is not to become God, and if you do, *don't tell anyone*.

Baba doesn't say he's God, but he seems to imply it, which is in bad taste. People like Baba succeed as God because other people spend a lot of time staring at them trying to figure out if they are God; if, after enough time, they're still sufficiently mysterious, people get sick of trying to figure it out and say, "All right, he's God!" just to get it over with. But most people are not this mysterious. Look at your brother-in-law; he may seem mysterious for about fifteen minutes, but you figure out after a while that what motivates him is sex, a fear of terriers, and the desire to play golf.

IT'S SUCH A world of icebergs and soldering irons we live in. "So you're searching," people keep telling me. I don't want to answer,

"No, I'm not *searching*." It sounds conceited. But it struck me yesterday that people look at me like "Poor guy, he hasn't found anything to believe in yet" — as if it were a search that comes to a preordained end, and that end is belief, whether in the Torah or the guru. Nobody thinks it's a labor, like science, with no end.

MAY 4

Dear Sparrow,

I suppose you got the letter I wrote from Delhi. I've been traveling for two weeks with a camel driver, but it turned out he only wanted my money. When he found out I've been telling him the truth — that I'm not rich — he decided to go back to his camel. It's not so much the rejection that hurts — I was already thinking we'd part soon — but the disillusionment is hard to take. It's so appealing to see peasants as simple and pure. Now I feel like a fool.

Benares is the pits; maybe it's my state of mind, but it's such an ugly city. I'm taking the bus tomorrow to Katmandu. Hope to rent a room out in the countryside and write a lot. Maybe spend some time in Bodhnath, where there are a lot of Buddhists. (Westerners study there.)

I've not had a letter from you since early March. I feel completely out of touch with you. I will try to define myself frankly: I still feel we have unfinished business, and I still have curiosity about how it would go with us. I also feel fear and foreboding about us being together. But we've both changed, so I know it would be an interesting challenge.

My job right now is to follow my impulses through as many experiences as I can handle, to see what works and what doesn't. I'm trying to figure out what kind of life I want to make. So far I've decided I don't want to spend most of my life in cities — actually, I don't want to spend any of my life in cities, but I'm not sure I can make a living in the country. Cities deaden something in me that I don't want to live without. Also I've determined that, ideally, I should have about two hours a day of physical labor. I just spent ten days at a work camp in the hills of Bihar, digging holes for yams and papaya trees. Four hours a day of it was too much, but two would've been just right. And I loved drawing water from the well.

By now I'm low on money. I have to work soon, and I don't want to go back to New York. I'm thinking of going through Nepal and Tibet to China and flying from Hong Kong through Japan. I will see how far I can get without asking my parents for money. I'll have to eventually, although I was hoping to avoid it.

If you don't hate me by now, you could come up to Nepal. I wouldn't come back to India to meet you unless you'd be willing to lend me some money. I'm getting curious about Calcutta, and I'd also like to go back to the village in Bihar and live there for a while without the pressure of a work schedule.

I have no idea how you feel about me by now, or how these words will fall on your eyes. (Such a serious letter.)

I met the brother of the man who shot Gandhi. He spent twenty-six years in prison as a co-conspirator. I was impressed by his idealism, but shocked by how much he hates Gandhi. Gandhi still evokes strong passions here, both for and against. In America you don't hear much about the people who hate him, about all the deaths they hold him responsible for.

I hope you're well, hope you like Israel — I keep meeting people who rave about it. Please write to Poste Restante, Katmandu, Nepal.

Love,
Jeanne

JUNE 1
Jeanne,

I was getting comfortable with the thought of having broken up with you — I like being alone, and it's an easier state for travel. When I got your letter I was depressed — *Now I have to worry about this again* — but also excited by the great hope that my book would have a happy ending. (Of course I hate you — my current wish is you hadn't told me about the camel driver — but I'm just beginning also to forgive you. So much of your crime is not accepting my arbitrary and insincere definition of you.)

We're not going to spend our lives together. I *like* cities, and though I associate countryside and marriage with the idyllic future, I find myself being a Manhattan type I've always despised — the kind who retreats to Maine to buck up for the return to Eighty-sixth Street.

It's easiest for me to consider us saying goodbye.

I expect to be in Bombay by July 1, possibly earlier, but maybe the boat goes to Madras. Who the hell knows? (*Can* a boat go to Madras?) I'll write when this has settled down a little (no information in Israel about Arab shipping).

I'm surprised travelers (whom as a group I dislike) rave about here. It's not "exotic." It's a strangely naive and unselfish place.

Love,

Mr. Sparrow

IT IS PEOPLE *without* beliefs who start all the beliefs. Did Buddha *believe* in anything? Even John DeLorean and Teddy Roosevelt didn't. Belief is a *comfort*, like a beauty parlor. It's something you should reward yourself with. But if you go every day, it's selfish.

My problem is I spent four years in Gainesville, Florida, believing everything I came in contact with. At one time I believed in astrology, healing with crystals, feminism, Ken Keyes, Edgar Cayce, *I'm OK, You're OK*, millet, the Farm, No Nukes, and *The Works of John Cage*. When I returned to New York City in 1978, my beliefs began slipping away like silverware from a Ramada Inn. At this point I'm down to reason and jazz — the *last* beliefs.

And reason, why have faith in that? Mostly, I'm just embarrassed by all the junk I've believed, culminating in the letter I wrote my sister advising her to cure cancer patients with white-light meditation.

Reason is like every belief in that it knocks out all the others — but its virtue is that it isn't consoling.

I'm going to India to see Baba for the same reason I saw Miles Davis. I like to see the classics of the era before they pass away, or before *I* do. I've been "following" Baba for twelve years: he was in jail for murder; his wife left him and said he was a fraud; all my friends think he's God — it's an Agatha Christie mystery I want to try my hand at. Also, I want to offer my life to

God. You reach a point in your life — I'm thirty-three, the number of revolutions per minute a record makes — where you want to justify things. You've worked fourteen years, you've been to college, you've danced at the Danceteria, Hurrah, the Mudd Club, CBGB's, the Pyramid, the Ritz, the Cat Club, the Underground, the . . . what's it called? Pavilion? No, Palladium. You've been to every one but the Surf Club and that video one. I'm not saying, "You haven't found fulfillment." You *have*. You've written poems, books, essays; you've been published; you've been unpublished. You have a band that changes its name every time it plays. You have friends who'll lend you their computers. You've *done* things for people, too. You've fed the bums in Denver once. You've taught meditation to paralyzed people at the Rusk Institute. You've marched against all the wars. You've worked with the retarded for six years. Now, well, because you've lived so *long*, because you were born too *young*, you just want to press your forehead to the ground and offer it all up.

O. Henry has this story "A Man about Town." This fellow's swank, hangs out at the Plaza Hotel, changes his clothes four times a day, goes to his club. One day his friends look up, and he's gone. *Has he gotten some girl pregnant?* they wonder, within the confines of O. Henry's polite prose. He leaves their minds.

Years later, two of this set are traveling in remote Switzerland. "You will be the first Americans to set foot in this monastery," their guide tells them. On a stairway they glimpse — can it be? — their friend, the man about town. They get special permission from the abbot to meet with their friend for ten minutes. Embarrassed greetings; "Charles, you had everything!" The final question: "What made you leave?" And as the bells of the church chime, he lifts up his robe in wonder: "I've found it — the perfect garment!"

And I guess that's why I do yoga. I've found the perfect garment — not the beliefs, but the silence and the bare room.

JUNE 15
Mom and Dad,
I was going to say Cairo looks like a run-down version of

New York, except that *New York* looks like a run-down version of New York. My Australian friend on the bus says that what you notice about the Third World is its smell, which is true, so far — but rather than smelling like camel shit, Cairo smells like *must*, particularly this room at the Lotus Hotel, with its grandmotherly desk and wardrobe.

The Israeli guide told us, "Don't eat in holes in the wall; eat in civilized places," and we all trooped over to the hole in the wall across the street. The falafel was less than a nickel.

I went into a department store and watched TV. A guy whose pants were too big was walking around trying to keep them from falling down. The same things are funny in North Africa.

Love,
Sparrow

MEN WALK ARM in arm with men, women with women here — which is as it should be — but it's weird to see a lady in full veil flirting.

What people forget, when they talk about the wicked Modern Age, is how much cars alleviate the suffering of horses.

THE SPHINX IS so beautiful, even though Napoleon blew off her nose. She has all these *ledges*, and a Mona Lisa look, and she gives the Pyramids a purpose. By themselves they look like postcards.

I sat against the wall behind the twenty-three stone pillars ("each weighing fifteen tons," a tour guide said). I worried I was meditating in a room once full of idols. Was I being the ultimate amoralist? Suppose Dachau had a good vibration?

But I left feeling I'd met the Sphinx — a woman who is also a lion — and one of the Wonders of the World. Funny to walk by her until you reach the right angle and it clicks: *This is the cover of* Collier's Encyclopedia.

WHO WOULD'VE GUESSED they smoke Camels in Egypt?

A SUMMARY SO far: In Cairo, my first night, I went to six all-night travel agencies and said: "I'm looking for a boat to India." "We don't know" and "Try Mena Tours" were the only two answers, in order of popularity. At Mena Tours, they said, "Forget Alexandria, try Port Suez." At Port Suez I got bitten by a dog, improved my ping-pong (here "pinga-ponga"), met Madgy, my actual Egyptian friend, who took me to Ali Yusef at Aswan Shipping, who said, "Try Port Said, nothing here."

Came to Port Said, visited four shipping agencies. Three said, "Impossible! It is against the law to take on passengers. The companies will be fined ten times what they would make from you," but one guy at Damanhour welcomed me immediately: "Where to, Madras or Calcutta? Nothing today. Come back tomorrow!"

I showed him my Indian visa. "This expires in two weeks," I said. "How long does the boat take?"

"Two weeks. You better fix up your visa."

Three days in Cairo, "fixing up" my visa. Finally, back at Damanhour, the manager said, "Go to the Indian consulate and ask what they suggest. I will pray for you."

TODAY AT THE consulati Hindi, the consul said, "You will find a boat to the Far East at Alexandria." It turns out there's no Soviet embassy here, so I can't ask about my new fantasy. I heard there's a boat from Alexandria to Russia, and a railroad to China-Tibet-Nepal-India.

I'M SO GLAD I'm in this mosque listening to an incomprehensible lecture rather than seeing *Staying Alive* with John Travolta at the cinema downtown.

AMERICA IS A good country," a kid says to me.
    "You think so?"
    "Yes."
    "Why? Why do you think so?"
    "Money."
    "What else?"
    "Money."

Another kid adds, "Very high homes."

AMERICAN EMBASSY: I never thought I'd be happy to see a *deodorant cake* in a urinal.

•

I RETURNED TO sex in Cairo. My second day I jerked off in the bathroom, thinking of a woman who was bent over, as if in prayer. And I felt big *circles* in my brain because I hadn't ejaculated in six weeks — felt my head wobble off my neck.

Then, the next night, a big blond woman (German?) was especially fond of me at the hostel, and I imagined her in my room. When she turned away from me and allowed me to put my hands on her full breasts . . . Was it the shiver from her that made the cream come out of me?

And why do I always enter women from behind? Is it their *faces* I fear? Or do I have the desire to subjugate, like Rosanna Arquette's husband in the film last night, who screamed, "Surrender, Dorothy!"? But in some ways there is more intimacy when the back is turned. One turns one's back on someone one *trusts*.

(And when I'd go in Jeanne's behind, and she'd clutch my hand, it was like holding hands in a storm.)

Perhaps it's this very intimacy I'm lacking, the *from-behind* intimacy, with Ananda Marga. I always face its front, and it's a little like saluting the flag.

THE ALEXANDRIA FINE Arts Museum was a delight worth postponing. A lady hunchback followed me from room to room so I didn't steal anything.

•

I'M BEGINNING TO suspect I have the only Bic pen in Turkey.

I have eleven days to make Baba. It looks unlikely, barring miracles.

WHAT ABOUT TRAVELING to Syria, Jordan, and Saudi Arabia?" I ask.

"We don't recommend going to Syria right now. There are anarchists on the southern border of Turkey."

"Kurds?"

Pause. "Yes, Kurds. Did you see today's paper?"

"No."

She shows me the full-color picture on the cover: corpses with their intestines hanging out.

OH, BEFORE I tell you anything else," my mother said when I called, "I want to tell you about Jeanne."

"Yes?"

"She's in Poughkeepsie."

"Poughkeepsie?"

"She has malaria. Remember we heard that she was sick in India? Well, she wasn't getting any better, so her parents paid for her to come home, and they found out it was malaria. Can you hear me? I'm shouting here."

"Yes."

"And she wrote me a letter and also called me from the hospital. Do you want me to get the letter? It's in the other room."

"No, it's very expensive."

But she went anyway and couldn't find the letter.

They're all going to Cape Cod for the month of August.

ISLAM CONTINUES TO impress me. It's the most recent of the great religions (excepting Mormonism), and it seems more advanced: more monotheistic than Christianity, less busy than Judaism. It's the way religion should be: you wash your feet, take off your shoes, bow a lot, stand on a rug with other men. It's the most physical religion, the most athletic, and the best for daydreaming. My attraction to it comes from its present role as the world's demon. Then there's that first moment, entering the courtyard to the Mohammed Ali Mosque in Cairo, seeing this conspicuous emptiness. Ah, a temple to emptiness. What a good joke!

JULY 14

Jeanne,

Well, well, well. I was happy to hear you're in the New York area, though I'd prefer you *didn't* have malaria (is it contagious?) because I'd like to try again with you. But if you don't want to (my mother said she had a letter from you, but couldn't find it — so I'm not sure of your tone), I'm afraid I'll be a hypocrite and ask to be your friend. I'm too curious about what you think now of the Group, or whether you'll ever become really political, and your poems. . . . I'd like a big Whole Story letter from you. As for me, I suppose I *have* changed since London (though it seems to me I never change). A lemon salesman's seated himself next to me in the Turkish Central Park. But I *want* a girlfriend. My worry is you want six boyfriends in the East Village — and I don't want to be one of your six boyfriends. Maybe you could hurry and do it now, if health permits.

Away went the lemon salesman — I hope not from my picking my nose. Istanbul keeps reminding me of the Bronx — the ugly Pelham Parkway part. I hate places where people come up to you and say, "Istanbul good?" I have not yet had the courage to say, "No, it reminds me of Pelham Parkway." I don't like how the people *look* is the problem. That policeman mustache is one of the things I hate about *America*. And they put meat on pizza — that is difficult to forgive.

Love,
Sparrow

■

I COULDN'T SLEEP last night on the train from Bombay to Calcutta. They turned out the lights, so I walked to the end of the train car, where the light was on by the latrines, to read.

Another fellow also stood in the light, a young guy, and he asked why I'm in India. I lamely said I studied yoga. Then he pointed proudly to his chest and said, "*I* am a Brahman." He showed me the *string* around his chest. "I *practice* yoga."

This fellow had an insolent bearing. "Look at my eyes," he said. "You see the glow. This glow is from *yoga!*"

I had noticed his eyes, in fact, which didn't seem to glow so much as *gleam*. (Later, he asked me for a cigarette, then proudly said he smoked and drank and ate meat and "fucked.")

"I say my mantra over and over," he told me. I forgot to

mention he had one arm without a hand, and the one hand he did have was missing two fingers and was covered with jewelry, as if to compensate. "Your mantra goes round and round like a turbine." And he rotated his handless arm around his handed arm. "And it gives you tremendous power." He pointed between his eyes. "If you concentrate it here, you have the power to move the world!" (He talked like a villain in a Marvel comic.)

Now I blazed *my* eyes and brought them to within an inch of his: "I've been doing meditation for twelve years."

He shrank back. "Then you must have psychic powers!"

This slowed me down. "I renounced my psychic powers! A yogi must renounce his psychic powers, mustn't he?"

"No, a yogi uses his psychic powers for God! Only for the will of God!"

"Do you have psychic powers?"

"Yes, I do."

"What are they?"

He looked slyly and humbly down. "I will not tell."

THE THING ABOUT the Third World is people are always kneeling and doing things you can't quite figure out, like tying sticks together, or tearing cloths into pieces.

Here's an example: a guy with a little mat out on the street, with three or four dead lizards and little bottles of dye.

And here's a sheep tied to a fire hydrant.

ON TO THE Critique of Pure Tourism. It struck me in Marmoris that a person exposed to tourism for any length of time would get a more negative view of human life than someone in a concentration camp. At Auschwitz one witnessed all kinds of unexpected kindness, but on the island of Rhodes, *no one* sacrifices their tan for someone else's.

◼

I GOT TO Tiljala, site of Ananda Marga headquarters, and a deep black fellow in a white T-shirt was getting into a motorized rickshaw: "I'm going to the *jagrti* — why don't we share?" The *jagrti* (the central office) is this gray cement dump behind a gate. Then he led me into the Nairobi-sector office: "I'm just changing my clothes," the black fellow said, "then I'm going to go back and see Baba."

Already?

And soon I was there, watching him lumber into his car. The line is "If you're not devoted, he'll just look like a little old man," but he definitely looks like a guru. He's bald, or has a shaved head, with thick eyeglasses and a kingly appearance. He seems to dwell inside himself and never to look at anyone. He looks a bit like God, particularly the way his head sits on his body almost like a hat.

Baba looks like God, but it seemed unlikely that God would notice me: "Hey there, Sparrow."

I did get a little kick, like alcohol gives, just after he left — a kind of knock on the head. Maybe that's his way of noticing us.

I forgot to offer my life to the Lord! I realized when his car had gone.

(He does look like an aged Jack Benny.)

When he came back, we were singing some of the songs he'd written. My eyes were closed, and I felt that riding sense you get in meditation, like a cowboy rocking over the prairie.

Baba's not a physical entity, it seemed to me. You can see him better if you don't look. This time I remembered and offered my life to the Lord, a bit abashed that I couldn't think of better words than that.

You know how you have this mental sense of the space inside yourself — as if you live in a dark hollow made by your dimensional silhouette, only at the top of it? It seemed to me that the top came off, and there was a spiral stairway going up and up into light ascending. But no one was standing on the stairway.

December 1990 – February 1991

THE MYTH
OF THERAPY

AN INTERVIEW WITH JAMES HILLMAN

Sy Safransky

A couple of years ago, I got a letter from Annie Gottlieb, a talented writer and devoted subscriber. "Have you ever read anything by James Hillman?" she asked. "He just rips up every one of the culture's unexamined assumptions. Right now, in a therapy-besotted world, he's on an antitherapy jag. He says we've withdrawn soul into the consulting room, into the boundaries of the individual skin, out of the neighborhood and the environment. He has especially harsh things to say about the 'child archetype' and our reverence for the 'inner child of the past' and its supposedly formative traumas, all of which keep us helpless adult children rather than empowered, political citizens. We look back instead of looking around."

And, she promised, "He'll make you kinder to your dark side."
As someone given to noodling around in the dark, I was intrigued.

"He has a way, for one thing, of uplifting and dignifying your most wretched obsessions. He's not in favor of transcendence but of transparency, not getting out, but seeing through, so your neurosis becomes a beautiful stained-glass window depicting a universal myth."

I followed Annie's advice and started reading Hillman. A brilliant and provocative thinker, he isn't always easy to understand. He can be a quirky writer, quick to throw out ideas and references. But what ideas! Reading Hillman is like stepping off a bus into the clamorous, exotic, slightly menacing streets of a foreign city. You're asked to leave behind fantasies of growth and self-improvement; to search the narrow, twisting alleys for better questions, not answers; to be prepared for trouble.

I don't always agree with him, but I respect the range and depth of his thinking, and value his willingness to challenge just about any theory, including his own. As Michael Ventura put it in the LA Weekly, *"Hillman knows that the work of thought is one of the most ancient and useful activities of humankind. To generate thought is to create life, liveliness, community. Consensus isn't important. What's important is how the generative power of our thought makes life vivid and burns out the dead brush, dead habits, dead institutions."*

As a younger man, Hillman studied with Carl Jung and was director of studies at Zurich's Jung Institute — though today he's regarded as a post-Jungian or a renegade Jungian or not a Jungian at all. In addition to being a therapist, Hillman runs a small, innovative publishing house called Spring Publications, and has held professorships at Yale, the University of Chicago, Syracuse University, and the University of Dallas. Lately, he's also been involved in the "mythopoetic" vanguard of the men's movement, conducting workshops with poet Robert Bly and storyteller Michael Meade. Hillman's the author of nearly a dozen books, including Re-Visioning Psychology *(HarperPerennial)*, The Dream and the Underworld *(HarperCollins)*, Healing Fiction, Inter Views *(both Spring Publications)*, *and the anthology* A Blue Fire: Selected Writings by James Hillman *(HarperPerennial)*.

I interviewed Hillman last fall [in 1990], when he was in North Carolina. Though nervous about meeting someone whom Robert Bly calls "the most lively and original psychologist we have had in

America since William James," I was put at ease immediately by his
friendly, gracious manner. A tall man, well into his sixties, who
manages to seem dignified and mischievous at the same time, Hillman
said he was feeling tired. He stretched out on the bed, I pulled up a
chair, and we began.

Hillman: You don't mind my lying down?

Safransky: This is perfect: the therapist lying down. The antitherapy therapist. You've criticized modern psychology for giving feelings too much emphasis. You've said we've had a hundred years of analysis, and people are getting more and more sensitive, and the world is getting worse and worse.

Hillman: I don't think feeling has been given too much attention. What one feels is very important, but how do we connect therapy's concerns about feeling with the disorder of the world, especially the political world? As this preoccupation with feeling has grown, our sense of political engagement has dropped off. How does therapy make the connection between the exploration and refinement of feeling, which is its job, and the political world — which it doesn't think is its job?

Therapy has become a kind of individualistic, self-improvement philosophy, a romantic ideology that suggests each person can become fuller, better, wiser, richer, more effective. I believe we have now two ideologies that run the country. One is economics, and the other is therapy. These are the basic, bottom-line beliefs that we return to in our private moments — these are what keep us going.

Safransky: When you say "the country," don't you mean those people who share certain cultural and intellectual attitudes? The insights of therapy don't seem mainstream.

Hillman: The insights of therapy *are* part of the mainstream. We have mental-health clinics all over the nation, in every city and county. And they all produce pamphlets about how to deal with the problems of addiction, battered wives, childhood disorders. There are therapists throughout the country, and they're very important people, because they pick up the refuse of the economic-political system. Someone has to pick these people up, and therapy does it. But therapy operates with an ideology, an individualistic, must-learn-to-cope ideology. The individual has to learn how to *cope*, and the therapist helps that person stay *in control*. This ideology is based on the idea of individual growth and potential. Most schools of therapy share the idea that there's an inner world that can be made to expand and grow, and that people are living short of their possibilities, and that they need help to — what shall we call it? — *fulfill* their potential.

Safransky: Still, it seems to me that those who run the country aren't more "sensitive," but instead deny their own woundedness. It's hard to see how the increased popularity of therapy has led to a degraded politics.

Hillman: I won't insist on a cause-and-effect relation. But I do believe there's a correlation, for two reasons. The first concerns the child archetype in therapy, an archetype which tends to depoliticize the client. Once one is engaged in feeling abused, in feeling victimized, one also feels powerless and seeks to locate blame outside of oneself. The client is concerned with the past, with what happened to him or her, with one's own individual growth. Yet the child is apolitical per se, is not a political being. Second, the class that first bestowed power on those who rule is composed of the white American suburbanites, who also happen to be the people in therapy. I may be wrong about this; I don't know the statistics. But there have been fewer and fewer people voting since the Nixon-Kennedy race. The withdrawal from the political arena of the "better" people, the more intelligent, the more sensitive people has allowed those now in power to gain power in the first place.

Safransky: I'm sure these criticisms don't endear you to the therapeutic community.

Hillman: No. That's part of what I mean about ideology. When you are situated within an ideology, it becomes very difficult to take criticism. These days there are two things — aside from religion — that are difficult to criticize deeply: our capitalist system and our therapeutic system. They share a common emphasis on the individual.

Safransky: Isn't the emphasis on the individual at the heart of the American experience? There's always been a mythology here of rugged individualism.

Hillman: There are many who have located the roots of the therapeutic movement in the individualism embraced by nineteenth-century modernism, in which everyone is the author of his or her intentions and is responsible for his or her own life. *Own. Own* is a very big word in therapy; you *own* your life, as if there's a self — an individual, enclosed self — within a skin. That's individualism. That's the philosophy of therapy. I question that. The self could be redefined, given a social definition, a communal definition.

Safransky: You're suggesting that this emphasis in therapy makes the self into a "private property."

Hillman: Right. You *own* your emotions; you *own* your feelings. In the last twenty years, philosophers and literary critics — particularly in France — have argued that there is no author, there is no central identity to the self, there is no self. They think we are a product of discourse, of language. I would say that we are also the product of a social network.

Safransky: The Buddhists have been saying for thousands of years that there is no self.

Hillman: But that isn't the way it seems to have worked with those Americans who go into Buddhism. They seem to work very hard at *self-*control, through meditation. There's a good deal of criticism of such practices even from within that tradition. What is the criticism? What are people noticing? Not just that gurus sexually abuse their people or buy Cadillacs. That isn't the issue. The individualism to which we fall prey — that's the issue.

Safransky: Can you say more about the child archetype, and the emphasis in psychology on going back to the past in order to learn who we are in the present?

Hillman: We have a biographical sense of psychology. We don't have as strong a social sense, or spiritual sense — we're less inclined to regard the soul as going somewhere, and less likely to define wherever it ends up as the most important part of our biography, the most important part of our existence.

I believe the soul is always attuned to the geographical, or ecological, world. Where you are is as important as where you came from. What you do every day is as important to the soul, to the revelation of the soul, as what your parents did to you, or what you were like when you were five or ten. We don't generally subscribe to such notions, not really; instead, we emphasize the notion of individual career, personal biography. This notion is faulty because it's too singular to begin with. We could fault this model of the self even further. But it's hard to sit here and imagine other models. Do you see what I mean? It's hard to shift to an emphasis on the end of life, or to the social, geographical context of life, to the "you" who is what you do, to the "you" that you create with every move. Now, that would be a Zen thing, wouldn't it? Every move you make, every bite you eat, every word you say is inventing yourself. We think the soul is already made by what happened early on, and we're always trying to fix it, to adjust it. But suppose I'm making it now, as I talk?

Safransky: Yet who you are, talking, is also made up of who you were.

Hillman: That's what we think. But can we conceive of the possibility that I change what I once was by what I say now? That I am no longer what I was? Perhaps what I was is only a fantasy, just as the time of time past is a fantasy.

Now, in a Judeo-Christian culture, that is tough thinking. Because our Judeo-Christian culture believes absolutely in the reality of history. We believe in it to such an extent that we send archaeologists to Palestine to find remnants of the historical Jesus. But Jesus is powerful not because he was in Palestine two thousand years ago, but because he's a living figure in the psyche. We don't have to dig up a remnant to show that he's powerful. But our culture is very historically minded. There are other cultures which are not historically minded at all. They're much more concerned with whether or not the trees are in good shape and are speaking to you. Much more concerned. Or whether the river has changed course: *that's* something to worry about. My goodness, if the fish turn belly up, that is far more important to my soul than what my mother did to me when I was four.

Can we imagine that way? What I'm trying to do is simply

imagine in this way, rather than make a literal statement that the fish are more important than my mother. Because if we don't begin to imagine in this way, the ecological problems are not going to change. We're still going to do something solely because it's good for us personally. James Lovelock said that the great problem of the moment concerns the destruction of the rain forest, while the depletion of the ozone layer is a relatively minor problem — though he's one of the scientists responsible for having discovered the hole in the ozone. We're obsessed with ozone depletion because we're so afraid of skin cancer and other direct effects on us from the hole in the ozone. We're still looking at ecological problems from an anthropocentric, individualistic, narrowly human point of view. We're not concerned with ecology, and we can't be until we change our notion of what an individual is. That's psychology's job.

Safransky: To some people, changing society and working on oneself aren't mutually exclusive.

Hillman: Freud argued that the self is truly noncommunal, fundamentally individual. Jung said that we are each makeweights in the scales — that what you do in your psychological life tips the balance of the world one way or another. The pervasive therapeutic ideology today urges a similar point: if I *really* straighten myself out — the rainmaker fantasy — if I *really* put myself in order, then the world —

Safransky: What's the rainmaker fantasy?

Hillman: It's the old, mystical idea that once the rainmaker puts himself in order, the rain falls. It's the shamanistic idea that unless I'm in order, I can't put anything else in order. It's also an idea basic to modern therapeutic practice: How are you going to help the world if you're not in order? You're just going to be acting out; you're going to be out in the street, making trouble. First get inside yourself, find out who you are, get yourself straightened out, and *then* go out into the world; *then* you can be useful. Understand, I'm arguing the therapeutic point of view now: Put all the architects, the politicians, the scientists, the doctors into therapy, where they'll find themselves, get in touch with their feelings, become better people. *Then* they can go out and help the world.

We've held to that view, but I don't think that's it; I don't think it works. I wish it did, but I don't think it does.

Safransky: You've written that pathology is not a medical problem to be cured, but the soul's way of working on itself. I was curious how that perspective extends to the question of addiction.

Hillman: *Addiction* is one of the big words of our time. Do you think addiction is located intrapsychically? Is the problem located inside me? Consider bulimia, the eating disorder. Now, I think an eating disorder is a food disorder. I think there's disorder in the food, in our relation to substances, so that we become addicted to them. We could say the addiction is a *symptom*; a symptom is always a compromise between an appropriate relation to a substance and a sick relation to a substance. What's important in an addiction is the value of the substance, the value of something external to me, on which I depend totally. It's this that the addiction recognizes: there is something outside of me with which I must be in touch. Whether it involves codependency — I'm talking here of a love object, of someone to whom I'm addicted in a relationship — or addiction to a substance, the result is the same: my psyche can't live without this other.

That's a big statement for the soul to make: *I must have it.* I must be in touch with that other thing, whether it's the person, the alcohol, or shopping, which is the second-biggest leisure activity in the U.S. TV is the first. Since we now have TV shopping, they're getting hard to tell apart.

To my mind, these are all ways of saying that, somehow, these things out there are carrying life for me. They animate it. We have a problem with the world of things; we have a problem with being dependent on the not-me. And we don't want to recognize that. We don't, because our ideology depends completely on the doctrine of individualism, a doctrine which assures me that I am a free agent, engaged in free enterprise; that I am on my own. I'm John Wayne. I'm Gary Cooper. I'm Rambo. I am a self-contained person. Or I'm a self-centered, abused victim/survivor. Yet I'm addicted to everything. It's breaking down my entire sense of who I am. *I can't live without you.* That's what an addict says, what every love-addicted person says: I can't go on without you. That's putting a *huge* value on that other person or thing. The way I see it, there's something instructive going on in the addiction.

Safransky: So is the way out *through*? Does one honor the addiction?

Hillman: No. I think the way out is contained in something Eric Hoffer said: "You can never get enough of what you don't really want." You don't really want the alcohol. If you can find out what you really want, if you can find your true desire, then you've got the answer to addiction.

Safransky: To what extent do you feel the twelve-step groups recognize this?

Hillman: They partly recognize it. They channel the desire toward something spiritual. But these "support groups" bother me, too. When you were a child, if you lived in a city, your father probably went out on Tuesday night to a ward meeting with the Democrats or the Republicans, to some meeting dealing with politics. Now we go out because we're fat; we go on a Tuesday night to meet other fat people. On Wednesday night we go out because our parents abused us; Thursday, because we drink too much. We meet single-issue people. We meet through our symptoms.

It's a new way of organizing the political world, the communal world: in terms of pathology. For everyone to sit around a room because they're fat   I don't know if that's a way civilization can continue. I want to meet with people who are fat, and black, and green, and white, and exhibitionists, and Republicans. That's what a democracy is about.

Safransky: I understand your point, but maybe you feel this way because you're not struggling with being fat or with having been an abused child.

Hillman: But why? Why is it that I have so reduced my struggle — the struggle of life, the very engagement that is life — to the fact that I am obese or that I fall in love too much? You see what I mean?

Safransky: I'm trying to see it from the point of view of someone in such a group.

Hillman: I think that group of overeaters could begin to realize what goes on in school lunches, and what goes on in advertisements for potato chips. There are acutely political dimensions here, dimensions that this group could work to identify. There has to be some imagination on their part, some effort, if they are going to see that their problem is not just something inside their own skin.

There's also the matter of the cell physiology, the physiological problems of obesity. There are lots of things. But all of them, all such points of view, tend to narrow the problem and in this way keep it from the communal. And I want it to go on into the communal. There's a fundamental political task, as Aristotle noted: "Men are by nature political animals." That's very important. Suppose we begin seeing ourselves not as patients, but as citizens. Then what would therapy be like? Suppose the man or woman coming to you as the therapist is, above all else, a citizen. Then you're going to have to think about these people a little differently; they're no longer just cases. I'm not sure what this leads to, but it points to a fundamental shift in emphasis.

Safransky: You're rather an uncompromising critic of spiritual movements and everything called "New Age." You once suggested that meditation was a fascistic activity, that people who meditate are as uncaring as psychopathic killers.

Hillman: I did once remark that meditation, in today's world, was obscene. To go into a room and sit on the floor and meditate on a straw mat with a little incense going is an obscene act. Now, what do I mean, what was I saying, for God's sake, aside from shooting off my mouth? I was saying that the world is in a terrible, sad state, but all we're concerned with is trying to get ourselves in order.

I remember hearing a student say something once that threw me into a real tizzy. He said we should meditate and let computers take care of world problems. They could do it much better than humans. I mean, he was really spiritually detached from the world.

Safransky: It sounds like he was also emotionally detached, but something called "spirituality" gets the rap.

Hillman: Your question is very legitimate. I don't want to be locked into an antimeditation position. I think every consumer

— for that is what we actually are — needs a lot of neutral time, a lot of turnover time: idleness, fantasies, images, reflections, emptiness; not necessarily disciplined meditation. But when meditation becomes a spiritual goal, and then the method to achieve a spiritual goal — that's what worries me.

Safransky: And the goal you're suspicious of is transcendence.

Hillman: Yes. The quest to flee the so-called trivia of the lower order seems misguided. Personal hang-ups, fighting with the man or woman you live with, worrying about your dreams — this is the *soul's* order.

Safransky: What if the goal is merely a few minutes of calm?

Hillman: If that's the goal, what's the difference between meditation and having a nice drink? Or going to the hairdresser and sitting for an hour and flipping through a magazine? Or writing a long letter, a love letter? Do you realize what we're *not* doing in this culture? Having an evening's conversation with people; that can be so relaxing. Moving one's images, moving one's soul; I think we've locked on to meditation as the main method of settling down.

It's better to go into the world half-cocked than not to go into the world at all. I know when something's wrong. And I can say, "This is outrageous. This is insulting. This is a violation. And it's wrong." I don't know what we should do about it; my protest is absolutely empty. But I believe in that empty protest.

You see, one of the ways you get trapped into *not* going into the world is when people — usually in positions of power — say, "Oh, yeah, wise guy, what would *you* do about it? What would *you* do about the Persian Gulf crisis?" I don't *know* what I'd do. I don't know. But I know when I feel something is wrong, and I trust that sense of outrage, that sense of insult. And so, empty protest is a valid way of expressing feeling, politically. Remember, that's where we began: how do you connect feeling with politics? Well, one of the ways is through that empty protest. You don't know what's right, but you know what's wrong.

April 1991

MIRACLE AT
CANYON DE CHELLY

Deena Metzger

Eight or nine years ago, an American Indian came to me in a dream. He taught me a simple rain dance. When I awakened, I practiced the dance — and then watched it rain. For a few days afterward, I could not refrain from telling the dream, even teaching the dance to friends and students. That was the winter of the floods in Topanga and Los Angeles. Though I was not so arrogant as to feel responsible, I came to suspect that I had betrayed a sacred trust.

Last summer I was traveling in the Southwest, trying to understand the landscape, the trees, the blue hills, the odd visual marriage of desert and miniature, twisted groves — piñon, juniper, and cottonwood — which characterizes that area. Such a landscape can break someone whose imagination has been formed by the art and vistas of the East Coast or Europe, who has been taught that beauty is limited to the dense green forest, the pastoral meadow, or the precise curve of the ocean's crashing waves.

I was disturbed by this landscape: the bare, harsh, rough hills; the open, splayed plain of brush; the palette of beige, pink, and brown rock; the dusty sage — everything muted by powder, dust, and heat, modulated by the patina of stone. It did not conform to any principles of beauty that I'd been given, but it insisted that it was beautiful and that I be willing to claim it as mine. I tried to reject it. "You are ugly land," I said, but it insisted itself upon me, demanded I take it in. *This is so ugly*, I thought, but then again knew it was mine. Ugly and mine and therefore suddenly beautiful. Ugly as long as it was outside of myself, and beautiful as soon as I allowed it to penetrate my heart. I came to understand that as much as I loved northern forests, redwood groves, English gardens, or flamboyant jungles, this harsh, dry plain, this mosaic of sand, disintegrating rock, unremitting sun, this fierce and dry beauty of desert — these were, always had been, would continue to be what I had been given, would always be mine.

I was traveling with my lover, Michael. He, too, had come to me in a shape or form that did not meet my expectations or fantasies. For a year, I had confronted myself with an ongoing question: "How do you know the beggar at the door isn't an angel?" The poor land I was looking at was also an angel, and it broke me to recognize this: to see that things are not what they seem; to see beneath the surface; to see the luminous in its tattered disguise. In a continuing progression, my heart opened to the land as it had to Michael: I took it in; I became afraid; I forgot the love; my heart shut; I was broken open again; I loved. . . . This dynamic became the bond between us. We did not give each other ease, but we were always alert and therefore enlivened.

Michael and I traveled to the Lightning Fields near Quemado, New Mexico. Because of the unusual amount of rain here, an environmental artist had set up a hundred rods over an area of a square mile to attract lightning. With four other people, we were driven to a small cabin, where we waited overnight, hoping to see an unusual display. To pass the hours before dawn, I told Michael about the dream. I speculated about doing the dance, but something stopped me. I said, "When I had that dream I was disrespectful to it. Magic is never to be used lightly, if it is to be used at all." Nothing in the dream had admonished me to be silent, but I had known then, and I certainly knew now, that only the most enlightened people have a right to use the power to make rain. Only they can know when the need is great enough and the permission to proceed is clearly given. If a Native American does a rain dance, she is already profoundly related to the land. I aspire to that relationship, but I have not yet achieved it.

I apologized for teasing that I might do this dance for our mere entertainment. Afterward, I spent some time repenting my earlier breach of faith. The rains did not come, and the next day we planned to travel north toward Mesa Verde.

The following morning, when we were setting out, Michael said he suddenly felt that we should change our plans and go immediately to Canyon de Chelly. I hesitated; Mesa Verde was the only place on our itinerary that neither of us had ever visited. But he was adamant, so we heeded the call.

We doubled back and drove through Navajo country, looking all around us at the storms and lightning we had not seen at the Lightning Fields. We watched the mysterious patterns of black rain falling in the distance, illuminated by pitchforks of light. When the roads became unmarked, then clay, we were a little concerned about the rain, because we had intended to take these back roads to the canyon. Just as we were approaching a rise, we came upon a Navajo man and asked him for directions. Was this the road to Canyon de Chelly? He responded by asking us if we would give him a ride to the camp where his people moved their homes and livestock for the summer. A bit surprised by his answer, we hesitated a moment, and then rearranged the luggage in the back seat to make room for him. He was a little vague as to whether this was the road we wanted. We came to the turnoff, and Michael drove down the very narrow lane to take the man to his door. We came upon a compound with a few houses and hogans, and the man insisted that we come inside to meet his mother.

When we entered, his mother was weaving a red, black, and white rug. She showed us the two rooms of the house, other rugs, photographs, and objects on the wall. She said, "I am so glad you are here. This is how we live. We want you to know this." She suggested that we might return to stay in the hogan if we didn't find a place to sleep.

While Michael got piñon nuts from the car to leave with them as a gift, I asked again for directions. She gave them vaguely, waving her hands casually, "Left, left, left, then right, right, right." She knew she wasn't being precise, and it amused them both. Then she looked directly into my eyes and said with a seriousness that could not be mistaken for road directions, "Don't worry. You will get where you are going."

We got confused at the first crossroads, and I told Michael what had transpired in that brief moment when he was getting the nuts. We proceeded as best we could until we again came upon a man walking along the road. Since we wanted to be alone, we hesitated before asking for directions, knowing that he would ask us for a ride; he did, saying that he'd been walking for ten miles and had another ten or so to go. "Is this the road?" we asked.

"Yes," he muttered unconvincingly. It must have rained profusely because, after a few minutes, the road turned to mud. The thick red clay was sleek as oil, the ruts four, five, six inches deep. The car could not hold the road, sliding in one direction or another; we went down an incline sideways, barely missing ditches, boulders, overturned logs. The Navajo man said nothing. My terror of heights returned. When we reached a place where the road went along a cliff, I demanded that we stop. I entreated Michael to abandon the car. "We'll buy another in Chinle," I said, as if cars were apples. Michael was determined to negotiate the road despite the clear danger, but fear overwhelmed me. I pulled on my cowboy boots and got out to walk. I could see lights some ten miles away. The Navajo said nothing, sitting still in the back of the car, serenely eating piñon nuts. Reluctant to leave me, Michael suggested that he would drive until he came to a place where I would not be afraid, and there he would wait.

And so we proceeded for about seven miles. I walked, then we drove together, then I walked a mile or so, they waited, we drove, and so on.

When Michael was alone with the man in the car, he asked him what the Navajo believe. The man said he could not tell Michael in white man's language, but it was something about worshiping the sun.

When we arrived at the man's house, he thanked us, took a package of piñon nuts, and left. Immediately after we turned the next bend, we were on a dry road that looked like it had never seen rain, and in moments we were on the paved highway to Canyon de Chelly.

It was almost sunset. Michael asked me to close my eyes, and he led me blind to the rim of the canyon.

I opened my eyes to the greatest natural beauty I'd ever seen. The light was just fading, but there was enough to make luminous the soft, deep, red-clay walls. One thousand feet below, a glistening snake of a red-clay river twisted through small fields of maize, which had been worked for more than a thousand years. The brush and junipers on the ledges, curling like the pubic hair of the Mother, and the cottonwoods beside the river all gleamed with a blue-green sheen, an underwater hue, as if phosphorescent with the light of another world. I stood on the edge, thinking that the Israelites must have felt like this when they first came upon the Promised Land. This was the sacred vision of paradise.

It was raining in the distance, and we could see black streaks against the sky. Where the sun was setting, the rain fell before it in a curtain of gold. I thought, *This is how so many sacred texts describe the appearance of God — a shower of gold light.* I was filled with awe. Then to the east, where the black fell against the darkening sky, wild explosions of lightning appeared, each stroke blasting the earth. This was another face of God. Then, as if this were not enough, there emerged a brilliant rainbow to the south, and beside it a gleaming, burning eye of platinum fire. Then came the magenta, the rose, and the wild colors of night. Every face of the Sky God appeared in all His glory, as if He were saying,

"Look at Me. Why have you raged against Me these last years?" I looked at the face of the Sky God pressed against the glorious body of the Earth Mother Goddess, and I knew — I knew — there are gods, and I thought, *I must not ever doubt again.*

Over the course of time, moments of beauty such as this have been enough to inspire that leap of imagination which asserts the actual presence of the divine. People have staked their lives — and the lives of others — on such visions. Such a moment could have allowed Noah to assume the rainbow he saw was a covenant between him and an immanent God. It is so simple and so difficult to believe. I vowed to myself to try to remember that, in this moment, I really did believe.

We had come to the rim at the very edge of time. Tired as we were, if we'd come even a half-hour earlier, we would not have stayed long enough to see the sky explode. If we'd come only a minute later, we would not have seen the unearthly light. Because we were there at that precise second, I could not doubt that we had been led most carefully, through a conjunction of simple events, to a moment of vision that would have been impossible to imagine.

As night fell, I remembered a sad tale my friend Steve had told me. When he was at Canyon de Chelly, he'd intended all day to leave a ring behind and, in the last minutes, he had forgotten, and the omission had tortured him for months. So I took the snake ring I loved, which Dianna had given me, and I gave it to the canyon for all of us.

The next day, we rose at dawn to see the sunrise and went down into the canyon. When we arrived at the edge, we were greeted by two goats, who nuzzled us affectionately, their bells breaking the silence. We fed them crackers that we'd intended to eat but had forgotten about. Ordinary as the moment was, it had a miraculous quality, as if the goats were there to invite us out of ourselves, back into the natural world.

Michael and I spent the day in solitude, together but not speaking, and as we walked in the *prima materia* of the shallow, warm river, a chant came to me with the rhythm of my steps: "In the dark red body of the Mother, at last." In the quiet body of the canyon, something in me opened, something I'd been taught to keep closed, so that learning to open it was the great mystery.

Michael spoke to me of his sense of revelation. He felt this beauty we were experiencing insisted that he devote his life to protecting the planet, to being her advocate. He said he would ask himself this question each day: "How shall I live my life to accord with such beauty?"

"With attention," I answered too quickly. "Attention, mindfulness, awareness." The words hardly served to describe the task of the willingness to see.

When I was meditating at the end of the day, a teaching came to me from the astonishing peace — the peace that passeth understanding — which was the gift of the canyon floor and the warm, primal red river of clay, where I had immersed my feet:

This beauty comes only out of great love.
This beauty comes out of a great heart.

I understood for the first time that love is the very nature of beauty. Skill, technology, aesthetics, or mind was not sufficient to create such beauty as confronted me; I could see the profound tenderness and caring in the hands of the Maker. It was as if I could feel the heart from which this beauty emerged; as if the impetus for this creation were a love which had been unbearable to contain and therefore had demanded to be manifested physically; as if the world were not a separate object, but the embodiment of a love so profound it absolutely required form; as if the love and form were not distinct from each other, but different faces of the same divine presence.

Now I felt I had to change my meditative practice so that after stillness, I focused upon the heart: no action taken, no word spoken, no writing initiated, no healing attempted before my heart was open; nothing that did not come out of love; not an instant to be lived without attending to the heart; nothing without the heart and nothing that did not serve the heart.

I also felt that I had been given questions to ponder:

Were the three Navajo people we met real beings?

What might have happened if we had refused either of the men a ride?

Was the second man a guide through the impassable barrier? Without him in the car, might we have gone over the side?

When the mother said, "Don't worry. You will get where you are going," what did she mean?

How much do we suffer in our lives, because we are not attentive to what is being asked?

Would I fall into doubt again, even though the revelation had come to me in the form and language I know best and trust most — in Story?

THE GODS SEEM to come to us in the forms we recognize. Had a flying saucer come, I would have laughed because I do not believe in the divinity of technology. But instead, the gods came to me in Story, the form to which I have devoted my adult life.

I believe that Story is a pattern, and that it gives coherence to our lives. For years, as a writer-therapist, I've worked with clients to help them discover the particular Stories they are living out. I've also had the hope that, at the moment of my death, I would come to know all the lesser Stories and the one great Story which have been my life.

The search for Story requires that we become aware of the essential elements that shape our experience. Story is a grid, an archetypal narrative, a divine scaffolding that organizes experience into meaningful form. What may otherwise appear to be unbearably random or chaotic can become coherent, even inspiring, when viewed as an intrinsic narrative consisting of elements like the Call, Setting Out, Meeting the Animal or Encountering Nature, Encountering the Guide, Purification, Sacrifice, Riddle, Making an Offering, Being Tested, Ordeal, the Barrier, the Monster, Blessing, Epiphany, and Return. Flailing around in the flotsam and jetsam of daily life, we may not be aware that we are engaged in a descent into the underworld or in an arduous attempt to find the entrance to paradise. Story is the template by which the most prosaic or mundane may be understood as the soul's search for the ephemeral but redeeming moment of vision. When I came to understand that there are mythic patterns in all of our lives, I knew that all of us, often unbeknownst to ourselves, are engaged in a drama of soul which we were told was reserved for gods, heroes, and saints. This understanding took me beyond psychology to the existence of spirit. I realized that Story is one bridge between the human realm and the divine.

Since that first insight, I've looked for Story everywhere. But I had never encountered it so blatantly before. At Canyon de Chelly, Story appeared as real as anything I'd ever encountered. The structure of those experiences was for me the absolute proof of the presence and existence of the gods.

At the end of the day, it seemed a good time and place to take vows. Michael and I sat together silently at the crest before the black-and-magenta sky. I vowed to try to remember. I vowed to devote myself to being a writer, lover, healer. I vowed to seek to do everything in my life from the heart.

When we were going back to the car after night had fallen, Michael invited me to sit with him again. We sat on the ground opposite each other, cross-legged. He was quiet for a while, then said he'd like to ask me two questions.

He asked me if I would marry him, spend the rest of my life with him, teach him what I know. He also asked if I would help him raise his daughter.

Before I answered him, I asked if he would be my partner on the path, take care of me as best he could, and help me with the work of my soul. I asked him if he would love me exceedingly well and let me love him exceedingly well. I said I'd come to realize how much I wanted to be loved and that, if I were to marry, it would be to find a place for my loving.

As we spoke, I was gripped with the desire to be married on the dark of the moon. The next week, when I looked at a lunar calendar, I saw that August 24, the night we said our vows on the rim of Canyon de Chelly, had been the dark of the moon and that the date we'd set for our wedding, December 20, would also be the dark of the moon.

We looked for rings on the Indian reservations as we traveled home. I had a ring in mind, a silver band with turquoise. We went from trading post to trading post without finding any two rings that were identical, though there were rings that resembled each other. But at the very last possible place, we came upon the only two identical silver bands we were to find. They were inlaid with turquoise, and they fit us each exactly.

January 1988

Somewhere Along The Line

Antler

What interested me most about gorillas
      when I first studied them
Was not that the males' penises are only
      two inches long,
But that gorillas shit and piss in their beds
      and don't leave to relieve themselves
          (though they build new beds every day),
Also they eat their feces, yes,
      they eat their turds.
And this made me realize that we
      (somewhere along the line)
Decided we wouldn't shit and piss in our beds.
We agreed we wouldn't eat our shit
      or drink our piss,
That we would wear clothes
      and not go naked in public
          and not shit or piss in public
      and not jack off in public,
Not fuck or suck in public,
Not stick our fingers up our rear ends
      and smell them
          (even in the privacy
      of our own homes),
Or on meeting another of our kind
      sniff each other's cock and balls
          and cunt and asshole like a dog
      but shake hands like a man
And rather than pissing and shitting to mark
      our territory
We invented money
And rather than gathering food from plants
      we'd work to plant them raise them sell them

And rather than killing animals fish birds
with our mouths and eating them
raw and bloody
We'd hire others of our kind
to kill them
and cut them up in little pieces
not with their mouths
but with sharp knives in their hands,
And somewhere we decided rather than live in trees
we'd kill them, cut them up in long pieces,
build houses and live
inside them while sitting in chairs
made from them and write poems
about them on paper made from them
with a pencil made from them
about how somewhere along the line
we decided to be different than
gorillas and monkeys because
our way of being was right
because we were better
than any other creature on Earth.

May 1992

 Cowards

Dan Howell

Maybe Dad was looking out the kitchen window, drinking a cup of coffee, or smoking. He must have seen something, because he went outside. I'm sure that my mother, my younger brother Steve, and I were in the kitchen; it was breakfast. I think I remember getting to the window in time to see Dad hurriedly close the doghouse door and signal. Anyway, I do remember all four of us standing in front of the doghouse.

It was morning and gray but not cold enough for coats. Brown winter grass filled our yard except near the basketball goal, where a big circle of permanent dirt had been trampled by a gang of boys — Steve and me and our eight best friends, all of whom lived on our oxbow street. Scamp's doghouse sat at the back of the yard, just in front of the new little hedge that blocked the creek from view. Behind the creek, the woods rose up much higher than the top of the basketball goal, like a deep green, leafy hill.

"A rabbit's in the doghouse," said Dad. He whistled for Scamp, who came with his customary speed, with an eagerness all of us liked, with a good-hearted purity of simple friendliness that sometimes seems conscious in certain dogs. Dad knelt to open the door, ushered Scamp inside, and shut it. All of us were silent, expectant.

No sound, no stirring or jostling. Nothing.

After a minute or so, Dad leaned over and opened the door.

Scamp just sat there, quietly gazing out. Beside him hunkered a medium-sized rabbit in its mottled brown-and-white winter coat, gazing out too. It wasn't trembling. Both animals stared at us. The rabbit's eyes were dark, shiny marbles. Scamp's eyes were light chestnut, human, grandmotherly.

Neither animal seemed at all excited by the forced proximity, as if they were old pals quietly hanging out together. So we had to laugh.

We kept on laughing as the rabbit stirred itself and hopped out nonchalantly, bumping along slowly past us and through the hedge to the shadowy woods, while Scamp continued to sit there, not even shifting his position. He wore a look of befuddlement, as if he knew we had expectations but could not figure out what they were, and so thought it best to be still — "Good dog" — until he could understand what was going on and respond appropriately.

A great dog. And that moment was great in its small way, the odd gentleness of it. Peaceable Kingdom: central Kentucky, the Bluegrass, in the fifties.

Only recently have I wondered why my father would have done that in the first place — set up an encounter that could have hurt or killed the rabbit, and even put Scamp at risk of injury or rabies. Maybe he wasn't thinking, it was an impulse, or maybe he meant it to be fun for us, or maybe he was simply curious, like a kid. I do know that for all his life growing up on small, leased farms in Dust Bowl Oklahoma, he hated the times when animals had to be slaughtered.

I GREW UP believing that I was a coward. In my adolescence I could think of nothing in my childhood or boyhood to convince me that I wasn't, and it was a long time before I changed my mind, although cowardice wasn't something I thought about much, except when circumstances forced me to, whenever I felt I should have fought but didn't.

Once, in downtown Lexington, I was sitting with Steve in the Ben Ali Theater, a worn but still lavish and rococo ex-vaudeville house, complete with filigreed box seats on each side, plush velvet curtains, and two balconies. Steve and I had gone to the movies by ourselves, so I must have been at least nine or ten. We were tucked quietly into the worn seats, absorbed by the action on-screen. Someone behind me started tapping my head. I waited a moment for the tapping to stop. It didn't. When I turned around to say something, a kid I'd never seen before spat directly into my face.

I can still feel it, how it almost slid across my mouth. I was stunned. I stared at the boy for a few seconds. Then I turned back around and wiped off the spit from my glasses and face and just sat there, without the faintest notion of what I should do, shocked and baffled, paralyzed. That I did nothing, nothing at all, was odd. Apparently the boy thought so too, because he didn't bother me again. Maybe he was waiting for me to do something, anything, and the longer I sat there without moving, the more nervous he became. Maybe.

It's not that I never fought. I had several fights with my friends, all of which I "won," all of which I cried after. Not from physical pain, but from the violence and sourness of our anger, from the strangeness of the urge to hurt someone I cared about. It always felt like some kind of violation, something deeply wrong.

STRANGERS WERE IN our woods one day, boys we had never seen before, about our age. Accurately or not, I remember them as seeming a little shabby, in a kind of disarray beyond that of boys outside at play, hinting at something threatening, like poverty or uncaring parents. I know they were "tough" in a way none of us were.

My brother and I and a couple of friends had gone into the woods for no particular reason, wandering. We knew all the paths, the best thickets for hiding, the sinkhole, the monkey vines. The place was ours by habit and usage — even today I can close my eyes and walk through it.

We discovered these other boys at the back of the woods, in the cleared-out circle of dirt surrounding the biggest tree, a massive old oak called the "treehouse tree" for the remnants of someone's treehouse ten feet up, still dangling. The boys — four or five of them — had been yelling at each other in play but fell into a wary silence as we approached. They assembled behind the largest boy and waited.

I don't remember anything more specific about the encounter, any words or gestures. Probably I indicated in some way our proprietary sense of the woods, but not forcefully. Whatever else happened, and in the absence of argument or any overt gestures, things finally came to the moment when I knew that I — the oldest and one of the biggest and the "chief" of our gang — would have to fight for our right to be there. I didn't.

By the time Steve and I reached the fringe of the woods and the creek at the edge of our backyard, I was crying — from surprise at the unexpected hostility, from the anger I hadn't expressed, from sheer frustration, from shame. We opened the back door and there was my father in the kitchen. He noticed my crying and asked me what had happened, and I told him.

Dad led us back into the woods. I don't remember any explanation, and in fact I can't retrieve a single word spoken during all of this. It's as if the whole episode took place as a dumb show, the images supplanting language.

Dad brought me forward, a hand gently on my shoulder, face to face with the boy I didn't want to fight; whatever he said, we understood that we had to. Maybe there was some feeling of a code being invoked, a tradition being followed. What I remember clearly is the unusual quiet that cloaked our little group, making a tentlike space of the opening in the woods where we had gathered; I remember the not entirely unpleasant sense of being enclosed, if not safe.

The boy and I faced off, and the others stood back a little.

About my size, the boy had dark hair and wore an old jacket. In my memory, his face is only a smudge, and I never found out his name. We began to "box," swinging at each other with tentative swipes and landing a few soft hits, nothing harder than the blows of play-fighting. Then, because the act may have felt sanctioned, or because I felt safe with my father there, or both, I swung hard and hit the boy sharply in the mouth. It hurt him. My father stepped in immediately and pushed us lightly apart. The "fight" was over. Those other boys left our woods and never came back.

For a long time, whenever I've thought about that episode, its sheer oddness has been tainted with just enough satisfaction to deepen the confusion I'm trying now to resolve. I think I understand what my father did, or thought he was doing. Part of it was an effort to teach a principle: "standing up for yourself." Possibly he knew already that this wasn't the first time I'd let myself be cowed. But beyond that, there was something about the setup, the scene, that reminds me of a gunfight in a western, a confrontation between two clearly defined representatives of right and wrong — and I don't believe I'm forcing that interpretation, even if my father was acting solely on impulse.

The American West — both the literal and the mythical — shaped my father. More than almost any other man I've known, he embodies and lives out (at least in his public presentation of himself) the rather worn and discredited ideal of the laconic Western male, "strong, silent." Uncomplaining, unfailingly patient in public and in private, he never lashed out at me or my brother as we were growing up (barring a few moderate spankings); even as we became more taxing in our teens, he struck — verbally — only a couple of times, memorable because they were so rare. All of our friends from the neighborhood and our maternal cousins and aunts and uncles recall him as gentle and calm ("recall" because they haven't seen him in decades since my parents divorced); they speak of his deep voice and commanding size, his pleasant, quiet, automatic authority. And maybe that authority was given extra weight, for me, by the sense I had even as a child that he was holding back, that his demeanor was more a product of his will than of his nature. It was something he had worked on, holding back anger, holding back the outward display of strong feelings, like a warrior. To conceal your emotions was to be a man.

BARE TO THE waist, his chest hair shaved away, my father was on his back, his chest sheathed in thick white bandaging where his breastbone had been split and his chest opened so a valve in his heart could be replaced with one made of plastic and metal. Wires from the electrodes attached to him ran to monitors, and several tubes issued from him, snaking from his nose and throat, from the back of his hand. The machines formed a semicircle behind him, at his head, and they combined with the spotlight effect of the lighting in the intensive-care unit to render the whole scene theatrical, in a way — as if staged for me.

I walked up to the gurney. Dad was awake but couldn't raise his head or speak. I asked how he was anyway, whispering, and tried to read his blinks and faint nods as I talked. His eyes seemed unusually clear. I reached out and put my hand in his and noticed with almost embarrassed surprise how suddenly my forty-year-old hand became as small as a child's would be in mine.

That Dad looked old was inevitable, laid out that way, silent, subjugated by the technology designed to maintain and protect him. He appeared helpless, and was. His stomach bulged up and his skin was pale and loose. Not much was left of the man whose appearance had once caused two little first-graders at my elementary school to stop dead in the hall, staring. My father was passing at a distance down the hall, heading for a visit with our principal (on official business as president of the PTA or Cubmaster of Pack 86), looking big and trim at thirty-five or so. This was the late fifties, and there was one famous popular figure he did resemble — in a suit, his curling dark hair swept back, with the same large face, the glasses. So I wasn't really surprised when one of the first-graders whispered earnestly to the other, "It's *Superman*!"

A FEW YEARS ago, my father and I were sitting in his TV room in Covina, California, watching an American and a Cuban boxer fight for a medal in the Pan-American Games. I remarked on their quickness.

"I never liked boxing," Dad said. "The only reason they're in that ring is to hurt each other. I don't understand the pleasure in watching that." I nodded, adding some comment. Then I lapsed into the silence that seems always to dominate our times together.

When my brother and I were young, we enjoyed traveling with Dad once in a while on his business trips to eastern Kentucky. More clearly than anything else, both of us remember those long stretches of time when we would all ride along in silence — a complete, comfortable silence, each of us in a separate reverie as the hills and woods slid by. That silence was never hostile or uneasy. It was almost a communion, a peaceable manifestation of our compatible souls.

But what was good about silence in those days is bad in these. Now that our father's life is closer to an end, both Steve and I feel pressed to be more open with him, and we don't know how. It's something we never learned. Too frequently, when I end a visit with my father, I'm sad and frustrated, once again having been unable to break through his reserve. Or my own. And what's worse is that I sense how badly he would like to be easy and demonstrative, to speak what he feels. He'll follow me outside to my car and we'll talk for a minute or two, awkwardly, and sometimes shake hands; then I'll drive away with ashes in my mouth, always looking back as if I might never see him again, which one day will be true.

My MOTHER HAS never been at all reticent about expressing her feelings. Spontaneous, emotive, even volatile, she could hardly be more unlike my father — to her detriment, I used to believe. Now I know better, and I'm grateful for the lessons none of us knew she was teaching.

As TROUBLE SURFACED between my parents in the sixties, and my brother and I drifted apart and into troubles of our own, Scamp spent less and less time with us, preferring to hang around our young cousins, who lived a few blocks away in our newest neighborhood. He liked playing with kids. He would still come by and visit us for a few days, then he would go away again and stay gone for a week or so. When Scamp was thirteen, he developed a liver disease, worsened quickly, and died. I remember seeing my brother (eighteen or nineteen at the time) almost staggering around in the front yard, in hazy bright afternoon light, crying. I was embarrassed that he would be so public in his grief, and I think he may have despised me for not showing my tears. I had begun to believe that things would hurt less if I didn't show pain, and for a while I lived that way. I know I couldn't help being the person I was, but I've found that very hard to forgive.

FOR YEARS I thought I had spent my boyhood in Eden — a place of beautiful woods and a kid-sized creek, with our big gang of "good" boys, our parents active in school and Scouts, our wonderful freedom to roam almost everywhere on our own, even downtown. I felt blessed. In fact, my old neighborhood seemed so golden that I began to suspect I was gilding my memory of it, until I discovered that my brother remembers it that way too, as do all our friends from those days.

But there was a snake in that garden — the snake of anger, one of the three "original sins" of Tibetan Buddhism (along with pig-greed and rooster-lust). I grew up believing that it was "bad" to show anger — and for me that belief ended up translating, now and then, into actions that I saw as cowardly. And of course there was the complicating influence of my father's general stoicism. Or maybe I simplify too much.

A number of years ago, my brother and I shared a small house outside Lexington. One evening after supper, we wound up, somehow, talking about anger. I told him about an incident in college when I was furious with my girlfriend; I was raging inside and felt explosive, even dangerous. But I didn't feel I should let that out, not by breaking anything, certainly not by any action toward Barbara. So I took the cigarette I was smoking and stubbed it out on one of my knuckles, grinding it. And my anger went away.

My brother was staring at me, smiling slightly. "How weird," he said. Leaning forward, he extended both arms until his fisted hands, knuckles up, were close to my face. "Do you see these scars?"

December 1992

Villanelle

Richard Hoffman

"You think you can walk right out?"
my father said. I did. I went
as far as I could from that house,

far from that town. I doubt
I could have traveled farther had I meant
to change my name, to walk right out

of the life I was born to, to applause
the world insists will make us confident
but never does. I trusted from that house

to this one, bridges burnt, roads out,
even demons wouldn't dare. But hellbent
quests are circles: storm right out,

a son hurt and rebellious,
and in half a lifetime, in bewilderment,
you come home to your own house

and your sullen son without
the evidence to prove you're innocent.
Some days I want to walk right out,
but can't, won't, must not leave this house.

April 1989

OF LINEAGE AND LOVE

Stephen T. Butterfield

My mother wanted more counter space in the kitchen. There seemed no way to accomplish this goal and still leave room for the washer and dryer. My father built a countertop that anyone could quickly disassemble in case the appliances had to be serviced.

"There are three ways to do a job," he said. "The right way, the wrong way, and the navy way."

His arms were hairy. He rolled his sleeves up past his bulging muscles. The armpits of his shirts were stained with sweat. My mother said he stank. I liked the way he smelled, except when he was glaring down at me, threatening, "You better straighten up and fly right, or I'll blister your ass" — which he would sometimes do. At those times I wanted to punch his false teeth down his throat.

He saw this in my eyes and smiled. "You can think anything you want, as long as you keep your trap shut."

Instead of punching him, I learned how to beat him at chess. I won my first authentic victory against him at the age of fourteen. After a battle of seven hours, as the dawn grayed the windows, he acknowledged that he had lost. He was too good a sport to chicken out by insisting it was past my bedtime. We laughed together and cooked bacon and eggs.

"Don't get too big for your britches," he warned. "The old man can still teach you a thing or two." By age twenty, I was beating him so consistently that he stopped playing with me.

His consciousness was formed out of the Depression and the Second World War. After graduating from high school in 1932, at seventeen, he went on the bum and rode freights for a year. He met my mother in Boston, married her in 1936, and supported her by setting pins in a bowling alley.

"I always found jobs," he said. "I wouldn't take no for an answer. Every morning I was out looking. You saw all these guys with their hats in their hands, saying, 'Oh, mister, please give me a job,' and this petty tyrant standing there sticking out his fat belly: 'Hmph! What makes you think you *deserve* a job?' I just walked up to him and said, 'I'm gonna give *you* a job. Your job is to find work for me. I'm gonna find it anyway, with you or without you, but if you want to make a little money, then I'm gonna hire you to save me some time. If you don't want the job, fine, I'll give it to somebody else.' I never let those bastards keep me under their thumbs.

"The way I learned the machine trade, I walked into a shop and said, 'I'm the best damn machinist around. I can fix anything you got, and the only way you're gonna find out how good I am is to hire me and put me to work.' Of course, I got fired after a couple of weeks because I didn't know my ass from a hole in the ground, but then I had two weeks of experience. On the next job I might last two months. If you really want to work, never give up. You want a job, find the guy with the money, and nag the shit out of him, every day, until you wear him down.

After a while he'll give you a job just so he won't have to listen to you anymore."

By the time the navy drafted him in July 1945, my father was nearly thirty, married with two children, and working in the defense industry. After the Japanese surrender, he went on a tour of active duty through the Caribbean and then suddenly showed up in the kitchen. I had forgotten what he looked like. I ran to his arms. He wrestled me, mussed my hair, and turned me upside down.

"Daddy, do you have to go back to war?"

"No, sonny, the war is over. The Japs were so scared when they heard I was coming, they gave up."

My father did not give up. "Life is a constant battle," he taught. My first history lessons about World War II came from his narratives at the kitchen table: "At Pearl Harbor the Japs caught us with our pants down, but we kicked their asses good at Leyte Gulf." I was brought up on this kind of talk. My brother and I staged wars with toy soldiers and made parachutes for them out of old sheets. My favorite rainy-day activity was drawing pictures of exploding bombs.

Bitten by the American Dream right after he got out of the navy, my father was determined to become a millionaire. He set up a printing business on the GI Bill and went bankrupt within two years. It was a devastating collapse that wiped out everything he had.

Our landlord told him that my brother and I were not allowed to climb the trees in front of our apartment. "Every kid has a right to climb a tree," my father fumed, having grown up in rural New Hampshire. He moved us into a winterized bungalow on the edge of a great swamp. "There," he said, "you can climb any damn tree you want." Then he locked the shop, came home, shut himself in the bedroom, and didn't come out for three days.

I was seven when his business failed. I remember a long, dark period of bitter fighting between my parents that went on far into the nights. Indeed, it went on for the rest of their marriage. I used to put a pillow over my head and cry.

One morning, my mother chased him out into the yard and threw every dish in the house at him. I never knew the cause of that quarrel, but there were always causes: she had been against his going into business; he was domineering, jealous, and unfaithful; she wanted to work outside the home; he insisted on being the sole provider. The dishes came flying out the door and bounced off his body as he stood there saying, "For Christ's sake, Emily, be reasonable, will you?" She packed her suitcases, called a cab, and left.

With a lump in my throat, I said, "Dad, is she gone for good?"

"Of course not," he said. "She'll be back by supper time. Help me pick up these dishes."

He was right. At supper time the cab brought her back to the door with all her bags.

Couples who fight a lot tend to use their children for a battleground. My parents often plunged my brother and me into conflicts of divided loyalties. I feared that I would have to choose between them, that they would abandon me, or turn on me, or that loving one would mean I had to reject the other. The pillow I put over my ears to shut out their anger grew into a habit of not listening.

When I was fourteen, he dragged my mother out into the yard and almost choked her to death; she was screaming for help. The neighbors called the police, and two cruisers came to the house and sat in the driveway for an hour, flashing their lights while the officers took statements. By that time, my "pillow" was so thick I'd slept through the whole battle.

His dream of making a fortune shattered, his third child born dead, and his credit ruined, my father went to work fixing Linotypes for a newspaper. Then he began telling his sons the stories of his boyhood.

"When I was sixteen," he said, "I tracked a buck in the snow for a whole day. I followed his tracks right out into the middle of an open field . . . and they disappeared! *Jesus*, I thought, *did he take off and fly?* Sonofabitchin' deer — he walked backward, in his own tracks, until he came to a thicket, and then jumped sideways, fifteen feet on the other side of that thicket, where you couldn't see his tracks, and took off. He didn't miss a single hoof print. You could walk up and down his trail a dozen times and never figure it out — unless you know deer. Dad found the new tracks for me the next day. He laughed at me over that one for years."

My father taught us to shoot, fish, fight, snowshoe, tap maple trees, cut wood with a bucksaw, and make our own Christmas presents. His mother had knitted afghans and sweaters, braided rugs from old clothes, and made shelf knickknacks out of twigs and thread spools. "We never had any money," he said. "To get a jackknife for Christmas, that was a major gift."

Having been defeated in business, and knowing he would be poor for a long time, he moved us to a place that resembled the woods and villages where he had been raised; then he drew from his past the stories, memories, myths, and customs that enabled him to keep his sense of humor and self-respect.

His stories gave me a sense that I, too, belonged to a family lineage, a tradition. I began building toy villages and forts to reenact the frontier life of colonial Massachusetts, without realizing how deeply personal was the meaning of this play. The great swamp that extended from our backyard was the ground where I explored the myth of the wilderness. He had forbidden me to hike there alone, and this transformed my explorations into a rite of passage.

"When you walk on the ice," he said, "if it starts to crack, lie down and spread your weight."

The following winter, lost in the woods, I started across a frozen river and heard it crack. I lay down at once to spread my weight; it saved my life.

Whatever his failures as a husband, he spent a great deal of time teaching and nurturing his sons. We took in children from the state, whom he helped to feed and toilet train, like his own. He was our provider, barber, first-aid medic, and nurse.

One measure of my trust in him was the fact that I allowed him to pick wax out of my ears with a hairpin. During this operation I held perfectly still while he probed the deepest part of my ear. "Good," he said, in the smooth, gentle voice he used to doctor our cats and dogs. "That's good, almost got it, hold still now. . . ." He would give me the lump of wax afterward for a souvenir.

My brother's arm was once burned so badly the doctors wanted to amputate it. My father said, "Like hell!" He worked on that arm night after night, week after week, changing the dressings, making my brother exercise the muscles, bend the joints, lift small weights, then larger ones, until at last the arm healed. It was scarred, but usable and whole.

Sometimes my father read poetry to my mother at night, after work. He liked meter and rhyme. At ten, I sat on the floor near his feet, enraptured by his strong, level voice reading ballads. He was a militant trade unionist, and the poems he most enjoyed were about struggle, courage, victory, and defeat. He used to gaze out the kitchen window and quote John Davidson's "Thirty Bob a Week," about earning a living as a worker:

> It's a naked child against a hungry wolf,
> It's playing bowls upon a splitting wreck,
> It's walking on a string across a gulf
> With millstones fore-and-aft about your neck;
> But the thing is daily done by many and many a one,
> And we fall, face forward, fighting, on the deck.

Watching television inspired him to express the same sentiments in more prosaic form: "There's a lot of money in entertainment. The only thing that doesn't pay is a decent job well done; but you have the satisfaction of knowing you did an honest day's work, as you drag your ass home with your pockets empty and fall into bed."

My mother's parents were Seventh-Day Adventists who taught me that the Bible was the literal word of God. My father would say, "If God wanted to reveal something to the human race, why would he choose a corrupt instrument like human speech?" Or: "If the Bible is the literal word of God, then how come God gave two different accounts of the Creation in Genesis? Didn't he know which was true?" Or he would write his questions in verse, which he read aloud to the family:

Spaceman, with your rocket bright,
As you cruise the stars so white,
Can you find the streets of gold,
Mentioned by the men of old?

If you can find the gate of Heaven,
Is there just a single one, or seven?

And so forth. His own view, culled from his reading of Spinoza, was that God is everything — which he would repeat with such dogmatic ferocity that I wanted to run out into the yard with my hands over my ears. Eventually I understood that he regarded philosophical discussions as a form of combat. You were supposed to stand up and slug it out with him. Then he would laugh and quote, "I may not agree with what you say, but I will defend to the death your right to say it," attributing this line variously to Voltaire and Thomas Paine.

My father's expanded notion of God showed me that it was possible to undertake one's own spiritual journey entirely outside the accepted conventions of religion. His version of Spinoza gave me a new way to think about sacredness. When I came to read Emerson, Thoreau, and Buddhist authors, many of their ideas were already familiar to me.

He was hungry for knowledge and carried books with him to work so he could read between jobs. Entirely self-taught, he was deeply suspicious of anybody who had a university degree. Intellectuals made him defensive. Yet he supported my efforts to get an education, caught between his pride in my academic success and his desire to cut me down to size.

"Tell me," he said, after I had become a doctoral candidate in English, "do any of these so-called modern poets write anything that a common working slob like me can understand? I don't know who they're writing for, but it isn't me. Why do they use symbols — why can't they just say what they mean? I've got a whole book upstairs *explaining* the work of T.S. Eliot! For Christ's sake, if he couldn't make himself understood in the first place, why the hell write a book about him?"

In the academic world, where the pressure to publish results in the production of endless and unnecessary scholarly books about books, my father's words often flash over me like the utterance of a prophet.

Eventually, my mother won me almost entirely to her side. I resented him for his abuse of her. Politics divided us during the Vietnam era — he supported the war; I wanted socialist revolution at home — but gradually I stopped caring enough about his opinions to disagree with them. He told the same stories over and over. I felt that he had betrayed my trust by lying to me during a family crisis, and his stories seemed like a smoke screen to hide his dishonesty. I found his company boring and hardly talked to him at all.

When he was old, I tried to introduce him to the Buddhist doctrine of emptiness; I thought it would ease any anxiety he might be having about the imminence of death.

"Ultimately," I began, "you never were."

"Maybe not," he said, peering over the rim of his glasses, "but I made a hell of a splash where I should have been."

That was the end of my Buddhist lecture.

I took him to the family grave plots to visit the stones of our dead relatives. This was a ritual my mother used to enjoy. We stood among the monuments for a while. "Well," he said, turning to go, "I don't suppose any of them are gonna get up and thank us for coming."

That was the end of our sentimental visits to the dead.

His major hobby was genealogy. Someone had once insulted his background by implying that he came from an inferior stock of poor white losers. His pride would not let him rest until he had won that argument. He traced our paternal lineage back fairly quickly to our earliest colonial ancestor, Benjamin Butterfield, who entered Puritan Massachusetts in 1636. Then, his hobby escalating into a full-scale obsession, my father spent twenty-five years tracing the lines of Benjamin's offspring into a bewildering maze of branches. A network of contacts across the country funneled data to him from vital records. He built up stacks of files on thirteen generations of Butterfields, containing thousands of names.

He tried incessantly to recruit me as his research assistant and ghostwriter. I said I was too busy. When he got started on the subject, I would yawn and go talk to my mother, who thought all Butterfield men, excepting her boys, were ignorant hicks with smelly feet, chips on their shoulders, and no respect for their betters. She was hard at work telling me stories of her Nova Scotia ancestry. "Our people," she said, "had true class."

"Your people," my father would shout at her, "were nothing but a bunch of goddamned Tories who went up to Halifax because the patriots ran them out of Boston. Read your history — you'll find out who's right." In this bizarre way, my parents were still fighting the American Revolution.

They were fighting while she was in the hospital, a month before she died. I was past forty by that time. I suggested to him that, after forty years of battle, it was time they made peace. "You mean," he answered, "when I'm surrounded by Indians, I should just hand over my scalp? No thanks."

"You see?" she said. "He'll never change. Always wanting to dominate. I should have left him. But he was your father. I couldn't take him away from you."

I grieved over her death for more than a year. I blamed him, and I blamed myself for not rescuing her. I could not touch him. I had not been able to communicate with him for a long time, but now his deterioration from Alzheimer's disease made even the attempt altogether hopeless.

My involvement in the Buddhist path had taught me to recognize the forms of samsara, the repeating cycles of passion and aggression that perpetuate human suffering. I had seen these repeated by my parents, and repeated them myself, until I was utterly nauseated by samsara and longed to wake from it. The Buddhist path also demanded that I cultivate respect and gratitude toward my parents. This is an uncompromising, relentless demand, a precondition for waking up. Without my parents, I would not even have obtained a human birth, and therefore would have no working basis to seek enlightenment. During one practice, I was taught to visualize my father on my right and my mother on my left while I prostrated, thousands of times, to images of Buddhist teachers. This practice makes it impossible to ignore your parents. Whatever you feel about them is vividly present in your mind. Often you remember and relive scenes from childhood, revealing symmetrical patterns that transcend time.

When I was a little boy, my father used to wake me up at night for the bathroom. If I couldn't go, he turned on the faucet. The sound of the water started me.

I woke up sick with nightmares and said there were monsters hanging on to my curtain. "They won't let me sleep." He told me to walk right up to them and punch them in the nose, and they would go away. I went back to my room ready to follow his advice, but they were gone.

Now he was a withered, senile old man who got confused and wandered off by himself. I once found him standing on a construction site, fumbling with a broken coat hanger.

"Dad, time to go home. Get in the car."

"Wait just a goddamned minute, till I finish the job." He stood there and peed in his pants. The monsters were not gone after all, however much he shook his fist.

I took him to the doctor and had to get a urine sample from him. I led him to the bathroom, unfastened his diapers, and instructed him to urinate into a specimen container. Nothing happened. This was an embarrassing moment. Having broken the biblical taboo against looking on my father's nakedness, I still had no sample. What to do?

I turned on the faucet.

Ironically, the person who brought me back in touch with my father's genealogical work was my oldest son. Every summer, my son stayed with my parents. While I was preoccupied with my divorces, academic career, union activity, and Buddhist education, my father was taking him upstairs and teaching him how to find and hold jobs, how to cope with an angry stepmother, and what family lore was contained in those notebooks and card files. The old man must have figured that if I did not care enough about the values of my own culture and family tradition to pass them on to my sons, he would make sure the job got done without me.

Eventually, I had to put my father into a nursing home. I stored his genealogical records in my attic and forgot about them. One day, my son asked me where they were. We dragged them out together.

"Gramp showed me all these books, Dad," he said. "Didn't you ever see these? Look, this is fascinating."

He rehearsed with me the lineage of our direct-line Butterfield forefathers, beginning with Benjamin, who settled in Chelmsford, Massachusetts. They had fought in every New England Indian war since 1675. They included Lieutenant John Butterfield, who enlisted with Rogers's Rangers during the Seven Years' War. In 1759, Lieutenant John fought on the Plains of Abraham at the capture of Quebec. His sons, John and Peter Butterfield, traded six-month enlistments in the Colonial militia, one brother going away to fight the British while the other stayed home to take care of their widowed mother. At the Battle of Bennington, John, by that time a seasoned veteran, led a charge on a Hessian wagon, captured several enemy soldiers, and saved a Hessian prisoner, who had already surrendered, from being shot by a panicked adolescent boy on the Colonial side.

This John had fifteen children. The tenth, Daniel, born in 1785, fathered the branch that leads to me. Daniel's photograph, taken in the 1850s, hangs in my study. "That's the grandson of the man who fought on the Plains of Abraham," said my son, looking at his face. "Doesn't it make you feel weird?"

I felt so weird that it was like waking up from a dream to find myself on fire. My sons and I went canoeing on Lake George in New York, to visit the places our ancestor would have seen. My father's work was a thread of continuity holding us together. We talked about him incessantly.

In this way, he had converted his isolation, anger, and injured pride into something of priceless value to us. I spent a few days in the Library of Congress, checking the accuracy of his research. He had been meticulous, using vital records to correct the errors of other genealogists as he went along. He preferred truth to legend, always. I found the names of our forefathers on the muster rolls for the French and Indian War and the Revolution. I read histories of the period so that I could make a chronology of their lives.

The domestic ignorance and aggression that plague the human species are well represented among these forefathers. My great-grandfather was a wife beater and child abuser who drove my grandfather, an eleven-year-old boy, out of the house to live on the streets. Great-grandfather was married and divorced twice; Grandfather was married three times and divorced twice.

Suddenly, I understood why my father should have regarded life as a battle.

While taking volume after volume off the shelves, looking up references, making notes and copies, I felt his presence over my shoulder, smiling and advising, "Check this, and this; I already did that; now you're on the right track." I returned home to Vermont at the end of the week with a briefcase full of new material, having visited Bennington while searching for the site where Revolutionary John had saved the Hessian prisoner. Some kind of expansion was happening to me that I did not fully understand, but I rode this current without resistance, curious as to where it would lead. I saw the forefathers lined up in a row, their faces creased and squinting from hard work, piling up stone walls, cutting trees, and watching for enemies from the woods. Their features changed slightly from generation to generation, but they all had the characteristic Butterfield glare. I felt lifted out of myself and brought before an archetype in colonial dress, named simply Father.

The night before my return, I had been reading a biography of Robert Rogers, looking for more clues that would enable me to reconstruct the life of Lieutenant John. In the 1750s, Rogers was a dependable and competent leader of Rangers, but he failed in all his business ventures. He died — depraved, drunk, impoverished, and friendless — in a London slum, destroyed by his enemies, his best hopes unrealized. The story overwhelmed me with more than the usual sadness at the futility of human ambition. His fate reminded me of my father: strong and competent in practical ways, but always girding his loins for battle, surrounded by enemies real and imaginary, and gradually undone by his own blindness.

I thought, *I must see my father right away*. I just knew there had been some kind of change in his condition. I would tell him, "I see it all now, all your struggle, your longing; I know why this work was so important to you." I felt certain that he would know what I had said, even though his brain could not make a coherent response.

By the time I got home, my phone had been ringing for hours. The tape on the message machine was loaded. While I was reading about Rogers's death, my father had died.

Now I understood. My surge of interest in Rogers and Lieutenant John had been my father's last attempt to communicate with me in this life. I had imagined sending him a message, but he had been the sender. The story of Rogers was only the medium.

There is no line of demarcation between the normal and the paranormal, or between the mind and the material world. Most communication from the dead probably comes through ordinary events, made significant by their timing and context.

My brother's connection with our father had been sharper than mine, closer in some ways, but with more of a sting. Religion figured prominently in many of their arguments. On the morning of the day our father died, my brother woke up, put on his pants, and was promptly stung by a hornet that had taken refuge in the pant leg. He flashed at once on the New Testament verse "O death, where is thy sting? O grave, where is thy victory?" He received the news late that night.

Soon I stood over my father's coffin and touched his lifeless, leathery corpse, faintly odorous with the smell of death. He was haggard and bony, like the rock ribs of the New England countryside where our ancestors have lived, loved, hated, worked, fought, died, and been buried for 350 years. I saw his body lowered into the ground, just as he had seen his father's body similarly disposed of, and his father had seen his father's before him.

As I sit in my office transferring my father's notebooks onto computer disks and preparing them for publication, I smile to think how he finally hooked me into this job. He did an amazing piece of work, valuable not only to me, but to anyone descended from the same roots. Using the numbering system he invented for keeping track of all these thousands of separate bloodlines, the genealogy of any American Butterfield on the tree can be traced in a matter of minutes.

I smile also to remember how uncomfortable he was with symbolism, even while carrying in his pocket the measuring rule that symbolized his demand for certainty. For him, any job with dimensions could be reduced to a solution. He could not measure the confusion of human life, which is far too messy to have such qualities as dimension and solution; but the product of his attempt — his devotion to his family, and his elegant and precise compilation of our ancestry — illustrates how we transform our confusion into love.

After he turned senile, I did not like taking on the responsibility for his care. I thought about euthanasia. I would not have killed him, but sometimes I wished him a speedy death. Now I can see how his long period of decline gave me a chance to feel close to him again. I needed that chance, in order to dissolve any lingering vestiges of alienation from myself. Sometimes I wanted to deny he was any relation to me, but it was impossible: staff members in the nursing home always knew I was his son. "You look just like him," they said. They reminded me that my father is woven into the very bones of my face and the color of my eyes.

Alzheimer's disease gradually took all the fight out of him. In the end, he was as gentle as a toothless old cat, smiling and weak, fumbling with his chair. I did not resent him for anything. Whatever had divided us came unwound along with the connections of his brain. If I had been able to dispose of him when he became inconvenient, or had been unwilling to be involved in his care, that release might never have occurred.

Despite their battles, both my parents did far more good

than harm. In my Buddhist path, I seek a greater vision than the teaching that "life is a constant battle," but no such vision is possible without first accepting the truth of who we are. My father gave me a vital part of that truth. Then he reached past me to make sure it would survive my indifference. He was right to do that. Because of him, both my children and I have a family tradition that we can use to help us transcend our small selves, just as he rejuvenated himself by returning to the woods.

My father was impelled by the same deep desire that impels me, and, I suspect, all others: the longing to continue, from age to age, not merely a blind propagation of genes, but a light of loving awareness, which is never swallowed by death, but only more fully revealed.

May 1991

Flesh Of My Flesh

Ona Siporin

The same year his wife died, Clarence
tripped off the tractor. Helplessly big, game
leg rigid as a plea in the air, he twisted
his head to watch the mower shuffle toward him
over the sweet timothy, singing its awesome melody.
It severed him midthigh. The storm of his own
blood twisting around him, he thought fast
and saved himself, dancing; a beetle on its back,
he jigged round and round, wild, to turn his head
downhill. Later, they told him what he had
already learned from the death of his wife:
when a limb is gone you still feel it, still
reach to touch it. So it was for him
in the armchair of the early dark
autumn afternoon.

August 1989

Heart Too Big

John C. Richards

Fourth Street starts in Gretna, Louisiana, and runs into Marrero, miming the curves of the Mississippi. In one six-mile stretch there were once more than a hundred bars. By 1985 half of them had closed, but the abandoned buildings served as a reminder of the boom years of the oil industry, when men slept above them during the day and drank inside them at night, spending what they earned in three weeks out of each month on the oil rigs in the Gulf of Mexico.

But by 1985 the money was gone, and the area was depressed and showed no signs of picking up again for a long time. The people who still lived there hung on in one way or another. If their incomes weren't dependent on oil, they had a head start,

---

but in any case there was simply less money to go around. Both Robert and Edsel had worked on the rigs, spending most of what they made, but managing to acquire a few things, like their instruments and places to live. Since the end of the oil boom, they'd relied on music and odd jobs to keep themselves and their families going.

I met Robert across the river in New Orleans, at the 500 Club on Bourbon Street, where we were listening to an Irish trio — two acoustic guitars and a mandolin. I don't even like that music, but I'd just finished a gig and was on my way home, so I looked in just for the hell of it as they started their last number. They were singing in those silly Irish accents that carve the air, and I thought, *Oh, shit, here we go*. It was one of those bar songs where they belt out things like "Round 'n' around 'n' around!" and get the audience to do it with them. But then the mandolin player started a solo, and five seconds into it he had me. He was just running up and down scales at first, but it was so fast and uncluttered that I had to watch. The other two picked up the tempo, and he left the scales and romped across his mandolin with obvious effort, but so obviously pulling it off. Redheaded and pale, this stereotype of an Irishman threw himself into his solo with so much joy that he silenced the audience for a moment, but they recovered and began to shout to drive him on. He folded his big linebacker's body over the instrument and ripped at the strings. Sweat ran off his face and onto the neck of the mandolin, then slid across the fret board and dropped to the floor. I found myself walking toward the stage, drawn to him and feeling a vague desire to let him know.

Robert was sitting in the middle of the audience, twice as close to the stage as I was. He was the first one to stand on his chair, yelling and flailing his arms as if looking for something to beat his fists on. When others joined him on their chairs he took no notice. He was on his chair because it was simply the thing to do at a moment like that; the situation called for it. The fact that he'd started something seemed lost on him.

Afterward I was standing in the street outside the bar drinking a vodka and tonic. Robert came out and walked straight at me, loosing an authentic rebel yell that lasted for at least six steps and never removing his eyes from mine throughout. He had a beard and coal black hair that nearly reached his waist — an imposing Southern Rasputin. He stuck his hand out, and we shook. I'd been chosen for something, it seemed, but all he said for the first thirty seconds or so was "God *damn*!" over and over, before introducing himself.

We ended up getting onstage together at a drop-in bar on Iberville Street, then at one on Magazine, and then drinking until 7:30 the next morning. He liked the way I played, and I liked the way he played, so we exchanged numbers and got together after that. He wanted me in Edsel's band to give it some kick, because they were losing the younger audiences. No one under forty wanted to hear Hank Williams and Johnny Cash, and the club owners were getting after Edsel to do something about it.

Robert introduced me to Edsel the following Friday. I didn't meet the drummer until our first gig in Burris on Saturday.

Edsel played a 1962 Gretsch with a tremolo bar, and he used medium-gauge strings that discouraged other guitarists from borrowing it. His strings were so tight that no one could bend them, but in his strong mechanic's hands they were pliable enough. Edsel didn't bend much, anyway.

He lived with Ellen Lewis and her daughter from a previous marriage in a tiny shotgun house tucked just inside a levee. With the river less than a hundred feet from the back porch, the house advertised its vulnerability, but it hadn't flooded once in the seven years they'd been there. Edsel never thought about it. The garage in Marrero where he worked intermittently was also in the shadow of the levee.

Unlike most of his friends, who'd been in the service earlier in their lives, he'd never left south Louisiana, and, at thirty-nine, probably never would. It was something that didn't occur to him. He had more pressing problems, the variety and assortment of which could have driven other people to collapse; but, like his part of the world, Edsel was always able to keep his head barely above water.

He could play four sets of music with a fifth of Crown Royal behind him, and he claimed that he played some of his best music on nights he couldn't remember. He relied on the word of his friends to remind him that he'd had the audience on the tables, women crying, and men firing pistols into the ceiling or wearing women's panties on their heads in gaudy celebration of the emotion he could wring from a song.

When I met Edsel, he was dealing with two problems and hadn't decided which was worse. First, Ellen had walked out on him, claiming that this time she definitely wasn't coming back. When Robert and I arrived, he was eating his way through the food she'd left for him in the refrigerator, and he shared it with us as if there were an endless supply. She'd been gone for four days, and had planned her exit with sufficient concern or pity to have left enough chicken, biscuits, pots of gravy, cornbread, beef, stewed tomatoes, okra, and red beans to last him until his sorrow cleared and he was able to make it to the grocery store. He hadn't gotten that far when we showed, though he had made it to the liquor store for a fifth of Crown.

He knew she would stay away longer than the two weeks she'd been gone the last time. That time she'd said she wasn't coming back; this time she'd said she *definitely* wasn't coming back. Edsel pointed out the distinction as if explaining the rules of a game he knew he was losing but had by no means given up. In fact, he was quite confident that, if Ellen's sister Ruth would let him talk to her, he'd be able to coax her into coming back. But Ruth, I gathered, was something of a protective monster who hoarded Ellen's affections and presumed to know what was best for her.

Edsel's looks invited wild speculations on what kind of woman would get near him. He stood five-foot-six and weighed more than two hundred pounds. His thick, stubby fingers were stained with nicotine, the nails lined with auto grease. His fat, round face pried itself open when he smiled. This he did often, revealing browned, broken teeth and a bleached tongue. The teeth were beyond discolored, would have horrified a dentist. Edsel wasn't the kind of man who went to dentists.

He served us on cracked plastic plates, the food steaming hot and astoundingly good. Ellen was renowned locally as a cook and could routinely match anything served across the river in New Orleans's best restaurants. We ate sitting on the sofa three abreast and washed the meal down with Dixie beer and Crown Royal. It was 10:30 in the morning, and Edsel had just gotten up to attack the second problem he had to deal with that day.

"The Fat Fucker showed up this morning," he said, mopping his plate with a biscuit. Robert nodded, indicating that he knew who Edsel was talking about but wasn't ready to speak until he'd concluded a particularly sensual moment with Ellen's cooking.

Edsel explained that the Fat Fucker was the sheriff, a man named Riley Jones, who'd pestered him for much of his adult life. They had an understanding built on rides in the squad car, handcuffs made extra tight on Edsel's wrists, and nights in jail. They understood that they hated each other and waged a constant battle, Jones trying to catch him at anything and Edsel fighting his war of evasion by hiding out at friends' homes and sticking to the back roads. This time the issue was child-support payments to Edsel's former wife, who happened to be Jones's sister. Jones had gotten to him about an hour before we arrived.

"He took my amp," said Edsel, nodding toward an empty space along one wall where his red Gretsch sat on its stand. The chalky paint in the area around where the amp had been was faded, but for a perfectly square shadow.

Robert's eyes grew large in wonder, emerging from his chinless head like an alien's eyes and making his face look top-heavy. "He took your amp?"

"He took my amp. The Fat Fucker took my amp. He said he had a buyer for it and went straight off to Slidell to sell it. Fat Fucker."

We went back to our plates with equal conviction. This was a real problem. I was there so we could rehearse for a job the following night in Burris. I didn't know any of Edsel's material, and he didn't know mine. My stuff was simple, and I suspected that his would be, too, but these guys tended to rearrange even the most standard Hank Williams songs, so I'd asked Robert for a rehearsal.

"What're we gonna do?" I said, trying to sound like I was only looking for information.

Edsel made one last pass through his remaining gravy with half a biscuit and bit into it thoughtfully.

"Kevin," he said, "we're gonna finish this bottle o' Crown. And then we're gonna go get me an amp."

He had five hundred dollars that he'd just gotten for pulling an engine out of a pickup and rebuilding it. He cleared our dishes away and put seventy dollars in my hand and another sixty in Robert's. I was drunk from the Crown and Dixie. I listened as they discussed the amplifier they'd seen in the window of Jason's Music.

"Thousand marked down to seven-fifty," said Robert. "That means he paid four."

"So I gotta get the cheap son of a bitch to take five."

Edsel tucked his shirt in and tightened his belt, preparing for action. He let the belt ride right across the farthest projection of his belly, displaying just how fat he was. Then he slicked his hair back and led us out to his truck, a 1961 Ford that ran like new. Driving carefully, he took back roads and came up on the rear of the store, where he edged the truck in next to Jason's green '74 Cadillac. Robert and I were to go to a coffee shop first. Edsel told us to give him forty-five minutes.

Jason's Music was one of those businesses that lingered inexplicably even when economic disasters wiped out everything around them. Gretna had suffered as much as any other part of south Louisiana, and the slide that accompanied the drop in oil prices had taken a Kentucky Fried Chicken, a convenience store, three bars, and a bakery from the three-block area around Jason's. In another city people might have figured the music store was a front, and that Jason Long sponsored gambling or drug operations and was hip-deep in the Mafia. But everyone here knew him, some since before he took the store over from his father and changed the name from Lester's to Jason's (resulting in two poisoned dogs and five years of complete silence between them), and they *saw* him inside the store every day, doing nothing at all, wearing the bad toupee that sat like a mound of shit on his head and selling guitar picks and music magazines to kids while dust gathered on his Hammond organs and Les Pauls. There was nothing to accuse him of, except trying to sell the same overpriced goods that could be had across the river for a third less.

Jason made just enough money to survive and would have been shocked if criticized for not doing more. He was a benign presence in the neighborhood, was known for his cheapness, and was visited nearly every day by two friends he'd had since high school. They gathered behind the glass counter on hard, uncomfortable chairs and talked. All three men were over sixty and had known each other so long and spent so much time together that they rarely had anything to say. Every so often Jason got up and made a sale, and nobody expected more of him.

I DRANK THREE cups of coffee and Robert had two, and by the time I was actually in the store and looking at the amp, I felt sober, but nervous. Edsel was leaning on the counter with his $370 in hand, looking like he'd been told of the death of a friend just seconds before. The amp was sitting on the floor beside him. I followed his instructions to the letter, buying an old issue of *Guitar Player* that I wanted anyway for an interview with Duane Allman, and taking my time about it.

"Kevin," Edsel said, his gaze cast at the floor. Jason was directly across the counter from him.

"Hi, Edsel. How you doin'?" I said.

"Not too good, Kevin. Mr. Long here says he's got to have at least six-fifty for this amp. I offered him cash, but I don't have but three hundred and seventy bucks at my disposal. And I gotta have this amp today."

"Yeah? So what're you sayin'? You need a loan?"

He waited, giving Jason ample time to take it in.

"I sure would like that amp."

Jason's eyes went back and forth between Edsel and me. I pulled out my wallet and peered into it, aware that my hands were trembling. I can perform music in front of audiences, but I'm no actor.

"I don't know. I could probably give you seventy."

"I sure would appreciate it. I got a gig tomorrow, so I could pay you back tomorrow night after I get paid. At least some of it."

I gave Edsel the money, and he added it to what he had and recounted the whole thing. He put his finger to his lips in an absolute burlesque of a man in thoughtful concentration. Jason's two friends sat a few feet away, blatantly enjoying the show.

"You know I can't do that, Edsel," said Jason.

"I sure would like that amp, Mr. Long."

"I believe you. But I can't do that."

Edsel squatted in front of the amplifier and played with its control knobs. He ran his fingers down the fabric that covered the speakers. The bell over the door jingled, but he never looked up.

Robert went to the cabinet near the register, and Jason slid over to pull out his scratched plastic cases of guitar picks, never taking his eyes off Edsel and the amplifier, as if fearing that both might disappear at any moment. Robert selected several picks, taking his time and testing their pliability between his fingers. Edsel got up and walked toward him.

"Robert, my friend," he said, "I gotta ask you somethin'. This might blow your mind, but I need to borrow some money."

Jason and his friends were now staring at Robert, who spun around and folded his arms across his chest smartly.

"Don't do this to me, Edsel."

"I'm serious. I wouldn't even ask if it wasn't serious. I gotta have this amplifier. It's marked down to seven-fifty from a thousand, and Mr. Long says he won't take less than six and a half."

"You got me at a bad time. Sorry."

"Come on, Robert. You know I'll pay you back."

"Yeah? When?"

"Don't do this to me, man. I'm askin' you."

"Come on, Edsel."

"I'm askin' you."

Jason cleared his throat and rotated a toothpick in his mouth.

"It ain't a good time, Edsel."

"Man, I wouldn't ask you if I didn't need it. I'm askin' you."

Robert turned a pick over several times and waited before replying.

"How much've you got?"

"I got four-forty."

Robert reached into his pocket and pulled out a thin stack of bills. He laid them out on the glass as he counted.

"I can give you sixty. No more."

Edsel took the money and repeated his labored counting, as if to verify for all of us what the new total would be. I was watching Jason, who stared at the bills as Edsel smoothed them flat on the glass countertop one at a time. His two friends leaned forward in their chairs. Jason turned the toothpick around again.

"Five hundred?" he said. "You got five hundred cash on you now?"

WHEN HE PLUGGED the amp in later that day at rehearsal, Edsel was whistling. He knew he'd pulled one off, but he was content to let Robert and me take care of the celebrating and the mocking of Jason Long. Only once did he join in. Squatting before his amp as he adjusted the settings, Edsel looked up, shook his head, and said, "Fat Fucker," softly to himself, as if thoroughly put out by a world in which he had to go to such lengths to make a living.

We drove down to Burris the next night in Edsel's truck and were met by Blair, the drummer, when we pulled into the lot at the Blessed Ark. It was a covered barn, formerly a church, that sat on a perfect spit of land surrounded by water on three sides.

Blair's wardrobe was pulled from different eras. He wore Beatle boots and striped blue-black bell-bottom pants. His belt was brown leather, with a tarnished brass Confederate flag for a buckle. The shirt was yellow, Mexican-style, with raised stitching along the buttons and a pack of Winstons in each front pocket. His earring brought him into the eighties, but hung amid spiraling ringlets of hair that reached past his shoulders and anchored him in the early seventies. He had a toothpick in his mouth; I never saw him without one, unless he was eating. He would drink, smoke, and kiss women with a toothpick in his mouth.

"Ed-*sel*," he said.

Edsel got out of his truck and appraised the building like a commander formulating his plan of attack.

"All right, Blair. Hey, meet Kevin. Kevin's a real fine guitar player from New York who's gonna help us out tonight."

We shook hands, and I told him I was from Philadelphia.

"Same thing, far as they know," said Robert, nodding toward the building.

"New York sounds better," said Edsel.

The Blessed Ark was built in 1908 and had sat unused for more than twenty years when two New Orleans policemen bought it in 1978. It had a gray tile floor and an old wood-and-brass bar that ran the length of the building. There was no stage, so we set up at one end of a space that had been cleared of tables. A beer cooler packed to overflowing with boiled crabs — a gift to the band — sat on a chair in the corner. Blair moved nonchalantly around the crabs, setting up drums and cymbals, adjusting his seat, smoking a cigarette, oblivious and comfortable, while I couldn't ignore the presence of their orange-and-white claws, lifeless black eyes, and desperate, broken feelers piled atop one another.

When we started there were only seven people in the club, but before the first set ended fifty-five minutes later it was packed. Every table had been taken, and there was an uninterrupted line of people at the bar. I felt that we weren't loud enough, but Edsel was running the show, and he didn't want to turn it up. His intensity matched the flow of the crowd. As they came in, he started calling more up-tempo songs and stamping his feet. He introduced me as being straight down from New York just for that gig and gave me two songs at the end of the set. During both of them he gracefully backed off, turning down his amp even further and just trying to play along to himself.

There were some rough spots at the beginnings and endings of songs, but overall I thought it had gone well. We met outside during the break, and the others seemed to think so, too. Edsel cut our critiques short when two women from Lafitte showed up. He introduced me as the real-fine guitar player from New York he'd told them about, and then abruptly steered them toward his truck. I went back into the bar and looked for the men's room.

It wasn't as bad as I'd expected. It was filthy, with dirty blue-green tile and no door on the one stall, but it didn't stink as much as it could have, and there weren't rolls of toilet paper in the toilet; I'd give it a moderate rating on the scale of bathroom horrors. As I stood at the urinal I heard the door swing open behind me, and a man wearing a crew cut and an army-fatigue shirt came in.

"Y'all soundin' good," he said.

"Thank you."

He leaned against the wall directly behind me, making no move to use the facilities. I stopped myself in midstream and moved over to the sink, where I could wash my hands and see him in the mirror.

"Yeah, re-e-c-a-a-l-l good."

" 'Preciate it."

"You the one from New York?"

"Yup."

"Yeah, New York City, man. What a place. No place like it anywhere else, that's for sure. You like blow jobs?"

I was getting a paper towel to dry my hands and tore it off very deliberately, with a quick, efficient, ripping sound. "Why?"

"Got a young lady right out back'll do you up just right."

"I don't think so."

"Twenty bucks."

"No thanks."

"Ten."

"No."

"Ten bucks?"

"No."

"You be cool about this?"

I nodded, and he extended his hand. We shook only after first slamming hands trying to do two different handshakes, and then he left. I dried my hands and went out right after him, but he had disappeared.

Our next set was better, and Edsel gave me four songs. He

was really enjoying himself, playing around the mike when he sang and pointing at me for my solos with a big windup of his hand. We did six straight Elvis songs.

At the break I joined Blair and Robert on the tailgate of Blair's truck, where he was rolling a joint. Blair was about as straightforward a drummer as I'd ever played with, content just to keep the beat and light on his bass drum. He preferred to stay on his snare and high-hat, beating out a steady rapping-and-hissing sound that, to his credit, never varied. He told me he liked my playing, and I told him I was having fun.

"Edsel sounds good tonight," I said.

"Yeah," he said, smiling and shaking his head. "Now."

"Wait till later?"

"Wait till later," they said in unison.

Blair had played with Edsel on and off for more than ten years. He said the whole night, right up until the moment we were home in our beds thinking about life, depended on how much Edsel drank.

"That's right," said Robert. "You might say we're his hostages. Short-term."

"Our lives are in his hands, it's true."

At this point Edsel was back in his truck with one of the women he'd been with on the last break.

"I been knowin' Edsel for a long time," said Blair, "and I love him, but I do *not* understand him. He's *always* fuckin' around, you know? His job and his wife and — have you met Ellen?"

I hadn't.

"Well, Ellen . . . Ellen's not your average . . . She's really special. She's smart and good-lookin', and she's got sense."

"Ellen's a priceless princess," said Robert; he was the lyricist for the band's original songs.

"She can cook the shit out of any food," said Blair, wistfully. "I know."

"Edsel's gonna lose her. She's gonna walk away and do lots better real fast sometime, and he'll never see her again. That's what's gonna happen, mark my words."

We smoked the joint and were joined by one of the policemen who owned the bar. He said it was nice to hear some "more rock-y stuff," and Robert said there was plenty more where that came from and clapped me on the back. I was having a good night, playing well and singing easily.

At 3 A.M. I drove Edsel's truck back to Marrero. He was off with the women from Lafitte and hadn't even said goodbye or gotten his money. He was drunk, but he'd held it together onstage, and Robert said we'd had a good night as far as that went. Blair gave me Edsel's keys and asked if I'd mind, then stuffed a folded wad of bills in my pocket.

"That's Edsel's, too," he said. "Leave it where he'll see it."

Edsel's place had accumulated a good deal of clutter in his five days of bachelorhood. There was a raw-maple coffee table he'd made himself, a sofa, and two overstuffed chairs in the front room. Each piece of furniture had something stacked on it — clothes, bottles, or the remains of a meal. The television set that dominated the room was the one modern-looking item, a huge state-of-the-art screen with a cable box on top and a remote channel changer.

Ellen hadn't gotten around to taking very many of her things. But she'd been back since we were there, and had left a note on the kitchen table. Scrawled in upright, schoolgirl handwriting, it read:

Dear Edsel,

I passed by to pick up a few things, and you weren't here. There's some iced tea and some more chicken in the fridge. Don't forget to pay the water bill, or you'll be cut off. 'Bye.

Ellen

I looked around the kitchen and went back into the living room, certain that she was hiding somewhere, watching me. The note reminded me that I didn't belong there. It was her home, her private place. Yet I couldn't resist looking further. I wanted to see more of the woman who had chosen this life.

Their bed was huge, a king-size mattress that filled the bedroom and faced a bureau-and-vanity set that had to be sixty years old. Over the bureau hung a round mirror with several snapshots taped to it. I picked out Edsel, Robert, and Blair, but there was no way of knowing which of the women was Ellen. In some of them the men were holding stringers of fish: fat and shirtless to reveal tattoos, squinting into the sun, in most cases posed too far to one side of the frame.

The closet door was open, and inside hung one dress, with several feet of empty space between it and Edsel's cartoonish fat man's clothes. The dress was purple with sequins and tiny, thin straps. It was a party dress, something she wore probably once a year and saw no reason to take with her. I imagined Edsel buying it for her and insisting that she put it on right away. He must have considered her sexy at one time, and she must have believed it and made love to him in that big bed. She had made a conscious decision to leave that dress behind.

Carla's room seemed barely lived in, a young girl's room perfect as a catalog photo. Sylvester Stallone's eyes challenged me from above a pink bedspread; there were pictures of Cher and Madonna, stuffed animals, and a yellow diary sitting on a child's desk. The tiny clasp on the diary was locked.

Returning to the kitchen, I took two pieces of chicken and a biscuit from the refrigerator. Even cold, Ellen's biscuits were marvelous. I ate over a paper towel and cleaned up after myself, then put Edsel's money on the table, near the note. I put the keys on top of the money and left, leaving the door unlocked so he could get back in.

AT THAT TIME I was working on Bourbon Street with another band during the day, and about a week later I saw Robert walk by the club while I was onstage. He waved, pulled a woman into the doorway, and pointed me out. She waved tentatively. She was tall and thin, with long blond hair.

"We're like brother and sister," he told me later. "Ellen takes care of me sometimes, and I listen to her when Edsel's givin' her a bad time."

"Whaddaya mean, she takes care of you?" I said.

"Lets me crash on their couch. Gives me a good home-cooked meal now and then."

"That's all?"

"We're . . . She's real good-lookin', yeah, but I wouldn't ever . . . We're good friends."

I looked away, ashamed I'd made him explain himself and unsure why I had.

After repeatedly attesting to his love for Edsel, Robert said he couldn't figure out what it was Ellen saw in him. It wasn't that Edsel was so bad, but that she could have done so much better.

"But she's country," he said, eyeing me. "This is what she's used to. She's from here. And Edsel's just like her brothers and her dad. She's kinda stuck with it, and she knows it. If you don't know it, it don't make no difference, but if you do, it's awful. Purely awful. That's what she's goin' through."

"Has she ever been anywhere else?"

"That's part of the problem. She hasn't, and she knows she's doin' herself short for that. She might just fit right in somewhere like Seattle or Milwaukee. She might be a real smooth lady in a place like that, but she's scared to try it."

He said he never actually suggested she leave. He just let her talk about what it might be like. Ellen had gotten a library card for the first time in her life and found books about Seattle and Milwaukee and read them in the library, never daring to bring one home for fear that Edsel would figure out what she was up to.

"Why Seattle and Milwaukee?"

"I was born in Seattle and lived there till I was eight, and I've told her about it. I don't know why Milwaukee."

BLAIR BOOKED US at a private party to be held in a bar in Algiers — as much food as we could eat and sixty dollars each. We kept trying to arrange a rehearsal, but it never worked out. Either I was working, or Edsel was working at the garage. He had a reputation as an expert mechanic, and he could make four hundred dollars a week when he wanted to.

Ellen still hadn't come back. Edsel was drinking, and she was dating other men. Blair said she'd never done that before. She was going out every night, taking her sorrow with her and accepting the invitations of men who'd eyed her for years, only to disappoint them in how far she would go. It was something she made herself do so that she could say she had. Being seen in public with other men was crucial; not only did it thrill her, but it silenced her sister Ruth and her friends who had urged her to leave Edsel for good. It delayed her decision and gave her time to think about just how badly he'd hurt her, to go to the library to read up on Seattle and Milwaukee, and to run over her list of his sins. These started with the third time she'd caught him with another woman, went on to his drinking, his unwillingness to work at anything but music, and then dissolved into things like his eating habits. Ellen knew how to handle Edsel and what to expect of him, even down to the other women. He had to be punished each time she caught him, and she was punishing him now. She regarded Edsel's behavior as justification for her thoughts of independence; she would never risk appearing so selfish as to want to leave the people she'd known since childhood without a very good reason.

WHEN I WALKED into the Red Barrel in Algiers with my guitar, I was greeted by a young waitress named Christine. She shook my hand and said she'd heard all about me, then demurely accepted kisses from both Robert and Edsel. She asked us what we wanted to drink before we'd even put our instruments down, then again once we had put them down. Blair told her to relax and go wait on her customers.

While we were setting up, Christine got into an argument with a customer, a huge man in jeans and a torn sleeveless shirt. He towered over her, clearly thinking about hitting her while she screamed in his face that if he ever, ever grabbed her like that again she'd kick his ass in front of all his friends. He stuttered a few times in a vaguely threatening way and sat down. Two bartenders came over to back Christine up, and the man growled at them, obviously more comfortable being tough with other men.

"Don't mess with her, Kevin," said Blair. He was right behind me, affixing a cymbal to its stand, so close to me that his voice seemed to come from inside my ear. "She beat my old lady up not once, but twice. In the same day. Anybody can do that can *fight*. I wouldn't even consider it without a pool cue or somethin' to hit her with."

Christine had moved on to another table, resting her tray on her hip and politely inquiring of an elderly couple what they would like to start off their evening.

That night, I finally got to see something that I'd heard about several times by then: men walking around in a bar with women's panties on their heads, and no one acting like there was anything unusual about it. When the time came, they either went out to their cars and fetched panties from their glove compartments or convinced a woman to go into the ladies' room and remove hers. The women did this willingly, laughing at the men, handing over the panties with exaggerated ceremony and enjoying what they clearly saw as the men's willingness to degrade themselves. It was a bizarre ritual, but really had nothing lewd about it. Most of the women wore jeans, and after they handed the panties over they went back to their conversations or dancing.

Midway through the first set, something came over Edsel, and his intensity picked up so fast it was alarming. The panties came out as soon as he got warmed up and started to reach the audience. The songs were slow, sad ballads, and he was drunk, but he never lost his grip. I barely played on these, just hid in the shadows between the lights and watched him. The Red Barrel was packed, and a solid cloud of smoke several feet thick hung over the crowd. Men with panties on their heads walked around talking or simply staring at the stage.

Eyes closed, Edsel stood absolutely still at the mike and sang in a soft, elegant voice that belied the fat man's shadow he was casting on the stage. It was some of the saddest, most moving singing I had ever heard. People were wiping away tears, seeming to hold their breath until he finished each number and they could clap and cheer, as much with relief as in appreciation. At the end of the set, I was grateful he hadn't called for any of my songs. I would have felt ridiculous.

Among the audience, I was a forgotten man, but Edsel was surrounded by people who were not so much talking to him or listening as just wanting to be near him. They held their drinks at their chests and watched as he smoked and drank from his glass of Crown, then his Budweiser. He nodded brusquely to the ones who complimented him, too much the man hard at work to be affected by their praise, too consumed by his emotions to want to talk. As we walked to the stage for the second set, Robert nodded toward Edsel and smiled at me.

"Ellen's here," he said.

The tempos the next time around were all up, and Edsel was tearing into songs instead of caressing them. Where once he'd given lyrical form to his sadness, he was now showing anyone who cared to watch that he could live with it, shouting out the words and ranging around the stage.

During my guitar solos Edsel was right at my side, yelling and stomping the floor around me. I have to admit, he drove me to a few reckless things that I pulled off in spite of myself. The men with panties on their heads were moving rapidly about the bar, some dancing, some just going quickly from one place to another, as if responding to a signal that Edsel was sending. We played well past the allotted time for our set and kept right on going. Feeling inspired and confident, I moved out to share the front of the stage with Edsel and sang as hard as I could. Behind me, Blair was pounding his drums with a drive I didn't know he possessed, and Robert was going out into the audience with his bass during almost every song.

We had to break eventually. Everyone in the place was at a fever pitch. When he snapped a string on his Gretsch, Edsel chose to end the set on that song. We finished to shouting and applause. I had several drinks thrust at me when I stepped offstage. At least a dozen people followed Blair outside to join him at his truck for a smoke. I sat down, and just as I noticed a cigarette burning in the ashtray in front of me and realized I'd taken someone's seat, Ellen came back from the bar. Her long hair slid off her shoulders and stopped abruptly above a slim waist and pronounced hips. One side of her mouth turned down crookedly, lagged behind the other, and rippled belatedly into her smile.

"Kevin, I'm pleased to meet you," she said. "I've heard so much, and if half of it's true . . ."

She sat next to me and pulled out a fresh cigarette, drawing the ashtray closer and offering the open pack to me. I declined, but in the same moment picked up the matches and lit one for her, regretting it immediately and imagining Edsel watching me from a few feet away.

Her sister Ruth was with her, sitting patiently with an elbow on the table and a cigarette perched above her ear, impervious to the smoke that drifted around her head and into her eyes. She complimented me politely on my playing and said that I was just what that band needed, because everybody had gotten so sick of Edsel's act they knew it inside and out. She said Edsel had worn everybody out with his cowboy music. Ellen stared vacantly while Ruth went on, clearly enjoying herself. Then a tall man wearing blue panties on his head and a flannel shirt came to the table and asked Ellen to dance to a song he'd just put on the jukebox. She put out her cigarette carefully, so she could relight it when she came back, and got up to dance.

He couldn't have picked a slower song and took advantage of the tempo to drape one arm around the small of her back and slide a finger through her belt loop. She pressed herself to him and laid her head on his chest. They had the dance floor to themselves. Ruth looked on with a pert smile.

I went outside and found Robert, who was leaning against the building trying to recover from a laughing spell. A short, bald, red-faced man was adding to a story about a shaving of red-hot steel in a machine shop that had found its way down the bib of someone's overalls and onto the head of his penis. Robert begged him to stop, saying he was serious — he couldn't breathe — and didn't want to hear any more. The man went back into the bar just as Christine came out with her cork tray and asked if we wanted anything to drink. Robert got a shot of Crown, and I had a Budweiser.

When Christine came back with the drinks, Robert asked her if that was the only job she had that night. She hit him in the head with her tray.

"I'm a good girl tonight," she said, "and I don't wanna hear about nothin'. That's it, case closed."

"You're not workin' overtime, then?"

"No."

"Tell me if you change your mind."

When she was gone, Robert told me he had visited Christine the week before behind the Blessed Ark, and that she had taken very good care of him.

"Ten dollars well spent," he said, like one smart shopper to another.

We were only in the club for about two minutes before we got back onstage, but in that time I saw Ellen kiss four men. Edsel was right there this time, so intent on ignoring it that he got onstage and proceeded to play the worst set of music I've ever been associated with. He was so drunk that the neck of his guitar kept slamming into his mike and making a horrible, metallic, banging sound that he stopped with a fat fist wrapped around the strings and a "Shhhh" spoken to his Gretsch.

Ellen danced with several men, spinning about the floor on the fast songs and clutching at her partners during the slow ones while Edsel suffered in front of everyone. He slobbered his way along, dropping his guitar picks then bending over awkwardly and running his hands across the stage to find them in the uncertain light. Blair retreated into his rapping-and-hissing steady beat, and Robert shot Edsel dirty looks. I turned my volume down and tried my best to blend in with the random chord changes Edsel was throwing out.

We'd lost the audience, but Edsel plowed on. The buzz of conversation filled every break in our sound, and the dance floor emptied. Only one man kept his panties on his head, too drunk to know better. Edsel called a song that he sang alone, one he usually did early in the evening but had saved that night. Glad to get away, Blair slipped from behind his drums and disappeared into the audience while Robert dropped his bass and ran offstage. When I took a seat next to Ellen, Edsel was alone in a blue spotlight that made him look old and tired. He strummed his guitar tentatively and started into a sad Hank Williams song, playing to an indifferent crowd.

Ruth was enjoying it, but for Ellen every moment hurt. She

turned to me just after Edsel started to sing, slipped her arm around my neck, and kissed me full on the mouth, holding it for several seconds and running her tongue across my lips. Then she pulled away and looked at me with tired, sad eyes. I blinked at her and she spoke, her breath hot and smelling of cigarettes.

"It's real simple, Kevin," she said. "You see that man?"

As if I hadn't, she grabbed a handful of my hair and pointed my head at the stage. Edsel was crying, his eyes closed and tears rolling across his fat cheeks and down his face. He was barely on key, but pushing ahead.

"That's my heart up there. He's terrible, and if he ever does it again I'm gonna beat him like a redheaded stepchild. But he's my heart. That's all there is to it. I've tried, but there's not a thing I can do about it."

And with that decision, in that moment, went any thoughts of Seattle or Milwaukee, of leaving, of other men or her sister's advice, of anything but getting back with Edsel and getting her daughter back in school and sleeping in her own bed. When the band finished and it was time to go, she argued with Ruth and told her to go away and leave her alone. Then Riley Jones showed up in his squad car, and she told him to get his fat ass back to Gretna before she tore his eyes out. We helped her put Edsel into the passenger seat of his truck, loaded his equipment into the back, and gave her his money and a box of boiled crabs. She'd been gone two weeks and one day.

February 1991

Jealousy

Lou Lipsitz

the man you went with
the man with hair that seemed
to lean toward the past

the man you went with
the man with so much
red-gold
left in his beard

the man you went with
who got you to laugh
so often at his other language
that seemed entirely
original

the man with the strange
clear eyes who just shook his head
gently

the man you went with
who taught you to listen
and forget

now I take you in my arms
and he is here with us
I look into his eyes
I reach out to touch his hands

I realize I've loved him
all my life.

November 1988

Last Year's Poverty Was Not Enough

Ashley Walker

The day hadn't begun well, but it was just another day in a long line of mean, anxious hours. Time mashed in on her like a couple of hands folded hard in prayer. You could call it a religious experience, or you could call it withdrawal. Either way it seemed like piss-poor pay for wanting to fly. She spent days and nights with the filaments in her legs dancing: kick, kick, kick. She kicked the sweaty sheets in the dance of deprivation. Then the phone rang and the voice asked her to come to lunch at Bar Tejas. She said yes because it was a chance: a sunny opening leading to normal people. She wanted a break. Even hell can be a rut.

Nights she lay on the couch, locked in spasms, the worst at about two in the morning, her legs and arms flailing like a dime-

---

Fiction

store toy played with too long. When her exhausted muscles quit, she'd read, and when the print started jumping, she'd listen to the symphony. She thought it might be Bach or Beethoven. The music had started on the third day without sleep, audible only to her; it played and played, accompanied by thin veils of fog blowing slyly past the corners of her eyes. Music and fog. Music and fog. Eventually, gray daylight beamed through the miniblinds and she'd shudder in the bathroom, apologizing to the thick air around her. *I'm sorry, I'm sorry*, she'd whisper, not knowing who she was talking to but thinking it was God. God never answered. Maybe he didn't care that for twenty-two years, Claire, who had begun as a sack of chemicals, had added more; she'd poured alcohol on top — sort of like a sundae — just trying to get it right.

Since she was sixteen, Claire had been trying to soothe the raw, screechy nerves that crisscrossed her body like silver freeways. She rubbed white powder into her gums, sipped on thick, frozen vodka; she was not present for much of her life, and it suited her fine. The problem was she'd black out. Other junkies and drunks envied her: *Man, I wish I could, wish I didn't know*, they said. *You're wrong*, she told them. She woke up in the middle of the week like a zonked-out Nancy Drew, gathering clues about what had happened: bruises on her legs, strange bills in the mail, a ding on the car. One time, when her lids creaked open, she saw her wrists already starting to crust with dark red scabs; an X-acto knife lay in the corner. *So that's how it's going to be*, she thought, *one half-assed attempt after another.* She came to another morning with a patch of sunlight about the size and shape of a kleenex next to her face, pills all over, and a big wet spot on the carpet she never did figure out.

She called AA. The man on the phone arranged to meet her before the meeting. Later, all she remembered was that she kept trying to pay everyone. She had dollar bills mixed in with the pill bottles at the bottom of her bag. She gave them out in handfuls. It was the last day. After that there were no days.

Her body said no, punishing her with tears, sweats, spasms, cramps, jerks. Fluid drenched the sheets as she jounced on the bed. When that stopped, her brain launched an out-of-focus home video of every failure, every lie, every heartbreak: her whole rotten life played out endlessly with bad actors. Claire got on the phone and called InterGroup. *Just talk to me*, she begged. *It doesn't have to make sense.* It didn't. Clutching the receiver, Claire listened to recipes, prayers, gossip, binges, and drunkalogues. *It's gonna be* OK, they all said.

Sitting in a kitchen chair, her teeth chattering like a baboon's, she watched her forty-one-year-old legs kick of their own accord. It made her thirsty. If she'd been able to drag herself to Red Colman's for a fifth, she'd have done it. But she kept seeing, in her mind's eye, the clerk screaming at the sight of her desperate, greenish face. So, sitting in a sweaty nightie with cigarette ashes down the front, she thought it was pretty damned amazing that the phone rang and Joe Avalon asked her to lunch.

It took 120 minutes just to take a shower and make up her face. It seemed like she had spent her life in bathrooms: puking, doing lines, lying on the floor feeling the cool tiles beneath her skull, praying next to the tub. She came undone like a loose knot and put herself back together in bathrooms. *Thank you, Jesus*, she said when she got on her eyeliner. Maybe she could have lunch, after all. She shuffled out of the bathroom on clay feet.

She sat on the bed, panting like a marathon runner, trying to keep her head blank and empty, like a big water-colored balloon: Just air. A perfect circle. Emptiness. But thoughts kept streaming into her mind, barbs on every one of them. She'd disappointed everyone, she decided. Disappointment isn't too bad a thing: just sad and gray like a wasted Sunday. All the things you were going to do and never did.

She got up slowly and walked to her closet. What outfit flatters withdrawal? She put on a black T-shirt with a flying television set emblazoned on the front, bluejeans, and silver loafers, all medium-sized. She was just an average-height, middle-aged woman looking the worse for a lot of wear.

Where had Joe Avalon come from? Faces gummed together in her mind like cheap candy in a box. Then she remembered: the temple. The part of her life she'd kept clean as blank typing paper: she never went to meditation stoned, never sat zazen after even one shot of vodka. It was like painting had been before she'd started smoking weed in the studio, doing what she'd said she'd never do: paint when she was wrecked. At the *zendo* she'd slurped golden almond tea, sat, waited for the gas flame of *kensho*, the great *I am*, the whatever that would allow her to stop like a watch run down. She was sitting in the front room of the temple, surrounded by paper cherry blossoms stuck in cheap vases, when she met Joe. He was big, with bad teeth and white hair, just come off the road after two days' driving. He clutched a jar of honey in his hand. The *roshi* told Claire he'd been sober for thirty years.

By that time, Claire knew, even with the works jammed up by cocaine and pot, that you could never predict in what forms angels might appear. Joe had studied Zen fourteen years to her twenty-five. She sat two cushions over from him in the *zendo*, heard him exhale loudly, felt bad when the *jikki* hissed at him, *Breathe quietly!* Maybe he'd given her the idea to call AA in the first place, in which case he wasn't Joe Avalon but the angel Michael: her salvation. Maybe it was time to be saved again. So Claire said yes into Joe's ear through the long, ropy telephone wires stretched across Dallas.

Time was what Joshu said it was — your skin. You didn't bob along on it, as if it were the Atlantic Ocean. Time was glued to you like your crazy family, your baby teeth, your freckles, your brittle bones. Time was slow because Claire was slow: a woman driving carefully through town, not quite sure how to get where she was going. Her T-shirt stuck to her shoulders and her pink eye shadow was smeared.

Squinting through the window of her three-year-old Mercury, she felt pure. Even though she was seeing fog and hearing an invisible symphony, Claire was shriven and holy. It was like going to the dentist on no sleep, too tired to care about a root canal. She didn't care how she'd seem to Joe. Psychic barnacles had been scoured off over the past few days. She was scraped down to an essence. Raw. And there, glory be, was the Bar Tejas shining in the October heat like a mirage, but real. She steered to a parking place in front of a brick wall where BUSH SUCKS was lettered in black spray paint. Claire paused to admire such an excess of concern. She didn't care who was president, had voted only once in her life.

Joe was in a middle booth by the window, grinning at her when she came into his line of vision. *Hey*, he mouthed at her: *Hey, hey, hey*. Inside, the bar was cool and dark, full of black shadows, the way a place should be after the hot sun. The bartender glanced at her, bored. She must be a stranger here, Claire decided. Maybe she hadn't been here before, written a note to a stranger, gone to a place she couldn't find again, arrived home in the clammy dawn not knowing how she'd gotten there. She felt gratitude, like tears, clog her throat. Two drag queens sat at a table, one of them weeping buckets of big black tears, his makeup a ruin. Claire walked straight over to Joe.

"So how are you?" Joe asked, grinning hugely.

"Not real great. I'm sobering up. Hurts like a mother."

"Haw, haw, haw!" Joe hollered, slapping his big hairy hand on the table like it was the best joke he'd heard all day. She grinned weakly.

"You'll be OK," Joe told her. "Haven't seen you at the temple lately, was why I called."

"That's the reason," Claire said, sliding into the booth to face him. "I couldn't walk until today. I had the shakes bad. I didn't know if I could move at all until you called. I guess I'm glad you did." Her shoulder blades, like bony little wings, punched into the wood behind her. The waitress, pretty and temporary as a gardenia, wandered over to them.

"You want something to drink?" she asked.

"Coupla coffees," Joe said. "And menus. I'll have one of your cigarettes," he told Claire, grabbing her pack.

He lit up a Marlboro, shook out the match, and looked at her with clear blue eyes. "We lost our monk," he said.

Claire gaped. "Really?" She remembered the monk padding around the temple in drugstore track shoes, his black robe crackling around him. He was a slightly homesick guy from New Mexico. Once, he told Claire that, when he joined the monastery, he'd brought his hair dryer with him. Every morning, a little before three o'clock, he'd gotten up to fix his hair, until the abbot found out. Claire had been charmed by such a small, sweet secret.

"How?" she asked. "Did the monastery call him back because we aren't really a temple?" She felt irrationally guilty. *Maybe if I'd gone more*, she thought. *Maybe if I'd brought in some new people*. She pictured the temple, a ramshackle white house in a bad neighborhood, and then let out a breath at her own craziness. Zen didn't try to get converts; it attracted. Zen monks, even masters, just hung out: sitting zazen, eating pickled vegetables, waiting for folks to show up. Not many had.

Joe glowered and thrummed his thick fingers on the table.

"Remember Anna?" he asked. "She moved into the temple before me. And I want to tell you, I voted against her. Everyone got on my case, but I did it. Anyhow, she made a pass at the monk, and he caught it, I guess you'd say."

"Holy smokes," Claire breathed. "Anna! She was the *soji*."

"I voted against that, too, even if it is a shit job. Now, tell me, how did she strike you?"

She tried to remember. Anna was thin and artless, a woman who wore black sweats and always had a friendly word for Claire. She envied Anna for being able to sit twice a day, endure Zen weekends, sit *sesshins* for eighteen hours at a stretch, sit, sit, sit until surely her legs were cramped and gnarled. But Claire had been suspicious, too. When she'd asked Anna how long she'd done Zen, Anna had batted her colorless lashes, poured Claire another spot of tea, and said, "Three weeks. I was doing bhakti yoga before that, but I like this better." It didn't make sense. Zen wasn't something you liked better than something else. One day, you just found yourself doing it because there was nothing else to do. Kind of like sobering up, Claire realized. But she mistrusted her own perceptions. Anna was living the life. Anna sat like a rock in the *zendo*. Her mind was probably as blank as an abandoned drive-in movie screen. Claire sat like a rock herself and thought about every guy she'd ever screwed.

Finally, she said to Joe, "I thought she was daffy. Something was going on with her. She was just waiting to happen."

Joe lit up like a jukebox. "Now, that's exactly right!" he

crowed. He leaned forward, and a whiff of Old Spice curled off him. "You know, the monk's a very confused young guy. He just got snared, and I suppose I'm being a chauvinist when I say, 'Hell hath no fury like a woman . . . whatever.' "

Claire, thinking of the past few days, said, "Hell hath fury whatever."

"Whatever. Anyhow, this has been going on quite some time. He feels just awful. Talked about quitting the monastery. I said to him, 'Just walk through it. Like the wind through the trees.' But he couldn't hear it. Anna called his mother."

Claire felt lightheaded, as though big eagle claws had just released her head, though she hadn't known they gripped her until now. She tried to picture the monk's mother talking anxiously on the phone in a dry New Mexico town, staring out the window as though she could see her son, small and black in the desert.

Joe took another cigarette and tapped the end on his thumbnail. "It gets worse," he told her. "Anna told his mother he'd raped her. Implied it, at least. *Then* she calls the master. Poor old guy. He's eighty-five, y'know. He's about ready to drop his body. Here's Anna calling with this crazed tale of rape in Dallas. What's he gonna do with that?" Joe looked at Claire like she might be able to tell him.

"My God," Claire said, "the world's just inside out. How's Sam taking it? He was the *jikki*, right? He and Anna were pretty tight."

"Little Zen pals," Joe said bitterly. His face darkened. "He stuck up for her. It's been pretty tense at the temple, I can tell you that. All of us walking around on eggshells." Claire noticed absently that tufts of white hair stuck out of the v-neck of Joe's aqua sweater.

"Whaddaya think?" he asked, while she was thinking what a nice old guy he was.

"I'm going to have the tortilla soup," she said, almost able to taste it. "I don't know, Joe. It's a big relief."

Joe regarded her, puzzled, almost angry. "How come? How come it's a relief?"

Claire sagged back in the booth, feeling her muscles letting go, her appetite returning, thinking that maybe, after all, she was going to be able to live out her life.

The waitress reappeared. "The tortilla soup," Claire told her. "Two orders. What I mean," she said to Joe, "is that all the time I was at the temple I was chewed up with guilt. I thought it was because of my drinking and doping, but it wasn't."

"Guilt?"

"Yeah. Zen guilt. That I couldn't be centered, in the now. I was fighting, always fighting. This whole deal is a weird blessing. It is. I thought I was the only one who didn't know how it should be." The air seemed darker and cooler around them — like a priest's breath, a benediction. Joe looked at her like he was about to say something, his mouth slightly open, but Claire rushed on: "I guess what I wanted never existed. It is what it is."

"Well, it shouldn't have happened," Joe said, flicking one thumbnail against another.

"But it did. It happened. With everyone trying for enlightenment. Maybe the right thing goes on all the time and no one knows it."

"I shouldn't have told you," he said, looking worried. "I was gossiping. But it's all been bottled up inside me. I've really been valuing nonattachment. Look where this gets you."

"It gets you where it gets you," Claire said, "and maybe that's the point. Just to go on."

The waitress brought the soup. Claire and Joe bent over the bowls hungrily: two people who'd said enough, who'd skidded close to a quarrel and found peace at the last minute. The time passed pleasantly, and for Claire, the succeeding hours were much better. Days were to come, one after the other, and at last, like a light in the window, Claire would feel a stirring in her heart she could name "hope."

SHE NEVER SAW much of Joe after that. Later, she'd think about him and wonder if he'd done his part and moved on. She stayed sober, and, as her head cleared, she realized she'd made a decision to save her own life. Others, newly sober, sometimes talked to her after meetings and asked her how she'd done it, as though she might have a big secret. They thought there had been a single instant when she'd packed it in. Maybe they were right, but she couldn't talk about it. There wasn't much of a story until that bright October day when, sitting in a restaurant, she'd bent over a spoonful of tortilla soup and seen her own face shining in it like a dime, like the morning sun. How could she say that some folks fell from grace and others rose to grace, and it was hard to tell which was which?

September 1992

My Mother Is Still Alive

Kathleen Lake

My mother taught me to worship beauty
And she hit me on the back of the legs with a shoe tree
My mother stirred up the dark
with her cigarette, loops of embers, we wrote my name
One time I came home and she was drunk and stirring
a soup with nothing but old water in it

My mother held me up so I could see
the tiptop of the tree at Rockefeller Center
She hissed at my friend Nancy and called her "that snake"
My mother gave me my own section of the garden
Once I cried and kicked all the pictures off the wall and she
            never came

A big cat killed my kitten and she wouldn't get me another
In the middle of the night she stood on a chair, painting
over the monster on my ceiling
She wrestled me and ripped out my new earrings
"The dark," she used to say, "is kind to tired people"

My mother didn't punish me for stealing all her money
Once she hired police to watch the house so I wouldn't fuck
"I want life on any terms at all" is what she told me
A turkey sat in the oven for more than six months

My mother called at 3 A.M. to wish me a happy solstice
While I had seizures she did a crossword puzzle "to stay calm . . ."
She let me stay up late to watch *Kidnapped*
Once she dragged me into the library by my hair

She left the house in clouds of lavender
I ran away and came back — there was nowhere else to go

Now I live by myself, I've had many lovers
I fall asleep with the lights on; at least I'm in bed

My mother believes God is merciful
I think he is too, and inexplicable
My mother yells at the devil, I bow to him
My mother is still alive and so am I

March 1993

Uniting The Opposites

An Interview With M.C. Richards

Sy Safransky

Talk about art by artists usually bores me. Perhaps that's because I'm more interested in what art has to say about life than about art. But with Mary Caroline Richards, such distinctions are moot: her art and her life are inextricably, elegantly entwined.

Richards — or M.C., as she is known — is a freelance potter, author, poet, and teacher. I remember my excitement years ago when I discovered her first book, Centering in Pottery, Poetry, and the Person (Wesleyan University Press). Here was a practical mystic, a stunningly original thinker who offered us the process of centering clay on a potter's wheel as a metaphor for centering the opposites in our lives:

"As you go out, you come in; you always come into the center, bringing the clay into center; you press down, squeeze up, press one hand into the other, bringing your material into center. . . . We bring our self into a centering function, which brings it into union with all other elements. This is love. This is destruction of ego, in that its partialities are sacrificed to wholeness. Then the miracle happens: when on center, the self feels different: one feels warm, . . . in touch, the power of life a substance like an air in which one lives and has one's being with all other things, drinking it in and giving it off, at the same time quiet and at rest within it."

It is an eloquent book, unpredictable and persuasive. Eschewing allegiance to any dogma, Richards is deeply religious without the trappings. To be truly creative, she insists, we must embrace life in all its glorious and painful contradictions. This means being cleareyed and unsentimental about ourselves. "The transformations that await us cost everything in the way of courage and sacrifice," she writes. "Let no one be deluded that a knowledge of the path can substitute for putting one foot in front of the other."

Richards is also the author of The Crossing Point *(out of print)*, a collection of her talks and writings, and Toward Wholeness: Rudolf Steiner Education in America *(Wesleyan University Press)*. She became interested in Steiner — the Austrian scientist and philosopher who founded the Waldorf schools — because of her belief that education shouldn't sacrifice the imagination or a sense of the sacred.

Today [in 1988], at seventy-one, Richards lives, teaches, and works at the Camphill Village for the Mentally Handicapped in Kimberton, Pennsylvania, which is based on Steiner's teachings. She was in North Carolina recently to give a workshop at Duke University, and we had a chance to talk briefly. Though tired, she endured my questions patiently, her answers thoughtful, her large, graceful hands moving through the air to make a point. What authority in those hands; what wild freedom. When she took my hands in hers to say goodbye, I felt as if we had hugged.

Safransky: So many people talk these days of being "centered." Twenty-five years ago, when you wrote *Centering*, what did you mean by that term?

Richards: "Centering" has become such jargon, connected

with terminal states like "bliss" and "self-realization" and "peace." People peer at their navels and ask themselves, "Am I centered?" When someone asks if I'm centered, I really don't know what to say.

I use "centering" as a verb, to mean a continual process of uniting the opposites. Centering, for me, is the discipline of bringing in rather than leaving out; of saying yes to what is most holy as well as to what is most unbearable. The severity of that, as a discipline, is not widely understood; what is more commonly understood by the word *centering* is something much more trendy — something, I think, that addresses only one aspect of reality.

If I have given any originality to the term *centering*, it's in the image of the potter's wheel. In order to center the clay on the wheel, you move it both down and up; you widen and narrow; you have both the expanding consciousness and the focus. This is part of its gloriousness and its mystery, too — that it has both of those qualities, of being inward and outward.

In *The Crossing Point*, I take those opposites and connect them with another image, that of the lemniscate — the geometric form of the figure eight. You have a figure eight which is not a line, but a plane: if you run your finger around the outside of the top loop, then through the crossing point, your finger ends up on the inside of the bottom loop. You have this marvelous law of the continuity of the opposites — the inner and outer — without any break.

I think it's important for us to know that we are both inward and outward; that we live both in the soul light within and the sun radiance without; that we are both earthly beings, with our feet on the ground, and beings of inspiration and imagination and weightlessness. We're both. That's our genius, and we must not be talked out of it. Of course, in the arts, one can particularly feel that. Anyone in the arts knows how much physical labor is involved and that you don't actually have the thing until it's made.

It takes a lot of work to bring dreams into physical expression. That's what makes the arts such a natural paradigm for what human beings are. We're artists, and life is an art. This connection between the vision and the practice is what makes it art; through the materials, the vision shines.

Safransky: How have your own ideas about centering evolved?

Richards: I am clearer, I hope, about the relationship of centering to evil. Previously, in describing centering as an embrace, as the discipline of bringing in and moving out, I had taken a sympathetic posture. But what about the antipathy? We also need to know how to resist.

I learned about antipathy through an illness. In the winter of 1975 I was in bed for seven weeks with respiratory problems. I was very weak, and I was not getting better. One day, I said aloud, "I have no resistance." I heard myself say it. It was a turning point.

In my own practice of centering, I had fallen into the error of an exaggerated sympathy, thinking that I should go to bed until I was well and listen to the message the illness could give me. I was thinking positively, but without any awareness that the illness could be the end of me if I didn't take some kind of action against it. Then I saw the importance of including in the centering a consciousness of resistance. It's a process of continually integrating the elements, so that you're also becoming conscious of the evil — that which is perilous, threatening to life.

Safransky: A devotion to art, to the inner life, may make sense for the individual. But what does it have to do with larger concerns, with human suffering? How does it apply to us as social beings, as political beings?

Richards: Well, we *are* political beings. We live in the world; we live in the world of politics. The inner life isn't separate from that. What is the goal of the creativity one feels — and wishes to develop and help others to develop? Is it just to make more and more pots or take more and more pictures? Is that what it's all about? I don't think so. I think that as we become more creative, we move toward a concern with social justice and compassion. That's the natural movement. We come, maybe through times of loneliness, toward experiencing the reality of another person. As we create, you might say, we are created. We move toward a deepened awareness of reality. Outwardly, we move toward social justice; inwardly, we move toward compassion.

Safransky: There's a good deal of religious imagery in your work. How would you describe your own spiritual inclinations?

Richards: In some respects, I've come to the place where the religious life for me means being attuned to the earth as a living, breathing being. When you watch the cycles of the year, you see there's actually an expansion and a contraction, and that there are creative energies of the earth — and destructive energies.

I now live in an agricultural community, Camphill Village, which is very earth-oriented. Now, there's another community nearby, based on the handcrafts. One might wonder why I didn't go there instead of doing this farming and gardening. Well, it's totally different energy. In the handcrafts community there is much more nervous, emotional energy. The rhythms of the agricultural community are quite different and are what I needed. I knew I had to go where my schooling could be continued. I needed to get my priorities clarified, and now I feel that I am connected in my daily life to the earth. Unless the earth can be healed, we won't have any place to put our art. There won't be an earth if we don't reconnect with its needs. It nourishes us, but it also needs to be nourished. It's sadly depleted.

So I find my religious life isn't a special thing. I guess an integration has taken place, so that at any given moment there's a paradox. There's the paradox of living alone and yet being connected; there's the paradox of being religious and yet unaffiliated with any institution; there's the paradox of creating art and yet being aware that there is artistry in everything.

There are moments when I like to drink deeply in ritual. I went recently to the Cathedral of St. John the Divine in New York City. I was so moved and filled by the music and the service, I asked myself, "Why do I think I can live without this?" It was so marvelous. Then there are times when I don't need that. I do use images from Christian worship, because I find them imaginatively real and moving. For example, I did seven clay low-relief plaques inspired by the imagery of what is known as Christ's "I am's": "I am the Light of the World," "I am the Door," "I am the Vine," "I am the Living Bread," "I am the Good Shepherd," "I am the Resurrection and the Life," "I am the Way, the Truth, and the Life." The plaques will hang in the Cathedral of St. John after Easter.

In the community where I live, there's a strong relationship to the spiritual world through the work of Rudolf Steiner. At one time, I studied Steiner and wrote a book about his theories of education. Steiner's work not only inspired me but enriched my life. I now use color in writing as a result of visiting the Waldorf schools and observing how children use color from the very beginning. At my workshops, we always work with color, color in relation to itself. To experience color for even ten minutes can be very refreshing. Spiritual awareness comes not through thinking about the angels, necessarily, but through something as basic as color — making some kind of contact with that reality we call color.

Consider words: more and more I have come to understand that words are nonverbal. Try asking yourself where the words are before you speak them. I think we make a mistake in trying to restrict a word into something that we can read about in a dictionary. A word is a being. Writing becomes a more intimate and inspired activity when you see you're writing out of your substance, out of the mystery.

Safransky: One of the paradoxes you mentioned earlier was living alone yet also being connected. Are you comfortable being alone?

Richards: Yes, I'm comfortable with it. I liked it better when my life included a partner, so that aloneness was balanced with mutuality and relationship. I think that was more to my liking. But that isn't the situation in my life at the moment — and it hasn't been for a while. I decided I would learn to live alone, in the sense of not being intimate with another person, although I've always had the warmth of friendships. Then there came a moment when I needed more community, so I came to live in the country village where I am now.

Life in my community suits me very well. I have both privacy and interconnectedness. I don't know if this is the last form that my life will take, but having lived by myself and developed strengths in myself, I approach community very differently now. When I was younger, I lived in two intentional communities where there was a kind of dependency on each other. But when that was shaken, the relationships seemed insecure, and people started leaving. When you leave a community, it can be like being a fair-weather friend. You feed on it and are sustained by it, but if it isn't able to give you what you want, you say, "Oh, this isn't a good place to be. I think I'll go somewhere else."

Safransky: Do you think that's why most marriages don't work?

Richards: Very often. I'd say that was true for me. I didn't understand that negative feelings between people are OK. I thought that if there was a negative feeling, you had to get divorced. I've learned a lot about conflict. I like to say that peace is an art of war. It's OK to be neither victorious nor defeated, to experience the continual energy of difference and conflict, which causes something in oneself to grow. As a potter, I use the image of the vessel. The vessel expands so it can contain more and more of life and reality. It was T.S. Eliot who said that human beings cannot bear very much reality. At first, that sounded to me like he was putting us down, but that wasn't really the case. I do think we have different capacities for reality in our lives. It's a good idea to be honest about how much clay you have to work with, so when you're making your vessel and stretching the clay, stretching it and stretching it, you know when you can't go any further, because the vessel will crack and split. You have to honor the potential of the material. Don't try to push it into a theoretical generosity of which it isn't capable.

Now when I come to a community, I'm able to bring more independence into it. I'm able to say that I'm "in community" wherever I am. I try to offer that to others. If you're wanting to build a community, you don't have to wait for some ideal situation. You can do it wherever you are. That's quite a different constellation of feelings from those I had when I was a lot younger. The aloneness and the connection are more like tides in your sea. You can feel separate tides. Sometimes you need to be by yourself, doing whatever you need to do — whether it's working on something or just resting or pondering. Sometimes you need to be with others: in the kitchen helping to prepare the meal, in the garden working, or doing an open-studio session. The tide flows in; the tide flows out. These needs play back and forth.

Safransky: Do you have children?

Richards: I was not able to bear children because of a physical condition. I tried hard, but it wasn't in the cards.

I did, however, enjoy a stepdaughter for five years when I was living and working at Black Mountain College. She and her father left, and I didn't see her again for thirty years. Finally, there came a point when I didn't want that estrangement to go on forever. It was important to heal that. I wanted to change my karma, you might say. So I found out where she was, and I wrote to her. We met. Of course I didn't look that much different: my hair was gray and my face was lined, but I still had the same shape and features. But she had changed from a skinny little twelve-year-old girl to a full-grown matron of forty-two. I just wanted to apologize for being such a lousy stepmother. But she told me I wasn't nearly as bad as I feared.

Safransky: Can you imagine being married again?

Richards: Yes. Oooh, I just had goose pimples all the way down to my ankles. Isn't that amazing — how one's body and soul are so connected?

I couldn't really think of it for many years after the end of my last relationship. That was in 1964, when I decided to break the pattern of dependency and learn to live alone.

Safransky: So, twenty-four years later, you feel ready to try it again?

Richards: Well, I allowed myself to be touched by a person not so long ago. It gave me a different sense of myself, which I like — I could feel that my heart is not hard or hardened.

It's strange: I don't really experience myself as old. I look in the mirror and see that I am a senior citizen, yet I don't feel particularly old. But there are stereotypes about what is suitable in terms of sex and romance. Every once in a while, I think, *Oh, maybe I'm too old!* When you have all those wrinkles and white hair, and your body isn't so beautiful, you think maybe you shouldn't allow your fantasies to continue. So I have that bit of momentary self-consciousness.

But it was nice to feel that release, to know that whatever needed to heal has healed. So I'm free to do some other foolish thing.

Safransky: What do you understand differently now about loving another person?

Richards: Your question brings to mind a Zen saying: "Now that I'm enlightened, I'm just as miserable as ever." For one thing, I understand that I still have to struggle with possessiveness, jealousy, and anxiety. And that I mustn't be embarrassed, horrified, and depressed when I feel fear as well as joy, and think, *Oh, why can't I just feel the joy?* You see how helpful the centering discipline can be in dealing with the polarities of our nature? I thought I would outgrow my needs for reassurance, and maybe I will — certainly they are not so extreme as in the past. I would really like to have the generosity of love to say, "Yes, my dear, you may do just as you like." But I'm not quite there yet. In the spirit of centering, I put my arms around the shoulders of my fear and sing my song. [Sings softly.] "Sit with me at table, oh my fear. Sit with me at table, oh my fear. Sit with me at table, oh my fear. Give yourself to me, oh my fear. And I will marry thee."

I also have a better sense of humor now, which I think makes a big difference. And I have more of an openness for letting the mind play the way it does, in a very wide field, instead of insisting to another person, "But what you *said* was . . ." As if the mind has no craziness. I think I'm much more prepared now for what is wild and nameless in a human being, along with what is elegant and proper and aspiring. I like that, even though it's full of risk.

March 1989

NIGHT OF DYING

Maureen Stanton

I had known all week that Keith would die that weekend. I knew he wanted me there when he died, not at work, or waiting at a red light, or picking up bread and milk. He waited for me.

We had moved from our small apartment to his parents' house the week before. Keith had returned to his childhood home to die.

His younger sister went to an amusement park in Ohio for the weekend. She knew, too.

I knew all day Saturday that Keith would die *that day*. It wasn't just my intuition, or the fact that I knew every inch of his lanky body. We functioned as one unit. Eighteen months before, when doctors had told Keith he had two months to live, we'd been separate — he with his cancer, I helping him fight it. But six months before he died, he quit the fight and quietly resigned himself to dying. This is the point at which we became inseparable.

I felt how the bones in his knees bore on each other as he lay on his side in bed, and I put a feather pillow between them to relieve this small discomfort. I knew exactly how to cook his eggs, removing the blotch in the white before stirring them up to batter French toast. I knew the pattern of his breathing, a slow, steady breath, a second of silence, then a slow, steady breath again. So it's natural I knew the day of his death.

It wasn't just my guts that told me. Months before, the hospice nurse had given me a list of clues. A checklist for dying, like you use when going on vacation to make sure you've packed everything. A checklist. This made me think of young men being drafted into the army. No allergies, no flat feet, no psychological imbalances? You made it. Congratulations, you can go to war. Keith was being drafted into God's army. That's what his mother believed.

The checklist included these signs:

An increase in pulse rate to 140. Keith's pulse rate had been a steady 120 in the month before he died. He slept ten to twelve hours a day and rose only to use the bathroom or to move from the couch to the recliner to the bed. Yet his heart was pumping furiously, like a runner putting it all into the last stretch. He was a quiet, motionless runner, speeding toward the end.

Cheyne-Stokes respiration. Who were Cheyne and Stokes anyway, whose namesake is the irregular, interrupted breathing of a dying person? I still don't know. After Keith died, I moved into a small house and was haunted by the old refrigerator. The refrigerator would give a huge shudder, then be completely silent. A low humming would develop, get louder and more strained, intensify, and then release a shudder. And the cycle would repeat, just like Keith's final breaths.

Clammy skin. His skin was cool and moist, like basement walls on a humid summer day. Despite this coolness, Keith was thirsty, desperately thirsty, all evening the night he died. We were giving him juice over crushed ice. His mother couldn't get it right. She'd bring a drink with big chunks of ice, and he would send it back. "Too big," he'd say. "Pulverize it." It was funny, his mother in the kitchen pulverizing ice. We smiled at the sound of the blender. His thirst couldn't be satisfied, though. I went to the nearest store and bought a large blueberry slush. He drank this, and his lips and tongue turned blue.

Bluish extremities. Keith's feet had been purplish for weeks, and his toes had a slightly blue tone. Little pieces of skin were peeling off his feet. It made me think of grated Romano cheese. It had the same texture and smell.

Incoherency and hallucinations. A few nights before he died, when we were lying in bed, Keith said, "What's that noise?"

I said, "What noise?"

"Listen," he said. "It sounds like someone turning pages."

"I can't hear it, Keith," I said. "Put your finger on my leg. When you hear the noise, tap your finger."

A few seconds passed, he tapped my thigh, then tapped again, and a third time. "I can't hear it." I became scared and tried to listen harder for any noise as he pressed his finger on my skin, but I just couldn't hear a thing in the dark night.

On the night he died, after I laid him in bed, he asked, "Who's frying something?"

"Nobody," I said.

"I smell something frying."

"I don't know what it is, Keith."

"What's frying?" he asked again.

These were the signs. At eleven o'clock his parents went to their room next to ours to go to bed: his mother without her glasses, bags under her eyes, blinking, in her daintily flowered nightgown that came to her toes; and his father still in his work clothes, an old pair of jeans that sagged to expose the cleavage of his behind and allow his belly to hang over his belt. "If you want to know what your husband will be like, take a look at his father," my mother always warned me.

I helped Keith up from his favorite chair, a brown La-Z-Boy recliner, to the bed. I fixed five pillows around him: one under his head; one between his skinny legs to keep his knees from jabbing into each other; one that he hugged to his chest; and two behind his back to keep him on his side, the position that allowed sleep. Sleep had become a major part of his life ever since he found out he had cancer. Toward the end, he was sleeping most of the day and night, waking every three or four hours for a short while to smoke a cigarette, have some juice, flick from channel to channel on the television.

Keith was restless in bed. He wanted to go back to the recliner. I helped him up, but halfway there his legs gave out, and he collapsed. I carried him the few remaining steps to the chair. Keith was six-foot-one. At that time, he weighed only about as much as I did, 120 pounds. As Keith shed pounds the last months of his life, I gained. I cooked and baked his favorite foods: peanut-butter oatmeal cookies, apple crisp, pancakes, mashed potatoes and gravy. I ate my food, the food he couldn't finish, and the food left in the pot. I ate an entire pan of Rice Krispies squares once, except for the one Keith ate. And when he wanted another one, I made another batch, gave him one, and ate the rest.

Perhaps this food gave me strength. Some strength from somewhere helped me carry Keith to the chair.

When his knees buckled, I cried out, "Oh, Keith!" I remember this clearly. His parents came running into the room, wrapping their robes around them.

"I think he's dying," I said.

Keith called out, "Mom. Dad."

"We're here, honey," his mother said.

He closed his eyes.

I was jealous about that for a long time. Why didn't he call out my name? I decided he didn't call my name because he *knew* I was there, as I always was, during the sickness, day and night, and before that, as his lover.

I bent to his ear and whispered, "Keith, I love you. Goodbye, Keith. I love you." I felt as if I was arming him with ruby slippers, that no matter where he was going, he could always come back to me, because I loved him. I kissed him, and he squeezed my finger.

He began to cry a little, moan softly. He complained his stomach hurt. He hadn't eaten in more than two weeks. He hadn't had a bowel movement in three weeks. He was connected to a morphine drip. A needle was planted in his thigh, the only place where he still had much flesh. I gave him an extra shot of morphine by pressing a button on his IV unit. But the unit was programmed to give him no more than one extra dose per hour — a "bolus," the nurses called it. At first, I thought they were saying *bonus*. If I could have, I would have sent it all down the tube into his vein.

His pain became intense. He held his abdomen and moaned louder and louder. Then he said, "I think I shit my pants." I looked and saw black fluid, thick like menstrual blood, coming out of his rectum. I massaged his abdomen. He moaned and moaned, until his breathing became moaning, and each exhale was a loud wail, a long, low moan. A baying, like an animal dying, so loud in that small house.

"Get me some warm washcloths," I ordered his father. I took off Keith's underwear. His cousin had given him blue and red and yellow polka-dot boxer shorts when he'd first gone into the hospital, a year and a half ago. He had been wearing them for the last two weeks. They were loose and comfortable, and there seemed to be no point in changing him each day. It was trouble for him to bathe. It hurt him. His teeth were gray and coated. He wouldn't let me bathe or clean him anymore. I understood. Why bother primping for the occasion? Might as well go into the ground dirty, earthy. That pair of polka-dot underwear stayed in the top drawer of my dresser alongside my underwear for almost three years after he died. And even then I didn't throw them away. I put them in a box with letters and pictures and a curl of his hair and a cassette of his voice — our voices: a fight we had taped long ago, before the cancer, to see who was twisting whose words around.

I placed a pad, like a diaper, under him. The nurses called these pads "chux." Black liquid, the color of old car oil, continued to pour out of him. The smell wasn't overwhelming, as waste usually smells. It smelled like guts, like cow livers, like internal organs. As fast as it was coming out, I was trying to clean him up, but it was coming too fast. I asked his father to get me a cup to catch it and a bucket to dump it in. His mother was rinsing washcloths and keeping me supplied with clean, warm towels.

Keith's eyes were rolling back into their sockets. He was still moaning loudly and steadily, almost rhythmically.

His father returned from the kitchen with a styrofoam cup. I tried to place it under Keith to catch the black shit pouring out. The cup broke. Keith's mother said, "For God's sake, Phil, get something that won't break." He left and returned with a small flowered teacup. She looked at him in disgust. I took the cup anyway. It worked.

The black fluid was slowing down now. But just about the time we got him cleaned up, wiped his sweaty face, and wrapped him in another diaper, more black fluid came out. There was so much. Each time it stopped I thought, *Surely, that's the end.* But it kept coming. It was like the sins of his life, thirty-one years of sins, pouring out of his body.

His moaning was still loud and regular. I got a syringe and some Valium, which he took four times a day. He wasn't due for a shot, but I gave it to him anyway. His mother didn't want me to. She had read the fine print on the bottles, where it warned that an overdose could lead to heart failure, could prove fatal. She was worried Keith was going to die of a heart attack.

"That would be a blessing," I said.

I poked the needle into the vial of Valium, and sucked the drug into the syringe. I jabbed the needle into a port in the tubing in Keith's leg and gave him not one but two shots of Valium. His breathing suddenly changed. It grew raspy and quick, and I thought, *His mother's right: I've killed him.*

But he continued breathing, and the moaning subsided to low, steady bursts of noise that sounded like a foghorn whose steady booming had slowed and become faint as its generator ran down. Keith's father was making coffee and drinking it out of a beat-up green metal thermos, as he always did, generously adding whiskey. His mother began filling syringes with Valium. She neatly arranged a supply sufficient for the next four days.

Keith was still moaning and holding his abdomen. I thought he must be stopped up and that was causing him pain. It had happened before. I put on the surgical rubber gloves that I had on hand for giving Keith enemas and changing dressings. I stuck two fingers up his rectum. I felt a blockage.

His mother looked at me and said, "I don't know how you can do that."

She sat by Keith's side and wiped his brow. I began to pull out hard masses of thick black sludge. More black liquid followed. I pulled out more hard balls of shit and massaged Keith's abdomen vigorously. His mother thought I was hurting him. Keith's moaning softened, and the flow of black liquid waned, then stopped altogether.

He was breathing restfully, shallowly, softly. I wiped his face, cleaned him up, and hoped we would sleep a bit. His mother was picking up towels and cleaning the mess. His dad lit a cigar, coughed up phlegm, and spit into his handkerchief.

Keith was sitting in his La-Z-Boy chair, clean, diapered, draped in a blanket, and resting his head on his favorite feather pillow. I sat on the bed. It was 5:02 A.M. I looked around the room. It was still a mess. There were half-filled glasses on the night table beside the bed, and plastic wrappers (from the chux), and rubber gloves and syringes on the aqua green carpet. I breathed and looked at Keith. His eyes were shut. He looked as if he was resting peacefully. But I noticed a strangeness, an absence of movement. His bony ribs were not rising or falling. I felt for his pulse, first on his wrist, and then by pressing my index finger and thumb hard to his jugular. Still. I called for his mom and dad. "Come in here," I said. "He's dead." I put my ear to his chest. There was no sound, not even a faint one. "Get a mirror," I said. His mother brought one. I put it under his nose. No mist appeared. I fell to my knees and began to cry. Keith's father said, "Son of a bitch," then punched the wall. His knuckles began to bleed, and he wept. His mother cried into a towel, and then we hugged.

For five hours, Keith died. Five hours, but really only one second, because he was actually alive, his heart beating, until that one second, that second I looked away, when I rested my eyes. He left without me, after all.

September 1991

This Life, This Word Unsaid

John Hodgen

Last year, first thing, the two women who moved in next door
    spent all weekend
putting up a post-holed stockade fence, then, smacking their
    hands, asked
if I'd foot the cost of weatherproofing my side.
Fair enough, yet, caught short, I demurred, and we have not
    spoken since,
half waves sometimes, eyes cast down as we lug the groceries in,
their side of the fence glowing in the afternoon light,
mine graying and green at the base.

They brought two dogs, Doberman, Dalmatian,
built a run, the loosing and howling now regular as rain, as sun,
morning barking means the mailman any second,
evenings rolling over, murmuring, *Jesus*, going back to sleep.
Last week skidding tires, the thump one knows as impact,
    assuredly,
the black-and-white blur gone right for the road like a
    Michigan hound,
no holding her, as if she had always waited for this,
and the women right behind, too late, crying, *No, no, no, no,*
left holding their broken dog like a double Pietà, the car
    pulling away,
stopping, then moving again, slowly, not a thing to be done,
not a word that would heal or matter in the least.
Tonight the quiet hanging in the yard, the Doberman nipping
    and snuffing at the air,
as if he wants to speak, but can't remember what to say.

And tonight my brother has gone to hell again, his stunted
    wife in a home,
his house up for sale, the sign on his lawn like a shame.

He wanders, he says, drifts from place to place,
like a father who has lost a child to the sea
and will not be comforted, pacing to and fro with the tides
    and debris.
I don't know what to say anymore, the daily news, the steady
    saw.
The words stand around unemployed in my throat, like shirts,
    frozen,
left out on a line, or dogs in a yard, the family gone away.
I am the quick brown fox's lazy dog. I lift my head. I look the
    other way.

And when my cousin with AIDS comes to Gloucester next
    month,
what words for him who will not last the year? What words
    after that,
when even the chrysanthemums might be moved enough to
    speak?

And this photo pressed into my hand today, me, long-haired,
    sixteen years ago,
Outward Bound, a little like Jesus in training, crossing on a
    rope
over watery water, calling to someone I can't even remember,
the words so lost by now they must have turned to shadows, or
    pieces of leaves
against a rusty fence, or souls broken up gone out to black
    holes.

And this, what pilots say most often before they crash.
Not *God* or *Oh, God.* One does not look for lord or love in
    such an angled sky.
Not *Father, forgive them.* Who could say that again,
having leapt from one's nails to the edge of the world?
Not *Mother*, more for soldiers, perhaps, hugging their helmets

or digging the earth gone wet with their weeping, their
    wounds.
No canticle at all, no litany.
Oh shit, we say, as if we've always been in it,
this ground, this stink, this awful place,
the earth coming up at us like the right hand of God,
us seeing it coming. *Oh shit*, we say,
oh shit, oh shit, oh shit.

February 1993

WHAT'S EATING ME: A MEMOIR

David Guy

I don't remember being fed in infancy — my memory doesn't extend quite that far — but I was seven when my younger brother was born, and I vividly remember his feedings. They were quite a production. The bottles and nipples had to be washed, scrubbed with special brushes, then sterilized in a big metal pot that boiled hard on the stove, the bottles rattling on a rack inside it. The formula had to be mixed; it had to be heated, but not too hot! I can still see my mother testing it on her wrist. Then, and only then, could the baby be fed.

My mother was very particular about stains on her clothes and slipcovers, so she always had a diaper or two draped in strategic places in case things got messy. You could never tell when the baby might open his mouth and, with an ease I always admired, empty the formula right back out again. What I remember most of all is the elaborate ritual my mother had for burping him. She would carefully drape a diaper over her shoulder, sit the baby up, then slowly — holding him well away from her body — lift him in the air and place him gingerly on her shoulder to give him little pats on the back. The baby always seemed awkward and heavy as I watched her do that. I felt sorry for my mother, having to shoulder such a burden. She held him as if he were a little time bomb.

A couple of years ago I was in New Hampshire visiting my older brother, and we spent some time with a friend of his who was a new mother. If there is such a thing as a polar opposite to my mother, it was this woman. She was tall, blond, and heavy, with large round breasts and a creamy complexion. She looked strong and healthy, at ease in her role as a mother.

She held her baby constantly, pressed him to her body. She wore a T-shirt and shorts and, if she wanted to feed her baby, just lifted her shirt and plugged him in. She didn't have to mix and warm her milk (and it wasn't called "formula"). She didn't wash or sterilize her breasts. There was no elaborate burping ritual. What I remember most vividly is that the baby was constantly pressed against her, as if he were a part of her. (He *was* a part of her.) My mother never held a baby that way. Even when she was feeding my brother, he always somehow rested on her arm, never melted into her body. In New Hampshire I finally said something to my brother about never having been treated that way when I was a baby. "No," my brother said, "our mother would have held us out there with a pair of tongs if she could have."

That was a poignant moment for me. Something I had always felt, but had not quite allowed to creep into consciousness, had been confirmed. The experience that my brother and I shared had had vast ramifications in our lives, ramifications that we knew only too well. No upbringing is perfect, of course. Being held too much must present its own problems. But that wasn't what we'd had to deal with. We were standing there staring at what might have been.

I KNOW THAT my mother loved me, but she didn't express her love physically. I would say that she couldn't. She would probably say she didn't think it appropriate. Because this was the absence of a good thing — not the presence of a bad one, like actual abuse — I was into my thirties before I realized how little physical affection my mother had given me. Even then, on my one or two visits home each year, we exchanged what passed for a hug — our bodies almost, but not quite, coming together; our hands touching lightly on neutral places, like the upper arm — and a brief, tepid kiss on the cheek. I could have counted on the fingers of one hand the number of places on my body that my mother had touched me. But I was so accustomed to that situation that I didn't think it strange. As a teenager, for instance, I didn't compare it to the showers of physical affection that my Italian friend got from every female member of his family. I thought *those* women were strange.

Someone might ask why, as an adult, I didn't hug my mother the way I wanted to be hugged. The answer is that I had repressed those wishes for so long that I no longer felt them. I didn't want to hug my mother hard, or kiss her on the mouth. The thought repulsed me. Furthermore, my older brother had come to the same realization a little earlier than I, and had tried to make his embraces with our mother more affectionate; he warned me not to try, that the resistance he had felt was more painful than any lack of physical affection.

I looked back on thirty years in which I couldn't remember a single passionate gesture from my mother. All I could recall was a kind of perfunctory touching: her hand on my shoulder, the offer of a cheek. I couldn't remember a different kind of hug when I was a child, or even when my father died, when I was sixteen. By that time, of course, I had already become that boy who didn't want to hug his mother. What bothered me in my thirties was that I couldn't remember when I *had* wanted to. I couldn't remember wanting to hug my mother.

I had always, however, remembered a particular moment from my childhood without knowing why. It did not seem the kind of traumatic once-in-a-lifetime incident that one usually remembers: I was standing in front of my mother while she sat at her dressing table. It was a quiet afternoon, and we were alone. I would guess that I was about six years old.

At first that was all I could remember. I couldn't see any significance in the moment. But the image came back to me so frequently, especially in therapy, that I knew there must be something to it. One day with my therapist, I sat in its presence for a while, and other things began to come back.

It was the afternoon of a school day. It was unusual — in a family of three, soon to be four, children — that no one else was around, and it may have been just this chance to be alone with my mother that made the day stick in my mind. Something had happened at school that had scared or worried me, that in some way had made me uneasy. My mother dismissed my concerns, as my parents tended to do. "You're not worried about *that*. You're not scared about *that*." They were trying to rid me of my fears, but they were failing to acknowledge that those fears were very real to me. It was terribly frustrating to be treated that way. The frustration manifested itself in my body as a restless, I-don't-know-what-to-do-with-myself feeling. I wanted to jump out of my skin. My mother probably gave me a little pat. She may have kissed my cheek. She assured me that everything would be all right.

Suddenly, at that moment in therapy, I knew what I had wanted from my mother. It had long since been made clear to me that this thing was forbidden; that was why, as an adult, I couldn't remember it, and that was why, as a child, I didn't know what to do with myself. I didn't want my mother to say something reassuring. I didn't want her to give me the answer to my problem; there probably was no answer to my problem. I wanted her to take me in her arms and hold me, comfort me, my whole body against her whole body. As I pictured her doing that — something previously unimaginable — I saw us fall back on the bed with our bodies together. We held each other tight. We held each other a long time.

Finally, I had remembered wanting to hug my mother.

Looking back, I recognized that embrace as the same embrace I had always sought from women, the way I always hugged a woman when we went to bed. I wanted that embrace more than any specifically sexual act. I wanted it as soon as we went to bed, and I wanted it to linger. It was what I had to have to feel all right.

WHEN YOU EXPERIENCE physical unease, that feeling of wanting to jump out of your skin, with no chance to get the comfort you need, your best strategy is probably to deaden it. Kill off the feeling. You are thereby cutting yourself off from your body, but numbness is preferable to pain. A person can stand only so much pain.

If you do kill off the feeling, it is likely that your body might change in some way: blow up into a blimp, or shrivel into nothing. You have lost contact with your stomach, so you have no idea what it needs or wants (what you think it wants is all the food in the world, to satisfy that longing that has nothing to do with food). If you get fat, the fat itself numbs you, so you feel your body less and less. The fat hugs and comforts you; you have solved your problem by hugging yourself. It is also the case that, when you are a child and don't get hugged, you take the blame on yourself. But if you can make yourself physically unattractive, then that can be the reason you're not getting hugged. You can even make yourself so big that people can't get their arms around you.

The stomach is the seat of the emotions. Sadness can feel like an emptiness, a gnawing, in the stomach. Anger boils up from the stomach. An excellent way to still these emotions is to fill your stomach with food. Just keep chewing and swallowing. Try not to think about it. (Please pass the noodles.) Get enough food down there and you won't know what you're feeling. Or at least your pain will have an obvious cause, and you won't have to acknowledge a deeper pain.

IN MY SEVENTH year, right around the time my younger brother was born, I suddenly got fat. I had been a positively skinny kid when I was younger, but by the third grade I was one of the three heaviest kids in the class. Thirty years later, I can still remember the day when the whole class got weighed. I was good friends with the other two fat kids; we had all known each other for as long as I could remember. We each weighed eighty-three pounds.

The family joke was that I had discovered Nestlé's Quik, a chocolate powder that you put in milk, and had gone absolutely wild, blown myself into a fat kid with this single product. (Mention of Nestlé's Quik still brings a roar of laughter from any member of my family.) I'm sure that no one food was responsible, but Nestlé's Quik isn't a bad symbol for what was wrong. It was a junk food that I stirred into milk to give it a different flavor. I wasn't seeking the taste or the nourishment of milk, but the sickeningly sweet taste of artificial chocolate. I wasn't drinking it out of a need for milk, but out of another need of which I wasn't aware. I was following the pattern of all senseless eating: if I can't have *that* (physical love from my mother), by God, I'm going to have *this* (the taste of chocolate). I would mix it up and drink it down quickly, the way a drinker pops into a bar for a short one. I would knock it back and dash outside.

It seems obvious that my brother's birth had plenty to do with all this, but I don't remember a particular loss of love at that time. I had felt the absence of physical love long before. Perhaps things just got worse. In the same way, I associate this new physical fact with an aspect of my personality that was probably there all along. (There was a fat person struggling to get out of my skinny person.) I am sure that, on the day when the three of us tipped the scales at eighty-three pounds each, the other boys scowled at the jokes that were made about them and pummeled somebody in the playground, but I laughed at the jokes and made some more at my own expense. I had become the kind of person who is automatically funny just by his appearance, and who makes himself agreeable by going along with the joke. The kind of person who eats — along with everything else — shit. I became a person who felt physically unattractive and spent a great deal of time dwelling on that idea, who thought his life would be much better — everybody would love him! — if he weren't fat. (He thereby had a pat excuse for why his life wasn't good.) I became a person defined by his physical appearance. If you had asked me to describe myself, I would have mentioned that first: I was fat.

Such a personality brings to mind a certain kind of comedian: Fatty Arbuckle, Oliver Hardy, Lou Costello — men who were funny partly because of their shape. I especially liked Costello, the pathetic little fat guy who always fucked up, who admitted ad nauseam how worthless he was, who was always being put down and always putting himself down. His only hope for affection was not that people would love him for his good qualities but that they would take pity on him, like a lost puppy or a helpless child. "The poor little guy!" his girlfriend used to say. She was a blond bombshell named Hillary Brook, and though she was nominally Costello's woman, she knew he was a pathetic sap and often admitted as much to Abbott (who was probably screwing her on the side). I loved the Abbott and Costello shows when I was young. Now that I can see what they meant to me, I can't watch them.

The amount I ate as I moved into adolescence — three or four times what I eat now — astounds me. It had nothing to do with appetite or physical need. It was habitual. It was partly, I think, a kind of comfort: if I got that amount of food, all was right with the world. It also had to do with a sense of injustice or outrage: if the world was going to treat me the way it had, it could damn well cough up five pieces of chicken at dinner.

More characteristic to me, and more fundamentally mysterious, is a particular eating habit I had. I have, as a writer, written about much that makes me antsy — I believe in writing about such difficult topics, largely *because* they are that way — but I have never written about anything that embarrasses me as much as this. When it came to mind the other day, as a detail from my life and a possible subject for my writing, I cried. It seemed so strange and sad. At other times it has seemed uproariously funny. I wrote about it briefly and humorously in one of my novels.

In our house in Pittsburgh, there was a downstairs den, a small room with two windows, an easy chair, and a portable television. It was my favorite place to be in the evening, especially alone (with my dreams). I would watch shows about what I took to be normal American life: *Ozzie and Harriet, Father Knows*

Best. I watched shows of teenagers listening to music and dancing. I wanted to be like them. I wanted to have fun with lots of friends, especially lots of girls. (I went to a boys' school, so girls were not friends. They were frightening alien creatures whom I saw only on social occasions. I longed to be close to them, but I didn't know how to do that. I didn't know what to say.) I wanted to be normal and average. For me that meant, above all, not being fat. And while I was watching those shows, dreaming that dream, I would go out to the kitchen, assemble an enormous chocolate sundae, and come back to the den and eat it.

I look back on that fact and am bewildered. I had undoubtedly had dessert, probably a duplicate of that chocolate sundae, not two hours before. I had put away a huge dinner. I couldn't have been, by any definition that makes sense, hungry. I ate the sundae quickly, with huge bites, and ate while I was watching television, so I didn't especially taste it. There I sat, watching those television shows, wishing I could be like the people on the screen, and simultaneously doing the very thing that kept me from being like them.

What really makes me sad about those sundaes is that I tried to keep them secret. I would tiptoe into the kitchen, open the freezer quietly, gingerly set down a bowl, slowly open the drawer to get a spoon. The fact was, sound traveled all over that house. Not only did the whole family know I was making a sundae, they could tell I was trying to sneak it. Perhaps they were thinking, *There he goes again, that poor, pathetic bastard. Another chocolate sundae. I don't know why he does it. Jesus, is that kid fat!* The bizarre thing was that, after all that secrecy, I would go back and leave the bowl in the sink. It was as if I didn't care who knew afterward, as long as I'd had my sundae. I was the classic criminal who couldn't help leaving the evidence behind.

Those sundaes were incredibly sweet — you might even say too sweet. They were loaded with calories. They were prepared and consumed in moments, without a thought. I didn't enjoy them. I ate them out of compulsion. I didn't notice eating them. I would only have noticed *not* eating them. Something would have been wrong. The evening wouldn't have been complete.

THE SECOND MAJOR physical change in my life took place in my first year at college. I remember that, a few days before my sixteenth birthday, at a checkup at the doctor's, I weighed 198 pounds. I have no doubt that, over the next couple of years, I crept over the 200-pound barrier, though my trained-down football weight was more like 190. But after my first year at college — a time when many people gain weight — I weighed 160. Recently, at my twenty-year high-school reunion, some people who hadn't seen me since we graduated were astonished at my appearance. They would have been just as surprised nineteen years ago. The changes they were seeing had all taken place in one year.

I didn't try to lose that weight. It just happened. The only time I have ever been called a liar to my face is when I tried to explain this to someone. (He had a weight problem himself and wanted to know my secret.) The summer before I went to college, I had a job as a tutor at my old school. I didn't like the breakfasts, and I got to sleep late if I didn't go, so I skipped them. I sometimes got hungry midmorning, but by lunchtime my appetite wasn't as large as usual — my stomach had shrunk, people would say — and I still wasn't hungry at dinner. It's also true that the food wasn't spectacularly good there. (I hadn't been a boarder when I went to that school. I'd shoveled down home cooking all four years.) I had stopped lifting weights, since I was no longer playing football. I had also started smoking cigarettes, which dulled my appetite and diminished my oral craving.

When I got to college, I maintained the same habits. I had noticed I was losing weight and wanted to continue. I was also selecting food in a cafeteria line and paying for it item by item, so I could no longer eat without thinking. I wasn't very likely to go through and ask for five pieces of fried chicken, two big helpings of mashed potatoes with gravy, two helpings of peas, and two huge pieces of cake with ice cream all over them. My food consumption fell to less than half of what it had been a year before. I hadn't gone on a diet. My life had changed.

For years I thought that those superficial details explained what happened to me, and in some ways I still think they do.

Why look deeper, when the surface explains so much? I ate less, and I lost weight. But it is also true that there was as significant an event in my seventeenth year as in my seventh. On that New Year's Day, my father died, and though people saw no immediate change in me, I do think there was a long-term change. My father's death changed who I was.

My mother, in a kind way, had always denied my weight problem — "David's not fat," I can still hear her saying — but my father acknowledged it and sometimes confronted me about it. He suggested I eat differently. On a few occasions he kidded me about my weight, which I found deeply humiliating. He had been overweight when he was young and also as an adult, though as he had gotten ill — he had leukemia — his weight had diminished. It was the common situation of a father seeing his own problem in his son and wanting to correct it.

But I also think that my father's attitude toward my weight was more complex than that. I look back on myself then and see a boy — he seems a different person — who was big and bland and quiet; who was shy and terribly self-conscious; who was physically strong but in other ways weak. His ego was weak. He didn't stand up for himself. He was also — I don't know how else to put this — sexually weak. He had an enormous yearning for physical affection, but felt so unattractive that not only could he not get it, he couldn't even ask for it. He had no right to it. He felt terribly vulnerable in that area.

It seems to me that, at some level, his father wanted to keep him that way. He wanted to keep him a bland, obedient boy (though strong, a kind of dray horse) who would do as he was told. Keeping him that way meant keeping him fat. Feed him enough fried chicken and cake with ice cream, and he would never be a problem. It was as if my father, despite all his annoyance with my excessive weight, wanted me to be that way. He wanted me to stay fat so he could stay annoyed with me. He wanted me to stay fat so I would stay a boy, so I would not become a man and challenge him.

I don't know what would have happened if my father, a big, strapping 220-pounder himself, had stayed healthy and lived. I have often explained to people that I never had the chance to rebel against my father because he died too soon, or because, when I was entering my adolescent rebellious phase, he was too sick and weak to be rebelled against; but in recent years I have wondered if that was really true. I think I was just cowed by the man. He had me by the balls. I wanted him to love me (he *did* love me), and I wanted to please him; I feared his anger and disapproval. Right around the time my father was dying, I was taking biology and chemistry, finding them very difficult, and beginning to think I didn't want to be "a doctor, like my father," as I had always said. The sciences were alien to me, but I was a hard worker and could probably have struggled through them. If he had lived, would I have remained a dutiful son, a solemn overweight man who continued to do the things he didn't like and to compensate himself with food? Would I have done mediocre work in a field I wasn't suited for but gotten into medical school because of my father's influence? Would I have joined my father's practice, become the third generation to enter the practice, another portly Dr. Guy with a wife and a big family, who drove a Buick and was an elder in the church? I probably would have. That possibility stands before me as what my overweight life would have been. Instead, my father died. I went to college, lost thirty pounds, let my hair grow, gave up the sciences for literature and writing, and somehow — in the way that all honest writers do — became disreputable, if only because I was trying to tell the truth. My fat self was a solid citizen, but my thin self was a little shady. Without fully knowing what it meant, I had decided to be thin.

IN BED, I am the most oral of lovers. I love sloppy, wet kisses, a tongue that wriggles and thrusts in my mouth; I love to chew a woman's lips, lick every inch of her body; I love to fill my mouth with her breasts, one and then the other. And my favorite erotic act is to eat a woman. Nothing else even comes close. This fact is so appropriate to my whole psychic history that the very thought of it makes me smile.

No calories!

IT HAS BEEN twenty years since I lost all that weight. Since then I have stayed between 165 and 175 pounds. When I was depressed after my marriage ended, my weight went down. More recently it has gone up. I would guess that my ideal weight falls somewhere in the middle of that range, though I don't like to worry about numbers. I have become something of an exercise nut, at first largely because I was afraid of how fat I'd get if I didn't do anything. Now, like many people who exercise, I have come to enjoy the activity itself. I like what my body can do, and I like the feeling of using it. I miss exercise when I don't do it.

Most people would think of me as healthy in regard to diet and food. I am not grossly overweight, and I don't go on eating binges; I don't suffer from any of the known eating disorders. That doesn't mean there isn't still a problem. I have simply learned a set of strategies for dealing with it. I have a general idea of how much food I can eat in a day and try to stick pretty close to that. It doesn't allow for much variation. If I am very hungry at an odd hour, I might have some fruit, but I generally don't eat between meals at all. At times, often late in the afternoon, I feel positively weak from hunger, but I don't act on that feeling. I don't trust it. There have been too many occasions in the past when it has really meant something else. In the same way, if partway through dinner I don't feel hungry anymore, I finish it anyway. Otherwise I might get hungry in the middle of the evening and not know what to do.

I was fat from age seven to seventeen, those vastly important years when we largely form our self-image, and I haven't been able to shake it. I still see myself as that five-foot-eight-inch, sixteen-year-old boy who weighed 198 pounds. There are periods when I worry constantly about getting fat. I think about it every minute of the day. It is the tape that is always playing in the back of my mind, especially when I've missed a day of exercise (even though I know exercise burns up only so many calories) or when I've eaten a piece of cake at an afternoon birthday party. It is as if I'm afraid that I'll eat the whole cake, or three cakes, or start eating cake every afternoon. It is as if a single piece of cake could make me gain back the thirty pounds that I lost twenty years ago. I am not free to do something that other people do without thinking. I am not free to vary from my routine.

It doesn't take a genius to see that my periods of feeling fat have more to do with how I feel about myself as a person than with how much I weigh. That doesn't make the problem any less real. I feel that imaginary weight on me. I look fatter in the mirror. My clothes seem to fit differently. I tighten up my stomach to make it feel smaller — I spent my whole youth tightening my stomach — and the rest of my body grows tense, becomes alien to me. In the summer, when I can't exercise as much because of the heat and my clothes are less flattering, I feel fat for three or four months and worry over every bite I eat.

In therapy once, I had a moment when I saw all I had made food into: a substitute for other things, especially for love and affection from my mother and other women. I didn't feel angry at myself, or at the women. I just felt sad. It all seemed futile, trying to use chocolate sundaes to fill a need for love. For that moment, food lost all its magic for me. It was just nourishment.

Sometimes I imagine a purely physical being, for whom food isn't tied up with other things, who eats what he wants when he wants to, who nourishes himself simply and well. Does such a person exist? I have no idea. Does such an animal exist? I think of my cat, once a scared and starving stray, who seems afraid of a time when she won't have food, when I won't feed her anymore. Not only does she go to her food bowl at every opportunity, she also wants me to fill it every time, even if there is already food in it. She apparently wants to keep establishing the connection between the food and me, to ensure that I will continue to feed her. It is enough if I just pick up the bowl and put it down again. My son and I call this "blessing her food."

At the end of the Japanese movie *Tampopo*, a prolonged, profound, and hilarious meditation on food, there is a scene, behind the credits, of a baby nursing at his mother's breast. It is a deeply moving scene, especially after all that has gone before. I saw the movie twice, and, both times, not a person in the theater moved while the credits were being shown. We sat and stared, fascinated. I realized then how beautiful a moment that was, how much I had always wanted to watch it, yet how many times I had politely averted my eyes from a nursing mother. I could have watched for hours: the baby's deep need as he sucked greedily, the way his whole body was involved in the feeding, the comfort he got from his mother's embrace, the way he opened his eyes now and then just to see her. *That's what we're trying to get back to*, I thought. *That's what all the fuss over food is really about.* And as I realized how much was going on in that scene, I understood that it wasn't strange that food was so complicated an issue for me. It wasn't strange that I had attached so much to it. What I wanted was the simplicity I was seeing on the screen. The problem wasn't that I had made things complicated. It was that I wanted them to be simple. I was wishing for something that would never be.

June 1988

Finding Out About Your Heart

Candace Perry

You find out just how bad a shape you're in with the first stress test. Not five minutes on a flat treadmill, and alarms start going off. You pay attention. The cardiologist says to take this medicine. You do what he says about the medicine, but not the cigarettes. Not yet.

You go back in two weeks and handle thirty minutes uphill. This time, there are no alarms. The doctor still wants to do that multisyllabic cardiac test. Same-day surgery, just a camera able to travel your arteries and take snapshots of your heart. When your daughter was young, you told her about little green men inside the radio who made the sound come out. You picture microscopic people, Japanese tourists with Minoltas, coursing through your bloodstream and photographing the damage.

The doc tells you this kind of chest pain is not uncommon for men your age. Men your age who play tennis twice a day, maybe not. But men your age who've smoked three packs a day for thirty years, sure. So it evens out. You're pissed at your college roommate who teased you into Marlboro Country. Like virginity lost, you can't go back.

You find out about the medicine you'll have to take if it's bad. You pretend to laugh when you ask the doc how a guy's supposed to give up sex and cigarettes in the same decade. He switches the talk to cholesterol levels, says let's not worry until we get the test results. Cross that bridge when we come to it. You know that if "we" have to cross that bridge, only you will be unable to fuck.

Not that sex is an everyday occurrence. Your wife complains to your daughter about it. Your daughter thinks that because she's a college senior everyone wants to hear her opinion. In her opinion, it is abnormal for a husband and wife to have no sexual relationship. In her opinion, you and her mother should be in marriage counseling. In her opinion, a man who says he plays tennis twice a day is really out cheating on his wife.

You do play twice a day. You don't need tennis for an alibi. There are other opportunities. The women you see are modern and don't require much. No cameras on the heart, just that flash of connection that happens sometimes. You phone one of the women, the married psychologist you think of as "nice." She lives enough area codes away that you can tell her everything. She is good at sympathy and easy chatter.

You talk about cities. New York, the city you love and hate, like family you know too well. The American cities that feel foreign: New Orleans, San Francisco, Seattle. (She doesn't agree about Seattle.) Then Canadian cities. How Toronto is like Chicago; Quebec like Paris. You say you'd like the two of you to go to Montreal together sometime. Sure, she says, sure, the same way your wife says sure when you tell her you'll take her on a cruise next year.

You turn the conversation back to your heart. It's nice how she's concerned. She says not to worry about the side effects of the medication; kissing and touching are enough. *For a woman*, you think. She understands how hard it must be to give up smoking. She asks if you know that nicotine is more addictive than heroin. She has never smoked, but tells you there are times even she craves a cigarette. Isn't that something.

You say you really wish you could quit. You tell her how your college roommate got you started. You say you have to quit; it's too hard on your heart. This is when you find out she's not so nice, when she says it doesn't really sound like you want to quit. You picture her sitting primly behind a big shrink desk, hair pinned into place. She says it sounds to her like you can't accept limitations. You tell her you have to hang up. She asks you to call her when you get the results. Sure, you say, sure.

You drive alone to the hospital. You don't allow your wife the waiting-room drama; you don't need it. A pretty nurse shaves your arm where the cardiac catheter will enter. She keeps calling you sir. When you say you bet she's a hell of a cook, she looks puzzled and says, Not particularly.

You remember when that line was a great starter. The stewardess in first class who lounged on your armrest and challenged, With a face and body like this, do I look like I need to cook? The women's-studies professor from Radcliffe who couldn't wait to have you over for stir-fry. Even the psychologist, offended by what she called a sexist come-on, warmed up when you confessed how well the line had worked with the stewardess. One of the first things you found out about women is that they like it when you tell them about women from your past. They mistake confession for intimacy.

You are waiting in Recovery. Your chest still bears little red rings left by the devices they hooked on you. You think of leeching cups, of how helpless men were in other times when their bodies began to fail. Your arm is bandaged and aches dully. A different nurse, older than you, asks if she can get you anything. You want to say, A cigarette and somebody to love, not necessarily in that order, but instead you reply, How about a new body? She laughs appreciatively. Don't we all, she says, when we start pushing fifty? You wonder whether she knows from looking at your face or your chart that you are forty-nine.

You are used to looking younger than your age. You were carded clear through your twenties; the other attorneys in the DA's office called you Babyface. Now, when you could use the youth, it suddenly leaves you. You can still work the arms and legs into shape, harden the biceps, calves, and thighs with enough tennis. But time is showing on your face, in the soft of your belly, in the failure of your heart.

The nurse tells you the doctor just called and is on his way. You make a joke about your pictures getting back fast from Quick Photo. She laughs. You wonder if the doctor told her anything.

You know all about death. You see it, you work it, you put it on trial. Other people's. But death isn't what's got you worried. It's consequences, limits. Up until this heart thing, you figured you would always be able to eat and drink and smoke and fuck as you pleased.

The door to Recovery swings open, and the doctor enters. He is younger than you and not practiced at keeping his expression neutral. After twenty-five years in the courtroom, you only have to look at the foreman to know a jury's mind. The doc's face tells you what he has found out about your heart.

You feel too sorry for yourself to hear him say, You're lucky to be alive. He is telling you what you must do and what you must stop doing to keep your heart going. You say, What's the point of giving up everything that makes life worth living just to keep alive? He says he's a cardiologist, not a psychologist.

Your psychologist friend always tells you to take care of yourself. She says you need to make it to old age because then you'll settle down and really love someone. You know she has herself in mind. The idea that you might finally connect — not necessarily with her, but with someone — comforted you. You were going to be different when you got old. The cardiologist is telling you that old is now.

You need Mrs. Brocato. She was your ninth-grade English teacher. You were tough in ninth grade, nobody to mess with. You wore a black imitation-leather jacket and combed your dark hair into a mean ducktail. Mrs. Brocato pulled you aside one day and said you were getting too old to pretend to be a hoodlum; soon enough you were going to become one unless you changed your ways. She said you could be somebody else; it wasn't too late. She told you exactly what you needed to do to change. That day you cut your hair, threw away the jacket. When you got into law school you called her with the news before you called your wife. You haven't thought of Mrs. Brocato in years; she died about the time you were studying for the bar exam.

You need somebody like her now. Somebody wise and understanding. Not your wife. Not your daughter, who thinks change is easy. Just because she stopped biting her fingernails, you were supposed to stop smoking. It takes a damn Ph.D. psychologist to figure out you don't want to quit a lifetime habit. Turns out the psychologist isn't so smart either. Here you are with an old man's heart and still not a clue how to settle down and love someone.

March 1991

LETTER TO MAXIM

Alison Luterman

The story of you is starting in me again. When I think of you, I see a road, a long gray stretch of lonely two-lane highway, a yellow stripe painted down its middle, a road in the middle of nowhere. You and I are standing by the side of the road hitchhiking, but no cars come. It is some obscure place in Canada: Thunder Bay. We are on our way cross-country, headed for the Canadian Rockies. And after that? Who knows.

Cold, overcast day, even though it's July. You wear a wool cap pulled down around your ears, a fisherman's sweater, bluejeans on your skinny legs. Under the cap, round, saucer-like brown eyes, uptilted nose, shiny brown hair. If it weren't for the astrologer's beard, you'd look like the little Dutch boy.

You're doing tricks to amuse me. Handstands. Now you juggle

potatoes and onions. You learned these tricks from a circus acrobat who taught a course at the university where you went for one semester. I say I could never learn to walk on my hands. You say of course I could. "There was a man of forty in my class, and he learned." It's never possible to argue with you.

No cars come, the clouds grow darker and lower. We haven't been caught in the rain at night — yet. We shoulder our backpacks and walk a mile back down the road to an all-night truck-stop diner. The big diesel rigs pull into the parking lot like dinosaurs at a watering trough.

We sit and order coffee, taking advantage of the free refills, tanking up on sugar and cream. We've been buying food at supermarkets and cooking it over makeshift campfires — a can of beans, an onion roasted on a peeled green stick.

This place smells deliciously of frying potatoes and meat. Men, alone or in twos or threes, sit quietly over coffee and food or wait blankly, cigarettes burning away between their fingers. Except for the waitress, I'm the only woman. I can't escape the eyes that assess me as I walk to the ladies' room carrying my backpack. I'm wearing jeans, a T-shirt, a sweater. My hair is wild and bright. I'm twenty-two years old. I have years to go until I'm safe.

In the ladies' room I take off my sweater and T-shirt and wash myself. My dark eyes in the mirror leap out of a sunburned face. I look like a little wild animal — a raccoon, or a ferret. Since the bathroom is empty, I strip off my pants, climb onto the sink counter, and wash my crotch. I have my period, and I'm tired of feeling sticky. I put on a limp pair of clean underwear from my pack, pull on my pants, walk back out. The men look at me again with their cold eyes. I think they think I'm a dirty hippie, a slut.

At the table you are drawing with a purple crayon all over the menu. You are making it into a letter to send to your younger brother, Paul, in France. He has never seen the world, doesn't even know it's out there. You didn't either, until you moved from your father's sheep farm in the provinces to Paris. Even then, you thought that was all there was, your working-class enclave where all night long you could hear Arab music blasting from the windows of Tunisian and Algerian seasonal workers, who endured with you the loneliness and snobbery of the most beautiful city in the world.

You didn't know about the world until you met Sylvie. She was nineteen. She had escaped her own family, out in the country where they could remember starving through several wars, to come to Paris and marry a Tunisian so he could get his papers. She lived in one room with him and worked in a home for old people. She liked old people; they were gentle. At home, she and Beshir had horrible fights. He raged against the racism in Paris, then against Parisians, then against the French in general. He hit her a couple of times, and she left.

I knew her, too; I loved her, too. Who wouldn't? She had soft brown hair and laughter like bells. She was hungry to see everything and know everything. She had befriended me, a gawky American college student thumbing my way through France with my backpack and purple socks. We slept together on gray beaches. She tried to organize me, scolded me about losing my washcloth, and I let her. She was the link that connected you and me.

"We're both dogs," you said to me during our first week in Canada, when we were still rapturously discovering all we had in common, "and Sylvie is a cat." That was true. She could sleep anywhere, all curled up, and wake with a stretch, looking and smelling fresh and perfectly content with herself. She liked eating little slivers of food — salade niçoise, the tip of an ice-cream cone. She was effortlessly tidy and self-contained and hated to be confined in any way. She swore she would never marry again. She said, "In friendship I'm as faithful as the moon, but not in love," and that was true, too. While you and I were trailing across Canada, arguing about who knew her better, whom she loved more, she had already pursued and left two American men, Boston Charles and Chicago John, and gone back to Paris for further adventures.

You look up from your menu-letter finally, and your eyes are soft and sad. But your letter is full of excitement, enthusiasm: we are having a wonderful time — "plein d'aventures"; he should come, he should do this, too, you lecture Paul. You're the older brother, advice giver.

I read it and think, *I'm not having a good time anymore.* I don't know when I stopped.

We drink coffee in silence. I look around the room speculatively. You are already hissing at me, "Try that one! He has an open face. For a Canadian." I hate this part; I hate that it's me who has to do it.

I have to pick a trucker to approach, preferably a man traveling alone, and ask him if he's heading west, if he's got room for two hitchhikers. I have to do this in such a low-key, unobtrusive way that the waitress won't kick us out for soliciting rides. I try to pick a man who is not too large, whom the two of us together might overpower if we had to, but almost all these truckers, even the short ones, seem to bulge with muscles and meat-eating bulk. You and I are scrawny from sleeping by roadsides and living on bread and cheese and coffee.

I go up to the one you pointed out and ask if he's heading west, if he'd like to take us. He shakes his head no and turns back to his plate while his companions look me up and down and my face burns. I go around the booths and try the other side of the restaurant. My voice is low, my manner polite and as reassuringly middle-class as I can make it. Will they somehow see beyond my filthy jeans and uncombed hair, see that I graduated *summa cum laude*, that I speak three languages, that I am a teacher? No, why should they? What does it matter? I'm stripped of those things out here, and it's good. I wanted to shed them, wanted to shed everything. What I hadn't counted on was that I'd still be left with the one thing I can never shed: being a woman, a young one. Their eyes reduce me to that — I reduce myself to that.

The third or fourth man I ask says he'll take us. Relieved, we follow him out to the parking lot with our packs. It's pelting an icy rain, and the sky is dark. The trucker's name is Doug. He's not too big; he's young, with dark hair and blue eyes, an Anglo. We're two or three days' travel past Montreal. The more Anglo the terrain gets, the more I take over the task of communicating with the outside world, since your English consists mostly of epithets and private plays on words.

Doug hoists our packs into the back of his truck. You scramble into the back seat of the big semi, and he offers me a hand as I clamber up. The smallest of warning bells goes off: is he just being helpful, or trying to touch me? Our safety depends on these warning bells. Without them I couldn't choose our rides so carefully or know when to cut them short. But I also have to override them constantly in order to keep traveling this way. I will sit up with Doug most of the night, assessing him, monitoring his every word and gesture as I listen to him talk about his wife or the girlfriend who just left him, constantly on the alert, constantly reassuring myself, while you whistle or sleep or write letters. It's my responsibility to worry about how safe we are with these strangers we depend on for rides, just as it's my job to pay for the rides by keeping them entertained, keeping them talking — everyone understands this. You climb into the little bed in back and immediately fall asleep while Doug tells me his life story.

It's a good story, though I can't remember it now. All the truckers' stories were good — and sad, too. I got to hear about a kind of life I never would have known otherwise, the quietness of hundreds of gray and blue miles unrolling in front of a high steering wheel, and all the time in the world for a man to remember the details of his child's face when he last walked through the door, or his girlfriend's back when she walked out.

At about two in the morning, my contact lenses are sticking to my eyes, and my chin is falling forward onto my chest. I tell Doug I'm going to climb in back and get some sleep, and he nods assent. You're waiting for me on the narrow mattress. One hard-muscled arm encircles my breasts, the other pulls my hips toward you. Softly, going with the motion of the truck, you push your penis against my backside. I can feel your stomach against me, your ribs, your hungry pelvis, your breath in my ear. I can tell the idea of doing it while we're in motion, two feet away from the driver, turns you on, but I don't like it. It seems rude, and Doug has turned out to be a nice guy after all. Also, I'm exhausted, so exhausted I manage to fall asleep squashed against you, but I'm woken an hour later when Doug pulls the truck over and climbs in back with us. I lie there, sandwiched between the two of you, Doug gently snoring from his all-night drive. When daylight finally comes, the rain has stopped and we slide out, handing our backpacks to each other in the bright, cold dawn.

You sling your pack over your shoulder and lead on, head down, thumb stuck out, not even bothering to look at me, or the road, or the few cars whizzing by. You're retreating more and more into yourself; your stories are becoming forced, didactic, the moral being always how one should really live life — not like the petite bourgeoisie, who stay protected within their locks and walls, but completely open, completely free. When did I become the example in your stories of how not to be? When did you start to use your knowledge of me against me?

I REMEMBER OUR first few weeks together, in the province of Quebec, picking strawberries in the chill June rains and the hot sun, moving along the rows on aching, green-stained knees. You got the runs from eating too much half-ripe fruit. The farmers who hired us watched us with suspicion; they didn't like you, and with good reason. The first day I made only fifteen dollars for hours of back-breaking work. You made twenty-three dollars, because you'd filled your baskets with green berries, with just a layer of red ones across the top. I think they knew no one could pick that much if they were doing it properly, but of course you chalked my misgivings up to my bourgeois mentality. The farmers were all exploiters, you said scornfully, waving your arm at the women with kerchiefs on their heads who were picking steadily and patiently.

We slept in an abandoned school bus, in the rain, on a mildewing mattress, our arms wrapped around each other. We slept in an old shack belonging to a farmer in the next town, a wooden structure with a tin roof that heated up like an oven in the middle of the day. I have a photo of myself in that shack, perched on a loft where we had wedged our sleeping bags. I'm wearing a halter top and rolling a cigarette of dried mint and tobacco, with a big grin on my face and my hair flying around like an electric halo.

I think those were happy times. It was new to me to feel so free. For six weeks I didn't call my parents, just sent them airy postcards from obscure small towns in Canada. I called Sylvie in France from pay phones, using the credit-card numbers of the Ku Klux Klan, the American Nazi Party, and the Moral Majority. We had long, luxurious, hundred-dollar transatlantic conversations, sometimes lasting up to two hours.

"Don't let him push you around," Sylvie advised me. "I worry about you. You are too . . ." She couldn't think of the right word. It might have been *boundaryless*, it might have been *naive*, it might have been *American*.

I called my friends and told them I was thinking of marrying you so you could get your papers. You had been refused a visa in the United States because you were a Communist. We'd almost gotten married in Montreal, but one had to establish a six-week residency first. I'd even sent away for my birth certificate. We thought maybe we'd do it when we reached Vancouver.

My friends all warned against it, especially John, a lawyer. "It's perjury," he said. "Immigration will separate you and ask you all kinds of slimy questions, like what was your first date and what's his mother's maiden name? You could get in big trouble."

I wasn't sure how I felt about marrying you. I never even felt like your girlfriend. At night we slept together, and during the day we hitchhiked or rode the buses, singing Jacques Brel songs off key, sharing chocolate bars and cigarettes. I learned about your childhood, the stone-cold silence of growing up among people with no language for feelings, only for hard work and poverty.

"A child must have someone to love," you told me. "And if there is no person around for a child to love, he will love an animal, and if there are no animals, he will find something. I loved my toolbox." An uncle taught you to lay tile; by the time you were seventeen you had quit school and moved to Paris, where you worked as a plumber, "up to my elbows in other people's shit." By the time you were twenty you had gotten enough money together to buy your own apartment. You could do anything, you boasted. Lay tile, fix plumbing, even build houses.

One should be able to do a little bit of anything. Not like me: I was brought up with books, only books. It was true, my education had been pretty useless, except for the languages — I couldn't even change a tire — but inside I knew I wouldn't be able to stand your incessant judgments for long. You talked on and on. You were passionate about children, about every aspect of their education and upbringing. A child should be exposed to the world, the whole world. A child should never be thwarted or told there is anything he cannot do. A child should be taught whatever he wants to learn, useful things, not what you get in school.

In Vancouver you insisted on swimming nude at a family beach. I sat fully clothed on the sand. The lifeguard came up to me and said you had to wear a suit. "I can't make him do anything," I said. You saw us talking and came leaping out of the salt spray, nude, elegant, coltlike, your penis flopping out of its dark bush, oblivious to all hostile stares. The lifeguard repeated his message. You pretended you didn't speak English and stared at me for assistance. I slid my underpants off beneath my skirt and gave them to you. You gleefully put them on — bikinis with red and blue sailboats — and went running and shouting back to the water. Another coup scored for life, against repression. It was all losing its charm for me.

In Vancouver my period was late, and I started to worry. You told me your girlfriend in Paris — not Sylvie — had gotten pregnant by you just to see if she could, and then she'd had an abortion. I never knew which of your stories to believe. One day, you put me on your shoulders and tried to demonstrate an acrobatic trick. I was unprepared to be thrown and landed heavily. A few hours later the bleeding started, intense and painful. I crawled into bed, moaning and rocking with pain, for two days.

I left you to save my life. I remember whispering that I hated you in a coffee shop in Vancouver, and you saying bitterly that you knew I did, that your mother had said the same thing to you. I cried that I had come to this point, telling somebody I hated him. My mother had said you could only hate Hitler.

After we decided to split up, something changed, and you became gentle again, almost the way you'd been in the beginning. We were staying at the home of a woman I'd met in France two years before, a Canadian named Alex. She'd become a lesbian in the interim and was angry and disappointed when I showed up on her doorstep with an obnoxious Frenchman. She took me aside and told me that even if I insisted on sleeping with men, she was sure I could do better if only I valued myself more. After a few days, she took off for her retreat up in the woods, where she was studying fish for a master's degree in biology. She was trying to prove that female fish are more cooperative and less competitive than male fish. We had the house in Vancouver to ourselves.

One night I lay tensely on my back in our hot, sweltering room. Our entire sex life had involved a lot of arguing, me trying to get you to do what I wanted, and you saying I shouldn't want those things. Your girlfriend in Paris had orgasms effortlessly. She could come if you just stroked her hair. American women were not really sensual. I was joyless, frigid, probably lesbian.

But this night, for a change, you were quiet. You lay on your side and with your fingertips gently stroked and touched my belly. I breathed deeply and let you do it. It was an effort for me to lie still, to just take it in. It was a gift you were giving me, your last, a goodbye gift to seal all we had been through together. You gave me an eternity of gentle touching, and when you were done you rolled over and went to sleep, all without a word. The next day you left to go pick fruit in the Okanogan Valley, and I never saw you again.

It was strange being without you. I was so used to your voice at my side. I had acquired strange habits, thinking in two languages, beginning an idea in English and completing it in French. Often when I talked, I groped for words or felt I should be translating. I had been translating everything constantly for you. Now there was a strange silence in me. I still had conversations in my head with you in French. I argued my case over and over, and then I argued your side. Neither of us ever won.

I no longer recognized myself in the mirror. I looked like a waif and felt transparent. You had cut off all my hair with fingernail scissors when it became tangled with burrs. I reeked with the bitterness of tobacco in my fingernails, under my arms, in my gums. I didn't flirt with the men around me. It was such a great relief not to have to defend myself anymore.

Two years later, I was with two of my best friends on the grounds of the DeCordova Museum in Lincoln, Massachusetts. It was a beautiful, clear September day, mild and sweet as a McCoon apple. We had just seen an outdoor concert and were taking a walk by the edge of a pond.

"Hang on, you guys. I have to pee," I said. I looked around for a spot where I wouldn't be seen and headed for a thicket of trees. It seemed like I had to pee every five minutes. My period was late, and my breasts were tender. I had scheduled a pregnancy test for later that week. I had met a man named Michael just a few weeks before, and we'd had sex on our first date, though I hadn't planned to. Still, after so many years of false pregnancy scares, I couldn't believe that I could really be pregnant.

When I came galloping back, Helen and Debbie were poking thoughtfully at the water with sticks. They looked at each other, and then Helen said, "We have to tell you something." Something in her face or voice made me sink down, and all three of us sat on the sand. Debbie said, "Maxim killed himself nine months ago. Almut told me. She heard it from Sylvie. Sylvie says she wanted to tell you but couldn't."

At home, I called Sylvie in Paris. She told me that for the last year and a half she hadn't seen much of you; you were hanging out with a group of friends she didn't like, doing drugs, and criticizing her when she tried to talk to you about it. Then, in January, in the coldest, darkest month of the year, you went up to a remote corner of your father's sheep farm, cut your wrists, and bled and froze to death. You left no note. It was a month before anyone found what was left of your body, after the animals had been at it. It was hard to make myself picture that, so that finally I could cry over you, the way you should be cried over.

Two days later, my pregnancy test came up positive. I lay stunned on the table while the doctor scraped and dug the baby out of me with long silver spoons. I welcomed the pain and the bleeding. At least something hurt the way it was supposed to.

There followed an incredible sadness of the body. I lay in bed and listened to music and let thoughts drift in and out. Sometimes I thought of you and the fierce drive with which you had approached life on the road — you didn't just approach it, you assaulted it. You wanted to seize it by the throat, squeeze it, make it give you something, wring it dry. I remembered you walking on your hands and juggling fruit, your harsh, insistent laugh, your rare moments of softness, sadness, or tenderness. I remembered your body, the long, elegant, narrow chest dented with muscles, the narrow waist, the small, tight butt. I remembered you laughing to cover your embarrassment the first time you undressed in front of me.

I remembered how desperate you were to escape, but you didn't know how, and I didn't know either. Always you were haunted by the specter of your mother, abandoned by her husband and worn by work to a nub of complaints. You dreamed of becoming a great singer-songwriter like Brel; you dreamed of being a world-famous juggler, juggling knives, juggling fire on the streets of every fabulous city in the world. You wanted a *life*, and you were so scared you wouldn't get one that you killed yourself.

Writing this a dozen years later, when your bones are dust and the girl who shared those adventures with you has also disappeared, I get out your photograph to remember how you looked. I'm shocked at how beautiful you were. Roguish, angelic, your face twinkles beneath the jaunty Irish fisherman's cap, and there's nothing to pity, nothing to rescue, nothing to hate. The future darkness is entirely unwritten there.

This makes me wonder about our faces now, how they shine with happiness or darken with doubt; I wonder which of these is real. Since you left, I've had times of such sweetness I could hardly take it all in, and times of overwhelming sadness, fear, and bitterness. I've worried about money, had headaches and birthday parties, noticed sunsets, cooked soup, and paid bills — all the ordinary things you rejected, all the details that have nailed me so securely to this life that I can fool myself daily into believing it will somehow never end.

Too late to talk to you now. I'm talking to myself in this moment. Things turned out better for me than I ever believed they would. I found someone who knows how to love, who has loved me steadfastly through many dark nights for seven years now. I made a decision lying there on the abortion table, right after I learned of your death. I decided that somehow I would change my life, change this senseless drifting from one desperate man to another. The face of my baby appeared to me then, and I said goodbye. I was trading a life for a life — mine. I have gone on. And things did get better. I made them so, not knowing how to do it, stumbling and cursing in the dark, falling down many times, getting up again. I did what you failed to do. I walked on.

August 1993

She Said, Can't We Just Be Friends?

R.T. Smith

After a week of sleeplessness
he dozed off at last
in the hammock and was
awakened by the sound of dead leaves
dancing. The next day
the collie bitch gave birth
to a sickly litter,
and he flung them one by one
against the stone wall,
always weeping. The dog
came round whining, her teats
thick with milk, and he said,
Yes, yes, that is how I feel.

September 1992

What It's Like

Dana Branscum

I t's like being in Miss Wheeler's class but wanting to play with the kids in Miss King's class. The thing is, they go to recess at 10:30 with the fifth-graders, while your class goes at 11:00 with the kindergartners. You face the hedges on your swing, throwing your head back so that the little brats on the playground dangle upside down into the huge, empty sky, and you pump your legs so that the world shifts and tilts alarmingly: you wish you were with the fifth-graders. On top of that, though Miss Wheeler is kind and soft-spoken, she has brown hair that hangs limply down to her shoulders, whereas Miss King has hair like white cotton candy piled high, and a coy smile that speaks of secrets as it creeps slowly across her pink, shellacked lips and into her

---

F i c t i o n

deep, green eyes lined in black.

The following year, in fifth grade, you get stuck in Miss Moore's class. Miss Butler, next door, wears orange powder on her cheeks, has feathered bangs, and hums to herself when she glides through the cafeteria with her tray. Miss Moore's pretty nice, but her hips seem to you inordinately wide. Your sister tells you that ladies are supposed to have big hips, that it's considered attractive, but you're not convinced. One Saturday morning, you study drawings of women in a party-jokes book that belongs to your parents. Because all these women have large, pointy breasts and large, rounded hips, you decide your sister was right. Years later, you will realize the drawings were sexist caricatures, women rearranged and distorted by male fantasy. You will understand that your teacher did, in truth, have inordinately wide hips; she was pretty nice, though.

In Miss Moore's class, you have your first boyfriend. You think he's interesting because he's a twin; he thinks you're wonderful because you always call him Mike, never Mark, because you are astute enough to notice the subtle curve in his cheeks that his gaunt-faced brother is missing. You let him carry your trombone and put his arm around your chair during filmstrips, but when he tries to kiss you on the sly, you get disgusted and announce loudly and indignantly that you are not eighteen years old.

In middle school, your science teacher is a man, scary and distant. He hides behind the lab's huge counters the way your father hides behind the newspaper. He talks to beakers and test tubes instead of to kids, and pretends he's important for making things bubble and smoke. But your speech teacher, Miss Gilbert, has freckles all over her face and on every part of her body that shows, and you think she smells like pepper. You love how she smells. Her aroma swirls around her like full, pleated skirts, in browns and blacks and warm maroons, rushing swift and soothing into your head, like a sip of your mother's wine. Sometimes you have to close your eyes: you love how Miss Gilbert smells.

At the end of the year, a girl in speech class — she has brown hair that ripples interminably down her back, like a waterfall in perpetual slow motion — writes in your autograph book that you're "cute as a button." Later, washing your hands without soap in the freezing-cold water in the girls' room, you stare at yourself through the graffiti on the mirror and smile engagingly. You decide to think about getting your bangs feathered. Though you have enough sense to hate the expression, you feel elated that this girl finds you cute as a button. You trust her judgment, because she is beautiful and because you'd love to bury your hands into her hair so that you could never find them to get them out again. You imagine her hair would feel nicer running through your fingers than sun-warmed sand or the bubbles in your bath or the smooth and cool adzuki beans in the white bins at the food co-op.

At eighteen, away at college, you take your first lover. You wonder why he treats you condescendingly when you're obviously more intelligent than he. You realize he's an asshole for "reading" pornography on your roommate's bed while you lie on yours writing poetry. You notice that you don't have large breasts or rounded hips; instead you have the long, straight awkwardness of a prepubescent boy at the peak of a growth spurt. You wear dusky plum powder on your cheeks, but it makes you feel false and trivial. In bed, you experience intense sensations you never imagined or discovered on your own. You are delighted by the small things: warm, moist breath inside your ear; hands tracing the lines of your face with the lightest touch, a touch the color of summer sky just brushed by clouds. But some of the acts involved in lovemaking — some basic acts quite inherent to sex — you find at best unpalatable. In the interest of being a good lover, you keep trying the unpleasant things. You tell yourself what your parents told you, that some tastes are acquired. You decide you don't like the expression "to make love" and start sprinkling your speech with the word *fuck*.

When you and your lover are separated for the summer, you feel a frustration that is embarrassingly disproportionate to the pleasure you're doing without. You cross your arms often, just to feel hands against your skin. You play with your hair tirelessly, not because it needs fixing, but in hopes of recalling that tingling in your scalp. Your awakened senses are demanding: they tug at your sleeve like a child, they whisper at and nudge you, they nag at you like a mother, like secret guilt, like work left undone.

At night, you lie awake in the dark, stretched diagonally across your endless mattress. You spread your hand across your belly. It's flat, smooth. One finger strays inside your navel — a silky piece of cloth in tight little knots. You circle it a few times. It has a nice texture. You think about your lover's muscled torso. Your hand slides up to your breasts for a perfunctory sweep from one to the other. You ascertain that, though they sink into your chest when you're on your back, though they seem to disappear altogether, they are still distinguishable from your chest and your ribs by their softness, an incomparable softness that wants not just to be stroked, but to be pushed against, to be pressed into, and your hand returns. One finger strays inside a nipple, which is concave, smooth, creamy as fudge that hasn't set, but then it stiffens, resists, pushes back against your touch. Your hand slides to the other breast, whose nipple buds instantly with contact. *Breasts are wonderful,* you think. You wonder if your lover minds that yours are small. You wonder what it would be like to have larger breasts, to touch them. Your hand slides back down, past your belly, into hairs — coarse hairs, soft hairs — that part as your fingers tunnel through. One finger strays into wetness, is pulled into wetness as by a current. And you begin to watch yourself from your lover's viewpoint; you imagine he is with you. A switch is thrown in your mind, and now another woman has replaced you, a woman from his magazines, and he's kissing her, licking her — she is cream against his tongue — he's kissing her breasts, her round, full breasts, meeting and parting and rolling against him.

Often, now, you borrow images from his magazines, all the while hating yourself. You are horrified, angry, confused. You hate those magazines and you hate those images that present themselves like an unbidden compulsion, like an uninvited guest you would ask to leave, but you can't find the right phrase, the right tone. You can't find a way to chase away those images that are becoming inextricably tied to your sexuality; they're as inevitable and unnerving as the tunnels you walk through in your dreams, looking for something every night, again and again. Sometimes you cry, wedged in that space between bed and wall, between pleasure and shame, and you rock back and forth between waking and sleeping.

Before the leaves turn colors and fall, you write your boyfriend a Dear John letter. You've realized that his contributions to your life were only sensual. Besides, you feel betrayed by him: all summer he's been sleeping with hundreds of women in your head, in your bed. Further, you're becoming confused by these scenes in which, as likely as not, he is missing altogether.

You read *The Women's Room* and become fascinated with Iso, the lesbian. Her character is so strong — so magnetic — that she seems real to you, but real with that intangible quality someone has when you've been introduced, and you see each other around, but you never once get the chance to sit down and talk. Further, you are puzzled by the women who make love as easily with other women as with men. The lesbian seems foreign to you, but the women who switch back and forth seem unreal.

You become acutely aware of your own feminine odor. You're startled to find yourself turned on by your own smells. You smile when you get your period and admire all the reds in your blood. When men flirt with you, you flirt back relentlessly but dart away before anything happens. This situation repeats itself with increasing predictability. It makes you feel incredibly attractive and incredibly empty.

You realize one day, all of a sudden, maybe while staring out the window to contemplate the gray snow along the curbs and sidewalks — how it rearranges itself like a sleeping cat, but never, never goes away — that you fantasize only about women, that you look only at women, that your psyche, your dreams, your favorite books all are filled with women. At a party, you pick up a man for a one-night stand and move in with him for two years. He's a nice man, an attentive lover, but his ideas seem to peter out just as they spark your interest, and, in bed, something is missing. You don't have an orgasm for an entire year, then you start thinking of women while he makes love to you. This makes sex much more pleasant. Now you're able to come, but you feel guilty as hell: you are a bad person. After he falls asleep, you cry softly and quietly, a silent siphoning of tears into pillow, drop by drop, or strung all together like streaming beads, a necklace unraveling without a sound into your dampening pillow.

You get your hair cut very short, so that it looks like a crew cut with a few strands trickling into your eyes and down your neck. You stare at yourself in the mirror and wonder if you've attained androgyny. You feel like a fourteen-year-old boy in drag. Your boyfriend thinks you look sophisticated — he says you remind him of Audrey Hepburn and starts calling you Tiffany, because he never gets anything straight. You start wearing only one earring, switching it back and forth from your left ear to your right, because you're not sure which side makes what statement. Sometimes you smile, thinking people might take you — *mis*take you — for a lesbian. One night when you're home alone getting ice for your third drink, you say the word out loud: "Lesbian." On your fourth drink, you practice walking with a boyish gait.

Sometimes your lover pulls you gently on top of him, and he is solid and warm beneath you, an island in the sun, and with his hands he envelops you from above, so that he is touching, claiming, soothing every inch of your skin — even places you can barely reach with your loofah in the shower. Every lonely, vulnerable pore he finds, and fills, and shields from the miles and miles of empty space the universe wants to cast around you. His palms, his thighs, his cheeks stroke your entire body, roughly and softly and artfully as a brush on canvas turning colors, turning wet, turning slippery with sweat neither his nor yours, a rich, scented oil that gathers on your flesh, on his flesh — wherever they connect — that gathers pure and gentle as morning dew, water on leaf, skin on skin, beads of moisture on glass that your finger is drawn to trace through. And then you feel whole, sexy, human; his fingers inside you seem real — they connect you with something universal, something ancient, something inside you deeper still. And you're carried by the humanness you share with your lover on this bed, this river, this lake, until he looks at you too intensely, with too much meaning; you realize that his eyes are looking into your eyes, not at them, that they are seeking in the green and black of your iris and pupil not only eyes, but your soul. You close your eyes — close them tightly — and keep them closed until morning, when you awaken dry and calm, separate and alone.

When you finish school and are offered a job in another city, another state, you go, relieved to be done with your lover. You tell yourself your next will be a woman. Maybe you're bisexual; maybe you should try sleeping with women, just to see. You have a few brief affairs with men, because they approach you and women don't. You realize that most men annoy you, and those who warm your heart don't wet your panties, unless you're stoned or drunk. When you see lesbians, you feel very uncomfortable: they are real lesbians and you are a fraud, a confused hetero-sexual who plays with women in her dreams, not in her bed. The prospect is laughable, inconceivable, insane. Besides, you wear eyeliner and carry a purse, and you have a cat named Dick. You get stoned and drunk a lot.

You get pregnant. After several weeks of anguished delibera-tion, you have an abortion. In the recovery room, you look at all the women while you eat your Lorna Doone shortbread cookies, and you wonder about the men these women fuck. You wonder if your aborted fetus was a girl. You wonder, getting up, if you'll make it to the john before you puke. Evenings, you lie on the floor between speakers, watching your ashtray fill, listening to Kate Bush. Her haunting voice draws you in like a lover's arms, pulls you into all the pain in all the world, validates the tears you cannot cry.

You see lesbians everywhere: in line at the bank, waiting at the crosswalks, squeezing organically grown peaches at the natural-foods grocery. This town is crawling with lesbians, and you know how to spot them. They're the ones without makeup, with "don't fuck with me" set on their faces, fingers without nails, unbound breasts, unshaven legs. They make eye contact with you as you pass, long after you meant to look away; they catch your gaze on them and hold it up to you, confronting you with the look you meant to hide. You turn your face, embarrassed. Or they turn away first, and you're disappointed, dismayed. But you're just a straight woman staring at lesbians, and they, of course, can tell that you're straight.

You see lesbians in couples, in groups, ignoring you. Their lips move, forming words you cannot hear and cannot fathom: their world is not your world. They call each other by name, talk, argue, wrestle; they love each other, meet each other for coffee, sleep in each other's beds, while you sleep alone with a recurring dream. They share ideologies, jokes, a vocabulary — all a mystery to you. You are like a child watching adults, bewildered by the pronounced differences between you, when you know that, fundamentally, you're alike. It's just a matter of learning codes, applying formulas, turning the lock to the right numbers, to the right, to the left, to the right again. You have a long way to go.

In the meantime, women keep talking to you about their men. And inside your head, the same tired questions are churning around like worn-out laundry. *Am I bisexual?* You don't know anyone who is. You're convinced bisexuals don't exist. *Am I gay?* You feel so alone with the idea — you're not like those women you see — that you're sure the answer must be no. And that *no* disappoints you so much that you're sure you must be, after all. But people you know think of you as straight and treat you accordingly. You don't know what to think.

You decide to give yourself a push in the direction of lesbianism by shoving your makeup in the trash. Along with the brushes and blushes and pencils and creams, you throw out your razor for good measure. You'll no longer look like a painted caricature of yourself, making a thousand readjustments on your face only to present it to the world with a thousand apologies still. Maybe you'll look like a lesbian now. Maybe as your pores begin to breathe, your thoughts will take on new life, your sexual self-concept will take deeper root.

You feel great. Without makeup, your face is clean and touchable. You rub your cheeks to gauge their unadulterated smoothness. When you rub your eyes, nothing smears. You never have to check your teeth for lipstick or wipe smudged mascara from under your eyes. You are free, natural, fresh — and unbelievably ugly.

You lean over the sink, steadying yourself against the cold white porcelain, nose to nose with the mirror. You've never seen anyone more plain or featureless. You marvel that the mirror can detect anything to reflect. Your nose is the only relief on the plane of your face; it sticks out pointedly, pointlessly, absurd as a beak, false as a carrot in a blob of snow. You have no eyes, only round, flat circles that blink at you as if dazed, as if dulled by all you lack. Your cheeks are voids, spaces defined only by their boundaries — a nose, temples, a jaw line — colorless flags demarking perimeters for a blank. You spend hours marveling at this version of yourself you've met only briefly on repeated occasions, at showers and face-washings. Now this is who you are; you must find your definition in this blank; you must present this nothing to others, who will never remember you, never pick you out of a crowd.

At work, a woman named Laura tortures you daily, making you laugh until you're weak and breathless. She stands straight-backed with the poise of a dancer while you are contorted with laughter, out of control. She is beautiful like a model, with an easy elegance that at first you found intimidating. Now the camaraderie between you is comfortable and warm, an obvious mutual pleasure. Still, unnamed barriers keep a distance between you. She seems averse to any degree of self-disclosure, and you are reticent by nature. You suspect growing closer will take time.

One day, Laura is visibly upset, showing an uncharacteristic vulnerability. You invite her home for wine and talk. You're relieved when she accepts easily, as if this were the natural next step in your friendship.

In your kitchen, Laura's tears choke off her words. You wait quietly, offering tissues she takes but doesn't use. You watch, fascinated, as the tears pour, drenching her face with melting blues and mauves and grays and blacks, until she looks like those grade-school watercolors that are always too wet, so that the paper begins to ripple and bloat and the paints, once separate and beautiful, begin to bleed and streak and mesh together. When her sobbing is stilled and her crying is spent, she dries her face, cursing and laughing at the colors smeared on her kleenex. She tells you she wishes she, too, were pretty enough to go without makeup. She says it so sincerely, you have to smile.

Calmed, Laura relates her story, her current pain. And in the story, though not at its center, is someone she calls "my lover," which she then shortens to "she," and finally names as Mary. "She's really wonderful," Laura adds as an aside. "You'll have to meet her." She continues unselfconsciously, as if every woman might have a lover named Mary, as likely as a cat named Fluffy or a dog named Fido. You compose your face and try to follow what she's saying because, after all, the point of this talk is Laura's sadness. But inside your stomach, nonetheless, a celebration is in progress, and all you can think is *She's a dyke, she's a dyke, she's a dyke,* and *If* she *can be one, so can I.* Miraculously, you manage to sit still and tune in to most of what she says.

Leaving, Laura hugs you and thanks you for listening and invites you to have dinner Friday night with her and Mary. You accept, outwardly pleased, inwardly elated. As an afterthought, when she's already out the door, she turns back to say that she somehow knew you'd be able to handle her being gay. You tell her that, oh, you have no problem with lesbianism.

January 1988

TRYING TO QUIT

Eleanor Glaze

My face and throat are swollen. This is the third sinus infection I've had this year. I ache, go to bed with fever and shivers. I take antibiotics, which make my face painfully sensitive. My ears feel like they're about to explode. My teeth hurt.

We seek pleasure, strive to avoid pain. Cigarettes give me more pain than pleasure, yet still I cling to them.

I smoke four or five before I get out of bed — the fifth is delicious with strong French coffee. I smoke cooking, planting, gardening, pulling crab grass. I smoke dancing, skating, riding. I smoke writing, reading, painting. I smoke in the bathtub. I smoke while making love. A cigarette is the last thing I reach for at night after I turn out the lamp, the first thing I reach for before

I'm fully awake. Show me someone more ridiculous than a jogger smoking. I can do five miles on the track, but only with cigarettes. Show me someone more dextrous and adroit than a swimmer on her back, floating, sucking on a cigarette, like a submarine with a periscope. If I am conscious, I am smoking.

Alden, my husband, has no intention of quitting. He gets bored with all my agonizing. Either worry or smoke, he says — not both. Worry will give you cancer, he says, quicker than smoking. He's been smoking years longer than I have, but not as intensely. Alden does not have to get up in the middle of a good movie for a cigarette. It's harder to quit smoking than to quit drinking, he says. For a Welsh-Irish poet who used to drink like Dylan Thomas, that is quite a statement, but Alden is incapable of any sort of greed, even spiritual: deliverance from one hangup per lifetime is enough for him. He sits serenely, smoking like a saint.

My addiction may have begun in the womb. Doubtless the air was thick-hazed at my conception. From the time I was born until my parents died, I watched them smoke. Both grandmothers smoked. My maternal grandmother's name (honest to God) was Gertrude Smoke. But no, I can't blame them. Born Caesarean, I sucked up blood during delivery. They had to swing me by the heels, whack and smack me, inject adrenalin into my heart to persuade me to breathe.

They should have offered me a cigarette.

All idols smoked in my youth. Remember the jaunty tilt of FDR's cigarette holder? Churchill's cigar? Remember Humphrey Bogart lighting up Lauren Bacall? It was romantic, intimate. They gazed at each other with smoldering eyes. They smoked, but I didn't. Not yet. I survived the death of Rowan (my first husband), poverty, life as a single working mother, all without a cigarette.

I got hooked at a writers' conference up north. A sophisticated Yankee editor of a famous magazine said she would seriously consider two stories if I could give her neater copies before she left the next morning. I bought my first pack of cigarettes, worked all day and most of the night. Got the stories to her just as she was leaving.

She kept them three years. Every six months I wrote her, politely requesting a decision. No answer. When my first published work appeared elsewhere, I wrote that if she did not return the stories I would charge her back rent. Home came the battered envelope without so much as a note, the manuscripts scribbled throughout with insults in red ink. I have no idea why she treated me so shamefully.

Meanwhile, back in Memphis, my home, I thought I could smoke now and then without getting hooked. Within a month, I was chain-smoking.

Five years passed. Five good smoking years with no physical or emotional complications. Then my father died of lung cancer.

THE FIRST TIME I tried to quit I went to a kind of Alcoholics Anonymous for smokers. (I'd been to AA meetings with Rowan's father, and with Alden. Compared to smokers, alcoholics are as tranquil as Zen meditators.) Never have I seen so many on the verge of hysteria: chewing gum, sucking candy, chewing their fingernails to bloody nubs, clinging to the backs of chairs, their eyes glazed, fixed in suffering. Some jerked and twitched.

We were each assigned a "smoking buddy," someone to call. My buddy never called. I never called her. Alcoholics talk for hours. Still, I did everything that was prescribed: walked miles, drank gallons of water, paced like a caged tiger. If only I had quit then. No one told me it could get worse.

At the last meeting a man confessed, "I haven't smoked in a year, and I still want a cigarette more than anything on earth."

WHEN ADVISED BY our leader not to eat spicy food, an old man leapt to his feet, tears on his face, hands quavering, fingers spread like the claws of a landing eagle. "You said we could have steak!" he screamed. Who knows what he meant? We were all about to go berserk. Our leader tried to soothe him. Someone asked our leader how long he had smoked.

"Never," he said. "I never smoked."

For one hideous moment I thought we were going to attack en masse, tear him to shreds. Hisses, boos, moans. A third of the smokers got up and walked out. I made it to the car before I reached for a cigarette.

In every other way, I am perfectly respectable. I love to cook. We eat healthfully and exotically. If promised immortality, maybe I could quit.

That I should persist in causing myself chronic pain confounds me. There's no way to predict one's addictions. Once, Alden and I smoked pot together. How silly, sucking on that little weed. I wanted a real cigarette. For twenty years I was given morphine for migraines. After dawn-to-midnight agony, morphine was bliss. But I never got hooked. Morphine is for agony, nothing less.

The second time I tried to quit, I added prayer to all the prescribed strategies. I do not believe in prayer, but by then I was willing to try anything.

The third time, I tried bourbon. (This was after Alden had quit drinking.) Normally I have a drink only while cooking — a dash of this, a zest of that, a spur-of-the-moment sauce with flair. That week I drank nearly a gallon of bourbon. I drank until I went to sleep. I woke, did a few things, drank, went to sleep.

I should have spared myself; I had scheduled a party for a friend that weekend. But the passion to quit smoking can strike at any time; it does not heed reason.

At the party, my best friend was trying to quit drinking; she was smoking like a lunatic. I was trying to quit smoking; I drank like I meant it. Around midnight, I said, "Drinking is so damned boring!" She promptly ground out her cigarette and grabbed a bottle. I poured my drink down the sink and grabbed a cigarette.

"This is it," she said, gulping.

"This is it," I said, inhaling deeply.

We stood facing each other, tears in our eyes.

You can't shame smokers by telling them their hair and clothing reek (smokers can't smell) or by reminding them that they burn holes in sofas and bedspreads. Those THANK YOU FOR NOT SMOKING signs only infuriate a smoker. I carry my own ashtray in my purse. Segregation in restaurants or airplanes does not bother me. If I could quit, I would still prefer to sit with the sinners.

To those health fanatics who fan the air, give us dirty looks, and would like to beat up on us, incarcerate us — with their self-righteous slogans and sermons about what we are doing to our bodies and *their* air — I would like to say, "We know all that!" There's a Zen saying that a fish swimming free has no concept of what is happening to a hooked fish, thrashing wildly, fighting for its life. We need help. A cure. An exorcism. For many, I fear there is none. Smoking to us is as natural as breathing. The next time you get the urge to lecture a smoker, try holding your breath for five minutes.

SLAUGHTERED WHALES. DOLPHINS drowned for tuna. Elephants murdered for ivory. Wolves gunned down from helicopters. Wild mustangs ground up for dog food. Baby seals bludgeoned to death in the snow. Foxes and minks with their hides ripped off. As a smoker, I identify with all of the above. All endangered species.

Meanwhile, let me confess — how I do love them. I think I would admit to any crime for a cigarette, no matter how atrocious. When trapped on the telephone with someone who won't hush, a cigarette. When late for an appointment, snarled in traffic, a cigarette. Waiting in long lines, in every caged frustration of so-called civilized life, I want a cigarette. Waking from nightmares, I reach for a cigarette. Worse still, with every peak of happiness or excitement — I want a cigarette.

Edna St. Vincent Millay once wrote of her body's ashes, "I'll be a bitter berry in your brewing yet." Oh, but when I am cremated, let me rise in the air as smoke; let someone enjoy the floating wisps of me.

Once, I had a funeral for a cigarette. Ceremoniously laid it in a clean black ashtray. Spoke to it. *It's you or me. It's down to that.* Got a pearl-tipped hatpin that had belonged to my Grandmother Smoke. Carefully stuck the hatpin through the center of the cigarette. Dead. Dead. Dead. Wrapped the cigarette in pink tissue. Then I put it in a little coffin. Not just any old matchbox, as I might use for a dead turtle or canary or mouse; no, it was an ornate box of turquoise, hand-painted with peach roses. Probably an antique worth money. Good. Part of my penance. In the backyard I buried it under a lavender rosebush between Ashley and Harriet, two deceased favorite cats. A stone for a marker. Out of every back window every day, I would remind myself that It was dead. Dumb with grief, I stumbled into the house. Threw all my cigarettes in the garbage. I had just buried my best friend. Cried until my blouse was wet. Cried and cried . . . until I was smoking one of Alden's cigarettes. Then I cried and laughed and smoked and cried.

YOU CAN'T THREATEN a smoker. You can't scare a smoker.

My father died of lung cancer at age fifty-two. It was a year-long nightmare. He suffered horribly. He was a builder. He never simply touched wood — he caressed it. He was French, a dreamer, easily excited. Between us, there was an intuitive communication deeper than words. He fought death every step of the way. Until the last few weeks, he was still making plans for all he wanted to build.

He was also an incurable, compulsive, and consummate gambler.

I haven't thought of his whistle in years. Never did he enter the house without that optimistic whistle: two notes, one high, one low. Last night I dreamed of him. Alden said I whistled in my sleep. Awake, I can't whistle.

Each time I've tried to quit, I've failed. Yet each time I've become more resourceful in trying to outwit the addiction. This time, not one word to a living soul, not even Alden.

We usually wake early, 3 or 4 A.M. Alden brings my coffee to me in bed. I so love that early silence, darkness, two cats and the dog cuddled near me on the bed; smoking, drinking coffee, planning work, pondering dreams. *Perhaps it's that inspiring red glow in the dark*, I thought, *that I'm hooked on*. So for ten days I prepared: I drank tea, lit a candle, and stared at the flame.

When the time to quit arrived, I started by counting minutes and progressed to an hour.

That was such an accomplishment I thought I deserved a whole gluttonous pack, but I primly limited myself to two. Through sheer blind fanaticism I made it to three hours. I drove to the mall (the weather was freezing) and walked inside. Four hours without a cigarette! A landmark!

I went to bed without a cigarette. It felt like some sort of adultery. The next morning, I had to have one. Instead, I got up, wrapped my wrists — both have been broken skating — and drove in the dark to the park to skate. The sun came up. Breathing, breathing. Clear, cold, clean, raw morning air. Exuberant, manic in triumph, I fell down. Rocking back and forth, both wrists hurting like living hell. A man in a truck stopped.

"Lady, are you OK?"

"Go on, go on, leave me alone. I'm trying not to smoke."

"Well, I think I'd rather smoke than have a bunch of broken bones."

"You don't know anything!"

He left. I got to my feet and skated.

I came home and gleefully mutilated some cigarettes. I cut them open with manicure scissors and watched their grainy, weedy, grassy, little brown guts spill out. *Ha! You're only plants, not demons.* I experienced an hour, then two, without acute craving. Reaching the peak of Mount Everest could not have so thrilled me. Freedom in sight!

Then it hit: an intense craving, a physical sensation deep inside my body, thrashing frantically. I thought I had been invaded by some alien entity. I thought I was going crazy. It scared me. It lasted maybe two minutes. I withstood it. Gradually, it subsided. An icy chill went through me. But that, I thought, was my last battle. I thought I had won.

A complete day, O Lord, O Lord, twenty-four hours without a cigarette. Listen, not an easy day either. Alden and I got into a terrible fight. Terrible. Alden in a rage is not someone any sane person would want to mess with. But I was not sane. I grabbed my longest, most vicious kitchen knife, pointed it at his throat. "One more word, Alden, and I'll kill you."

"You don't have to warn me," he said. "I know you French.

Gentlest people in the world, the French, up till the moment they cut your throat."

Then he hushed.

Sunday morning. Before I was fully awake, that thrashing again, spirit, soul, whatever, trying to tear out of my body. I put both hands to my center, held it down. The only thing on earth that would pacify it, make it stay in there and be quiet, was a cigarette. I turned on my stomach, held desperately to the bed. "If you get out," I whispered to whatever it was struggling deep inside me, "there's nowhere to go. You'll only get trapped in the bedsprings."

It got quiet.

I got up, dressed, got in the car, and drove without knowing where. I came to houses I had not seen since childhood. I found an old school, the low stone wall where once, when I was seven, a boy kissed me. I found the house where he grew up. That ugly little shotgun shack still standing. Amazing. The swing on the front porch where we used to sit. I drove to St. Mary's, where we danced in the basement when I was fourteen. I got the guard to unlock it, let me go in, look around. It was now a day nursery. When I got back in the car, stinging tears blurred my vision. It was bitterly cold.

Alone the next day, I caught myself drifting into thoughts of suicide, quietly wondering the best way to go about it. I wished I knew a kindly vet who would give me a shot, put me out of my misery. I say misery, but it was the serene, spaced-out simplicity of not wanting to live through that day. It was the worst fear I've ever experienced. Something collapsed. Something awoke.

I reached for a cigarette.

I'd never gotten that far before, far enough to realize that I smoke in order to live. And smoking will surely kill me. If I'm dead, I can't smoke.

I'm down now to the lowest in tar and nicotine, and still in physical distress. Cigarette manufacturers, Alden says, have to put something in cigarettes to replace the tar and nicotine, and they don't have to tell the government what it is. For all I know, it could be arsenic.

But that first cigarette — after nearly two days without one — I lit it, and every doomed cell in my body relaxed, expanded, glowed. Every doomed cell sang heavenly hosannas. I passed out from sheer ecstasy. Fell over on the sofa and slept for hours, a dreamless, depthless bliss.

So I smoke and shrug and suffer and sometimes feel wonderful and try to go on about my business.

And I say to myself, *Well, she burned to death. Moment by moment. One by one.*

February 1991

On Being Unable to Breathe

Stephen T. Butterfield

Something was drastically wrong with my lungs: every night they made sounds like a basketful of squealing kittens. I was always coughing, had pains under the sternum, and could not push a car or even run up a flight of stairs without gasping like an old melodeon full of holes. This condition came on slowly; no single daily or weekly change was ever big enough to scare me out of my habits. For three years after noticing these symptoms, I continued smoking pot.

It requires effort and a certain amount of ingenuity to practice ignorance and denial. The symptoms register clearly enough; we are not that stupid. But they are explained away, or the attention distracted from them. *It must be a chest cold. One more joint won't hurt. When I stop smoking, the cough will disap-*

pear. I can stop anytime. Tomorrow.

I decided to look into the problem after a long period of meditation practice, at a time when I was looking into my whole life: marriage, livelihood, everything. I gazed at the chest x-ray as though it belonged to someone else: thick white clouds in the center, where I felt the pain, and swirls, like mare's-tails, reaching deep down into the bottom lobes. Air capacity for a man my age and size should be 5.5 liters; I had 2.8. The next step would be a biopsy. Meanwhile I read up on the possibilities, ruling out the ones that killed you in less than four years; I had already survived those.

According to the medical profession, the cause of sarcoidosis is unknown; my doctor is not convinced that it is a direct result of inhaling smoke. He thinks it is hereditary, triggered by environmental insult of some kind. The word *karma* is not part of his vocabulary. I find it a particularly evocative word, a doorway into an increasingly subtle understanding of how the process patterns that we call "reality" fit together.

Lung disease runs in my family: asthma, emphysema, and bronchitis are common ailments. When I was a baby, I almost died of pneumonia, complicated by spinal meningitis; it was 1942, and the doctors said they had a new, untested drug they would like to try as a last resort — they called it penicillin. My grandparents prayed for me round the clock. At least one of those methods must have worked. Pneumonia came back twice in my twenties. Of course I should not have smoked anything, but I did. Karma is the influence of a fact that cannot be abolished, no matter how intense the regret.

Sarcoidosis is a progressive inflammatory swelling of the alveolar membranes, gradually scarring the tissue so that it no longer exchanges gas. Oxygen will not pass into the body through a scar. The only method of treatment known to Western science is prednisone, a corticosteroid drug which suppresses the inflammation; the drug does nothing, however, about the cause, which is still not understood. The side effects of prednisone are, alas, unpleasant: leg pains, weight gain, craving for sugar, dizziness, cataracts, liver spots, adrenal imbalances, weakening of the immune system, and worse, depending upon the dose. But if it is a question of being able to breathe, you take the medicine. It buys time.

How much time? More than if you had cancer, typically. I could wheeze out a sigh of partial relief and say, "I'm glad it's not lung cancer." Approximately one-third of the cases get better, one-third stay the same, and one-third get worse. Even with treatment, it can get worse. I have lived with it for about seven years. Right now it is in remission. Maybe I will have a "normal" lifespan, whatever that is. When I watch my father stack and restack old newspapers for hours, put cat food into a dish, then forget he is feeding the cat and sit down at the table with the dish in front of him, a "normal" lifespan does not seem so desirable a goal.

Being a Buddhist, I accept that nothing lasts, and that impermanence, suffering, and absence of solid reality are the three marks of existence. Saying this is one thing; living it is another. The actual presence of a chronic, disabling, possibly life-threatening disease is a relentless and vivid reminder of death. It wonderfully accelerates your spiritual journey.

We would like to avoid that kind of acceleration. Armies of joggers and physical-fitness buffs are out there right now, trying to strengthen their cardiovascular systems and increase their lung capacities to ward off the message delivered by the Buddha that day in the Deer Park of Benares twenty-five hundred years ago. Plenty of my readers could probably give me good advice on the diet and wholistic treatments I should try in order to cure myself and prolong my life. I would listen to this kind of advice, but what interests me the most is whether I can make use of the disease. Magazines are full of articles about this or that public figure who carried on a "battle for life" against cancer, AIDS, emphysema, kidney failure, or some other agent of transformation. The articles usually imply that these people are models of quiet heroism in the never-ending struggle not to go gentle into that good night. Illness and death are assumed to be very bad news, perhaps a punishment of some kind, boogeymen in the dark closet of deep, dark fears.

I would rather not make a knee-jerk reflex toward "battle."

There is a message in my body; do I have to go to war about it? Paraphrasing the words of Dylan Thomas, do I have to burn and rave and rage? Can I make another kind of response? What is this message all about; where does it come from; what does it say? I don't mean in a purely medical or scientific sense, although medicine and science are not to be disregarded. Disease is experienced, and perhaps originates, primarily in the mind. What is it doing *there*?

Let us begin by examining the effects, without pity or hope. Being unable to breathe introduces challenging modifications to your daily schedule.

In one of their movies, Sally Field wraps her arms around Burt Reynolds's neck and her legs around his hips; he carries her that way into the bedroom. It is a sweet and tender scene. To be able to walk around screwing your lover against walls and doors, or to bounce her on the bed while imagining that you are very powerful, taking charge — to play that scene requires air. Oxygen must be taken in through the alveoli and pumped to the muscles. Otherwise Burt and Sally (especially Burt, since he is doing most of the work) would gradually deflate, crumple, and fold up on the floor, and their faces might as well be painted on the surfaces of withering balloons.

In terms of life passages, I am entering the withered-balloon stage. Corticosteroids and sarcoidosis are not all that bad, even though they tend to make sex less interesting than seaweed. Now, if I am to imagine myself as powerful and taking charge, the power must come from somewhere other than physical strength.

Singing is punctuated with interesting silences at times, depending on the duration of the line to be sung. If the line is much over ten syllables, my voice simply disappears. "Whoops," the lungs wheeze. "Sorry, we are having technical difficulties." Then I might like to go cross-country skiing with my singing friends, but I don't; for them, it would be like having to take a giant snail out for a walk.

Household projects are scaled down drastically. I would like to dig flower beds and build an addition onto the living room. "Too bad," says the body. "Be grateful you can still walk across the yard." Ten years ago I swore that I would never own a riding lawn mower. Now I bless the person who invented it. If I had to make my living as a construction worker, I would be finished. I am in the same health class as victims of black lung.

When somebody's car gets stuck in the snow, there is not much I can do. Pushing or shoveling are out of the question. I become sharply mindful during bad weather, driving fully present, right on the dot all the time. I like that; going over the mountain to the store in January is almost an adventure.

But not being able to climb the mountain — ah, that really hurts. Years ago I could, and did, hike up almost all the peaks in New Hampshire and Vermont. I loved mountain air. I loved swishing through leaves, coming around a bend in the trail and catching sight of a distant peak — the lines and folds, the twist of the path winding up the ridge, the wooded plateaus and cliffs. It all looks like a model of a mountain on a big topographical relief map. Then, as you get close to the slopes, the features become huge, the folds turn into steep ravines full of complex detail, the nubs enlarge into crags high overhead, partially screened by foliage, waiting to be discovered, at which time they will dissolve again into gestalts of moss, bushes, talus, fissure, and ledge. The mountain always eludes comprehension; it is pure magic. From the peak in the evening, the whole valley sings and shimmers with the receding calls of birds, which keep winking on and off over the forest like thousands of bubbles manifested as sound. The greater the distance, the fainter the sound, and the vaster the space it fills.

In my twenties I dreamed of climbing someday in the Himalayas; I collected pictures of mountains and read about the great expeditions to the sky. But now, even as low as Katmandu, I would need an oxygen tank. In the Alps, taking the cog railway from Grindenwald to Jungfrau, I had to get off at nine thousand feet to keep from passing out.

Walking anywhere with friends, especially uphill, is an occasion for silence; I cannot walk, talk, and breathe at the same time. Every gram of oxygen must be used for locomotion; there is nothing left over. Superfluity must go. This becomes an amazing metaphor: in my life, in my mind, what is superfluous? Anger that freezes into resentment; jealousy; greed; gossip; ego-clinging; pretense; embarrassment; any form of fixation; running after pleasure; the discursive thought that maintains the story line of *me*. These things are very costly, in terms of the life energy that it takes to keep them going. They are what conversation is mostly about. I cannot take in enough oxygen to support them anymore, except by holding completely still and doing nothing else. When the oxygen is diminished below a certain point, you must choose, absolutely, between feeding all your mental bloodsuckers and taking care of your true business. You cannot afford to keep them around as pets. What an opening, what a discovery, follows from that simple realization: Could I *ever* afford it? Can anyone? What made me think that I could not let go of this expensive baggage before now?

It is more difficult to be speedy, about anything, when your supply of oxygen is exhausted simply by making a bed. When I had the energy for speed, I wasn't mindful of time. If I was late for an appointment, I might dash out the door, spin the wheels, and get stuck in the driveway. Then I'd have to dash in for a shovel and a bag of sand. If I left the book I needed upstairs, I would have to make an extra trip for it. After a whole day of this kind of waste, no wonder we feel drained and just want to lie down and complain about our bad day.

Now imagine that you are unable to dash. One trip up the stairs and you have to sit for a few minutes to pay the oxygen debt. Rushing anywhere, for any reason, leaves you gasping like a fish on a dock. You have to give yourself space and time to recover from the most trivial wasted effort. To get angry about having to move so slow just cranks up more waste. What are you going to do with the anger — throw something? That will make you gasp and pant all the more.

But giving yourself space and time is also giving yourself kindness: no pressure, no speed. Do I really need that trip up the stairs? When I am there, what am I forgetting; what can I take down with me so that I won't have to come back in two minutes?

If my car is stuck on the ice, how can I handle it to avoid physical expenditure? Take it easy. Look around.

Sitting calmly and looking around, I notice the lavender ripples of light on the snow in a field and the stubble of dead weeds coming up through the crust. The snow makes a separate system of rings around each stalk; no two systems are alike, but they all show the direction and patterns of the winds. How important is it that I go anywhere? The light on a group of stones — that is a masterpiece of art, made in the roadside ditch by nobody at all.

Once, when my car was stuck in a snowbank, the rear wheels buried up to the bumper, I had to measure each and every shovelful of snow, like an ant moving a mountain one grain at a time. It was a pleasant surprise to find that, in such a situation, even with a serious lung disease, I was not entirely helpless. There was plenty of time to see the tracks of the car, how they slid into the brook, how I had turned the wheel trying to bounce out of the rut and did not quite have the forward momentum to regain the road. What a precise reflection of my state of mind.

I walked slowly to a house. There was a bag of salt on the porch and a four-wheel-drive truck in the yard. The owner of the truck was doing carpentry upstairs. We talked about carpentry. He was glad to be of service and pulled out my car with a tow chain. The world is full of generosity. By having to ask for help, I tune in to that inexhaustible bank of kindness that is all-pervasive and unconditional and feels so good when it comes through us to someone else. Because of my need, his routine changed; maybe he took another step on the path.

We have little choice about anything, moving around as we do in a sleepy, anxious cloud of habit and conditioned response. When we slow down, that cloud settles, finally, and the details hidden within it begin to emerge with startling precision. I hold the kettle to the faucet; hear the water swirl in the bottom; place it on the stove, the little drops sizzling on the hot burner; stare out the window at the vortex of snow down in the valley, swirling over the trees. Finally, the steam whistles through the spout, and I pour a cup of tea. My thoughts flutter and swirl like water, like the snow. Having to slow down begins to seem less like a disability and more and more like a precious gift.

But I cannot delude myself that this is some kind of accomplishment, for I would dearly love to leap, like my cat, from the stairs to the floor; I would love to dance, run like a horse across the yard, play football, go out for a pass. The fact that slowing down is *choiceless* becomes part of the gift: taking credit for things just keeps stirring up that cloud. Since I cannot take credit, what really matters is the scent of the tea. The only choice we have anyway is to wake up.

A year before my mother died, she asked me what was wrong with my chest. She was the one person in the world who would want to hear whatever I had to say about it. As I narrated the details, she passed her hand over her face and named an old friend of hers who had died of my ailment. "I hope you don't croak before I do," she said. "It wouldn't be right."

I felt as though at long last I had graduated to her level. The generation gap vanished; I had finally grown up. I had something more serious to reveal to her than mumps and divorce. It was as though I now possessed an admission ticket to some kind of secret society — the society of those who have made friends with Yama, the Lord of Death. We would call this the Order of the Black Monk. The requirement for admission would be terminal or disabling sickness, battle experience, prolonged imprisonment and torture, waiting for execution, or working with the dead and dying on a regular basis. Attempted suicide would not qualify, since this act implies continued attachment to the illusion that you can escape.

While I had health, youth, food, friends, and comforts, I would hear about disasters and think, *I'm glad that did not happen to me.* Then I would play the "what-if" game: what if I were trapped in a plane going down; what if I had six months to live; what if I had six minutes? . . . But during this game there was always the separation between me and those who were "less fortunate." I think that the purpose of the game was to maintain that separation. The initiation rite into the Order of the Black

Monk is realizing that there is no separation anymore; you have been tagged; you are it.

In such an order, I am still a mere novice, for my mind remains cluttered with the detritus of hope and fear. But some kind of flip has taken place: from the viewpoint of the order, hope is irrelevant; fear is fulfilled and consequently dissolved. There is no need to maintain any kind of class system between more and less fortunate, happy and miserable; each experience has its own texture, and absolutely everything is part of the path. If a disease brings this kind of realization, then in what sense does it continue to be "disease"?

We think of disease as infirmity, disability, and tragedy, or reify it as "enemy" and project our aggression onto it, as though flesh could ever be preserved from decay. In looking at how great spiritual masters handle disease, other possibilities begin to appear. His Holiness the Sixteenth Gyalwang Karmapa, Tibetan Buddhist of the Kagyu lineage, who visited America twice in recent years, died of cancer in 1981. Right up to the final moments, he never uttered a word of complaint and was totally concerned for the welfare of those who waited on him. In a real sense, the cancer was not "his" cancer, any more than life itself was "his" life; because he was on this planet for us, whatever he suffered was part of that gift.

The late Chogyam Trungpa Rinpoche, my root guru and student of the Karmapa, was paralyzed on one side of his body throughout most of his teaching career in America. He limped, wore a special elevated shoe, spoke (and sang, comically) with one vocal cord, used a wheelchair, and endured intense pain without pity, hope, apology, or false heroics of any kind. His response to anyone in pain was not "Quit your whining and be like me," but limitless compassion and sense of humor. Affliction was his principal teaching tool. He transformed it into dignity and presence simply by the way he took his seat on the *vajra* throne. The concept of "disability" melts before such an example like a shadow in the sun.

The ignorance that results in karma is a kind of localized or attenuated intelligence, falling away from the luminous emptiness which is the ground of our being and into identification with structures of *this* and *that*. The unconscious components of personality are formed out of whatever we ignore. Having invented ourselves, for example, we forget that we have done that, and then we generate a story line to maintain our invention: other people become characters in our own melodrama; we define them as "good guys" or "bad guys" according to how they fit the labels and preconceptions of the story and project mental systems of aggression or seduction onto them, which might be interpreted as "relationships." Then we forget that we have a story line or a labeling system that keeps it going. The projections appear to us as an external "reality." Our emotional responses are shaped in terms of these ignored and forgotten systems; eventually our illnesses begin to be shaped by them as well. In this way, our physical and mental functioning becomes determined by skeins of psychic energy that we have frozen, from the very beginning, by diminishing luminous-emptiness-without-boundary into ego-form.

Perhaps this description can be understood more clearly by the image of looking at a light source, like a bulb on a Christmas tree, through half-closed eyes. The light is all over the place. It takes the shape of your lashes and spreads out to fit the contours of the lachrymal fluid. Some photons bounce and smear off the icy window and head for the stars at 186,000 miles per second. Others might be glimmering on the snow across the street, or entertaining an amateur physicist next door by making an Airy pattern on her wall. (Airy patterns are what scientists call the shapes of concentric light and dark areas made by waves.) There is no way to separate the light from the electricity, the eyes, the snow, the total environment. Likewise, there is no way to find the essential reality of the situation, the ultimate source, where the photons begin or end. This is luminous-emptiness-without-boundary.

When the eyes open, the streaks disappear, the fovea pins down the bulb, and the brain perceives it as a bulb on a Christmas tree. The localization of the light is an equally valid experience, if we know what we are doing; but typically that awareness is lost. We are heavily invested in losing it: a local reality solidifies the self and gives it something to relate with; a reality with no reference point turns the self into an Airy pattern on a wall that isn't there.

With loss of awareness, the total environment is taken for granted and forgotten; the mind wanders away into discursive thought. This is the ignorance, the attenuation of intelligence into structures of form. This creates the setting for the karma of disease.

The first stage in our habitual response to disease is to crank up more ignorance — that is, to deny anything is wrong. Denial is ignorance at a hysterical pitch; by the time you reach the point of denial, you know that there is something to deny, that the ignoring process is falling apart, and you are beginning to panic. You say, "No, it isn't true," but you have a pretty good idea that it is.

When the ignoring process falls apart, there is a possibility for luminous emptiness to reemerge. Disease always brings a friend: unconditional intelligence and compassion. It is like a door through which the friend may walk. Because you have this door, she can return. Chronic disease is like an unwanted baby you could grow to love, because it brings something out in you that you may not otherwise have felt.

According to Buddhist teaching, from ignorance we create neurotic tendencies in our stream of being that will ripen at some future date into painful results. The ripening process is karma. A painful result always removes the particular cause of the karma, although we can renew the cause by our failure to understand it.

My case is a clear demonstration of this process. Begin with a weakness in the lungs, which itself is a karmic result of some kind, carried in the chromosomes perhaps, or in what biologist Rupert Sheldrake would call the "morphogenetic field" — a good Western synonym for the Buddhist concept of a being-stream. Add now the attenuation of intelligence, which creates the belief in separation — *me* and *that* — which is the beginning of the neurosis characterized by grasping pleasure and avoiding pain. Viewing the source of pleasure as outside of myself but somehow necessary for the well-being of me, I begin to smoke. By smoking I induce euphoria and dispel boredom and fear. At this stage a full-blown conditioning process is underway: behavior, reinforcement, strengthened behavior. It is part of an addiction to pleasure. Additional psychic factors might be a general holding back, holding in, a tendency toward intellectual fixation and brutal self-criticism that is alleviated only by more smoking, and that slowly begins to manifest as a swelling of lung tissue leading toward complete suffocation.

The ultimate purpose of it all is to maintain ignorance — that is, to confirm ego. How does this happen? Boredom, loneliness, and fear unconfirm the notion of self: they are the first reassertion of luminous emptiness manifesting as a gap, a space with nothing to do, nowhere to stand, nothing to hope for. That experience brings on fundamental uncertainty; from the ego's point of view, it is a living death. You could go mad in such a space; you could shuffle from room to room haunted by guilt and failure. We call it "the pits," "the black hole," showing by our choice of terms that we know perfectly well it is a gap; it is open space, no boundaries, no definitions, no beliefs. Trying to avoid the gap, I set into motion the chain of causes leading to physical infirmity. I do it to myself as an individual; we do it collectively to each other as a social system.

My first step toward self-healing may seem almost masochistic to a non-Buddhist, but it makes good sense and has very far-reaching effects: I allow the disease to be there and make friends with it.

Tai Situ Rinpoche, in *Way to Go*, writes that we can be *grateful* for affliction, once we have understood that it is the ripening of karma and that its appearance removes the cause. This traditional Buddhist doctrine resembles the view that fever is the body's healthy response to the intrusion of harmful microbes; by means of the fever, the microbes are processed out. The Buddhist approach is not merely physical, but comprehends our totality over unlimited vistas of time. Affliction inspires wakefulness, which in turn removes ignorance, which is the ultimate cause of suffering.

It is important to distinguish between this approach and "positive thought." Both agree that illness is strongly connected with mind. The notion that illness begins in the mind has gained widespread credence in our culture, but what that means is still not properly understood. "Positive thinkers" generally perceive that sickness originates in, or is in some way supported by, negative mental attitudes — such as the desire to be punished or taken care of in a dependent role — and the suggested treatment might include looking in the mirror and telling yourself how wonderful you are, pacifying your anxiety with soothing cassette tapes, lying in a circle with a support group for mutual strokes, or forcing yourself to continue performing difficult physical tasks of self-help even while your strength is ebbing away.

But something is lacking in this kind of technique. There is still an underlying assumption that death is *terrible* and that we can talk ourselves out of it. What shall we do with patients who refuse to be "positive" — lecture them on their inadequacy? Or do we simply write them off, saying they cannot — will not — be helped? Negativity is real; it does not fade away because I listen to ocean waves and tell myself jokes.

For the most part, we want to understand disease only in order to get rid of it as quickly as possible, just like we want to get rid of the inconvenient tasks of caring for the old and the homeless and disposing of the dead — as though we could thereby reach some dreamland of health and happiness where people never get sick and die. Hearing that disease is caused by improper diet or unhealthy attitude, we might become health-food fanatics and search aggressively for the mental hang-ups that make us fat or cause our skin to break out in a rash. But our desire to reject negativity and cling to pleasure has been the problem from the very beginning.

As long as our goal is to hang on to something or to get rid of our own mortality, we are still only suppressing symptoms. "Healing" could become one more ambitious project by which we try to ignore the message of luminous emptiness: that there is no place to stand in the endless cascade. We thought that we had a cozy little observation balcony, but it is all Niagara Falls, no matter where we turn. Even the parking lot is being swept away. On the other hand, it might be fire, too; we are on fire every moment, dying and being born all the time, spreading out every-where. The work of ego is a mode of experiencing the heat and the color; within that mode, enlightenment is the ash.

Making friends with the disease leads to discoveries of the sort that I have described earlier: slowing down, shedding excess baggage, observing without struggle, deepening mindfulness, letting go of attachment to pleasure, feeling the texture of dis-comfort and pain, finding the roots of fear. Slowly the ground of ignorance is dispelled, like beginning to recognize a landscape in the very early dawn.

Ignorance is the environment of the whole karmic chain. If there is no ignorance, then the concept of "disease" becomes superfluous, as do the other "dis-" categories: discomfort, dis-continuity, disillusionment, disability, dissonance, disappoint-ment, disbelief. Some Buddhist masters wake themselves up fur-ther with a practice in which they invite all the demons of chaos and disaster to visit them. My little self says, "I am not at that point," but a braver, more expansive self answers, "Maybe that is what I have already done."

But whether I am brave or not, the "dis-ease" is here, how-ever it is defined. By using it as a path, and as a means to inspire someone else, I am hanging out with the masters. That is not such a bad result. If it means that I can hang out with the likes of Trungpa Rinpoche and His Holiness the Karmapa, then bring on the demons; they can sit on my shoulders while I type. (The mouse, getting drunk, bangs his tankard on the bar and declaims, "Bring on the cat!")

Another method to reject luminous emptiness and cling to the ground of ignorance is to make affliction *mean* something: to say, for example, that it ennobles the human spirit, toughens our courage to endure, takes away our sins — or even that we are being punished for our sins.

The second of the four stages of dying outlined by Elisabeth Kübler-Ross, in her classic study of the death process, is "bargain-ing" — the attempt to hang on to some kind of ground by giving up something else: "I will give all my money to the church if only I can get well." (The four stages are denial, bargaining, anger, and acceptance. These stages are all from the ego's point of view, and they are encountered also in meditation practice as we slowly give in to the experience of egolessness.)

Interpreting pain and loss as meaningful in a philosophical or religious context is another form of bargaining, a consolation prize: "My legs are gone, I have been tortured, my children went to a concentration camp, my tumor is inoperable, but there is a reason for everything: it is to test my faith; it is all for the greater glory of God or the Party; it brings humankind to a realization of existential despair; even from the jaws of defeat we can salvage a mustard seed of victory; we shall rise from the ashes, we shall be changed." Explanation is still the game of hope and fear: "If I submit to suffering, then maybe I will gain a higher truth."

The pressure of affliction tends to blow these answers away like chaff. A political prisoner being drowned in a bucket of sewage is interested in only one thing: the next breath. Sarcoidosis is not as severe, but the concern is the same. Beyond that, I want to open completely, without disguises, consolations, or illusions of any kind. By refusing to assign meaning to it, we stay with the emptiness of suffering, and thus begin to live in a reality that is luminous, limitless, unconditional, and immediate.

The cheerfulness that results from this renunciation of meaning cannot be destroyed, because it does not deny anything, and it does not have to be maintained. Misery and death are included and allowed. It is not "my" cheerfulness; it does not come from anywhere. You can let go of it and laugh. You could drink from skull cups and make trumpets from human bones.

Remaining with emptiness is also a gift of compassion — to myself and, perhaps, to others. Instead of giving the arrogance of Job's comfort, we can offer the witness of silence and the example of path. It may be a more generous and intimate experience, for both parties, to wait on someone who is disabled than to preach a sermon on self-reliance. Where action is called for, it comes best from an empty mind.

The basic terror of nihilistic despair seems to arise from sensing the truth of emptiness, and yet not living it. Can we open that final closet door? With the best intentions, our comforters say, "No — keep it closed."

Even if all the pain in the world means absolutely nothing, can we admit that and live? Can we still wake up? The answer is yes — with a smile. But not until we give in to that "nothing" and know it in our bones.

March 1988

The Cure

Kathleen Lake

The cure for pain is in the pain.
— Rumi

The cure for pain is in the pain.
As if an inside knife could air
the venom from a swollen vein.

Or does it mean, Increase the weight you bear
deliberately; speed forward into grief;
assist that thirsty alchemy, despair?

It may be saying, Drop the sheaf
of what you must have, like a useless broom
of frivolous weeds, whose life was brief.

You will not have it. Not till the room
you are is empty, blown clear in a shutterburst of rain,
and the wind plucks heedlessly a threadless loom.

It strips us clean as bone. In failure, we're laid plain
as whitewash, old as grain.
The cure for pain is in the pain . . .
and yet I cannot say goodbye again.

June 1992

THE UNHEALED LIFE

Yaël Bethiem

Sunlight slanting through my window touches the leaves of my plants. The light is luminous and magical, filling the room with warmth and promise. But notwithstanding the beauty of the moment, I am filled with longing — a longing to dance and run, to raise my arms in salutation to the sun, to wade breathlessly in ice-cold mountain streams, to feel the earth beneath me.

Instead I live in this one room. It is here I eat and sleep, here I teach, here I learn about love. It is here, from my bed, I reach out to the world, here I dream of a healthy body, the heavy, solid presence of strength. It is here I awaken each morning to a life very different from the one of my dreams.

My favorite cat slams against the door in an effort to rouse me to let him in. This is our ritual: he jumps onto the screen and careens like a wild man into the door, and I lie still and listen. The walk to the door looks very long. I lie quietly, waiting for the determination to rise. In the corner stands my walking stick, symbol of my "different" life. I am not sure if it is friend or foe.

Ankylosing spondylitis (AS) is a genetic disease affecting the immune system, causing it to attack the body as if it were an intruder. My father had this disease before me. Like me, he was forced to grapple with a life of disability and intractable pain. Although focused in the spine, AS affects the body as a whole, causing inflammation, fusion of the joints and spine, and inflammatory bowel disease. For me this has meant stays in the hospital and high doses of steroids to stop the bleeding. It has also meant a life of restriction and pain.

Sitting has become very difficult. Each day, I can manage about three hours in a chair. Consequently, "up time" is of great value. It is cherished, planned for, and jealously guarded. If people call when I'm sitting up, I ask them to call back later. Sitting is too important to be interrupted.

Essentially bedridden for three years, I have invested a great deal of energy in teaching my body to move again. I have worked for a year to be able to stand outside for ten minutes at a time beneath the trees. I live in a town full of hills and old Victorian houses. Outside my door, on one of the few flat spots in town, I practice walking. Under a stately oak tree that has become my friend, I struggle with the limitations of this body. At times, when I cannot walk, I stand in silence. Never before, when my body was well and I rushed to and fro in the world, did I see the beauty of the sky in such bold relief. I didn't understand the whisper of wind against my cheek. Nor did I realize the nourishment I received from the world of nature — not until it was denied me.

When I go outside each day, my cats rush to meet me; they know this is a special event. My time with the sky has been dearly earned, a shimmering victory that disappears like a mirage when held up against the easy movement of someone else's life.

For ten years I have worked to heal myself. I have eaten the right foods. I have cleansed my body and fasted with vigor and conviction. I have gone to healers. I have had the laying-on-of-hands. I have visualized, relaxed, and prayed. I have gone to pain clinics. I have analyzed my childhood and worked with my dreams.

I am deeply engaged in my life, but I am not healed. My mobility remains impaired. The pain remains constant. All my imagery and prayer has not been able to change anything.

One of the hardest parts of this experience has been my constant struggle with the philosophy that says we create everything in our lives. It follows, then, since I have not healed, that I have somehow missed the right thinking or the right beliefs. In other words, I have failed. This philosophy has been deeply hurtful. It has caused my friends to forgo their humanity for the sake of an idea. It has brought me to despair. But it has also brought me healing — by forcing me, again and again, to turn to God, looking for answers to my failure. In this anguished turning I have come to understand another truth: that the healing of the heart is more important than the healing of the body. It is the truth that says some lives are a greater message unhealed.

In the silence of this room, I have begun to understand the power of an unhealed life.

Even here, I am not alone. This silence is alive with the unfolding of other lives and with the turning and movement of the earth. Eventually, I began to sense my connection to the world's pain and my part in healing it. I realized that my transformation of pain into love was an act of service for humankind. By embracing my existence, I could bring courage to others to face their own pain and to acknowledge what it had to teach them.

In the pain of an unhealed life is the pain of the world's heart made manifest. I am sure our individual thoughts and needs affect our reality, but I am equally sure there is a larger picture. We are connected, to each other and to the world. An unhealed life is a statement of our need to work together to heal the whole. It is an opportunity to refrain from turning away, separating our reality from the reality of others.

According to the Gaia theory, our earth is a living being of which we are a part. Many people in many cultures believe this is true; they have felt the earth and know their lives are intimately connected to it. How then can anyone say, "I am well, and you are sick, and therefore your illness is not my problem"? If we are, as the theory goes, the consciousness or nerve cells of the earth, how can any one person's struggle not affect everyone? And if the earth is a living being that is hurting and polluted, would its nerve cells not be affected?

Is it possible that our illnesses and disabilities are not always the result of our beliefs or subconscious programming? Is it possible that sick or disabled people are like random nerve cells affected by the being of which they are a part? If this is true, then people who are not well should be given attention, respect, and support, for they are taking on a share of the world's distress.

I believe in personal responsibility. I believe we must examine ourselves deeply and constantly and increase our ability to see causes and effects. But I also believe that we must expand our vision beyond our self-centeredness. Perhaps, when one's back is hurting, it is not that one lacks support or cannot hold oneself up or whatever the reasoning is. Perhaps it is the earth whose back is hurting. Perhaps it can no longer hold *us* up. We may be the symptoms of its disease. We may be part of the working out of a design greater than our individual destinies.

The unhealed person calls attention to a need for healing and awareness. To disregard his or her reality is to disregard our connectedness.

The unhealed life is also a reminder of the pain and fears within each of us. It can help those who are well to look at the scared and uneasy places within themselves, to reevaluate the old programming about people who are "different," and to address the fear that whispers, "What if this happened to me?" Unexamined, these hidden reactions separate us from others unlike ourselves, from the conscious acceptance of the world as whole, and from our own existence.

There are other gifts of an unhealed life, other messages. Many people have told me that seeing how I live has given them courage to take the next step in their own lives, to face the hard places. I believe this is true — that the grace with which I embrace what I have chosen, or been given, can have meaning for others. I believe the unhealed life is a message that says, "Even here, love can grow."

I have been with those who have healed, and I have been with those who have not. There is no less beauty in the lives of those who have not healed. There is mystery; there are unanswered questions.

Those who have healed have their vision to impart. They have an offering that shines, radiant with possibilities. But often that dream is too far away to grasp. It is too far away from the next step that must be taken. The unhealed life is a bridge between terror and perfection. It is tangible because it is close enough to everyone's pain. It is hope and courage held out to everyone; it is the triumph of the human spirit. It acknowledges the uncertainty; it says there is love in it all.

I once had a spiritual teacher who drove nonstop halfway across the country to sleep one night in a motel room over a fault line where earthquake activity had been recorded. When I asked her why, she said she went to help hold the earth together. She said it was very painful physically, but it made her happy in her heart. I believe that those who are unhealed have chosen to take on, in their own way, the pain of the earth — and, through the courage of their lives, to transform it.

The idea that we are responsible for everything in our lives denies too many possibilities. Perhaps those who have not healed are responsible in a larger way, answering to a wider vision or an unnamed understanding. Reality may be bigger than we think. Perhaps we don't choose our reality as much as we would like to believe. Perhaps we choose it before we are born, when we can see clearly the needs of the world and the part we have to play.

Can we live with the mystery? Can we embrace the unanswered? Can we refrain from labeling others to maintain our own sense of safety? There are levels of our own being we have not even learned to touch, a wholeness we are barely able to glimpse.

To be willing to hear the whisperings of a different truth in the lives of those who have not "succeeded" in healing themselves is to take a step toward healing the world. Out of our willingness to live with this mystery, a new understanding can evolve. The message of the unhealed life is that it takes all of us, all of our caring, to bring this about.

January 1989

Combing

Veronica Patterson

How the chemicals that might heal you singe the hair inside.

How at the wig store you were angry with the clerks, because
they had no wig that was your hair.

How, when our generation came of age, hair was our
   exuberant *no*,
and *Hair* was our musical, and everyone had so much.

How we used hair unthinkingly for our own purposes.

How young girls in shining hair spend hours on a nuance of
   curl,
and that is youth: hours for a nuance.

How, falling gold in fairy tales, hair reveals the prince
or princess, reflects the kingdom to come.

How heads are shaved as punishment.

How Rapunzel made hair a staircase and a door.

How the woman in the story sold her hair to buy her husband
   a gift,
and then he bought her combs.

How you called yourself vain, but I say the strands of our hair
write our names.

I will bring you a broad-brimmed hat wreathed with fruits —
cherries, frosted purple grapes, peaches so small they never
   were; and

blossoms — daisies, roses, rue. No one would dream of bare
    land beneath
such abundance. You would live in its shade, private and
    imperturbable.

You would live.

December 1992

WHY I LIKE DEAD PEOPLE

Sallie Tisdale

Maud is eighty-six years old and weighs just that many pounds. She is nearly bald; her thin, fine white-gray hair has been rubbed nearly away by all her years in bed. At her age she is balding around her genitals as well, worn and loose where the catheter tube emerges. She is bare like a young girl, but the work of decades has left its erosion.

Maud had a stroke years ago, and then another, and another. She doesn't open her eyes, never speaks. She is fed with a big plastic syringe that the nurses slip past her resisting lips; the right amount of puréed chicken or spinach tickles the back of her tongue and makes her swallow, involuntarily.

Tonight I discover that Maud has cellulitis, an odd but common infection under the skin. Her right hip and buttock are red, swollen, hot; she has a temperature of 104 degrees. I call her doctor. He asks me, "If you were her granddaughter, would *you* want me to treat her?" "I'm not her granddaughter," I answer. "You know," he sighs, alone in his office, wanting to go home, "I promised that lady years ago that I wouldn't keep her alive like this." He pauses, and I wait. "I *promised* her." Eventually he orders an antibiotic, because of the slight chance Maud's infection could spread to another patient, and her temperature drops in the evening, and she goes on.

I AM OFTEN asked how I can stand my work, and I know that it is this very going on that my questioners mean. Not only the uninitiated, but other nurses and physicians often dislike this "gutter work" that I do: part-time charge nurse in an old, not very good, urban nursing home, working with the sickest patients, the ones who won't recover from an unfortunate age. Some of the nurses I work with are always looking for a "better" job, competing with thousands of other nurses for the hospital positions grown suddenly scarce in recent years — hospital jobs where patients come and go, quickly, and sometimes get well.

I feel a measure of peace here, a sense of belonging that is rare for me anywhere else. Partly it is because I know what to do, because I am competent here. Over the years, though, the ease that I've felt since my first job as a nurse's aide when I was eighteen has become layered with fondness, the way one grows used to a house and its little quirks, the slightly warped kitchen floor, the sighing upstairs window. Here, all is aslant, and I have to tilt my head a bit to see it clearly.

Coming in to begin a shift, I pass the activity room, crowded with humpbacked, white-haired people asleep in wheelchairs, facing a man playing "The Star-Spangled Banner" on a musical saw. In the corner, one upright, perfectly bald man spins slowly round and round in his chair, like a windup doll, bumping the wall at last and spinning back the other way. This is a scene of astonishing absurdity, and no one is paying any attention to it. We take it for granted, like the faint, lingering smell of urine tinged with kitchen steam and disinfectant. I leave the elevator on the third floor and step into furnace heat — July without air conditioning — and the queer conversation of the confused that will dog my steps all evening long. They give me this gift of skewed perspective like a gift of non-Euclidean sight, so that I become as willing to dip and bend with the motion of a damaged cortex as a tree in wind. I pass medicine room to room, and in each room the television is tuned to the same channel. For my four o'clock pass, it is *People's Court*, plaintiff and defendant, as I travel down the hall. At six we watch *Jeopardy!* "It's the only military medal that can be given to noncombatants," says the host, in Monte's room; then I pass Sylvia next door, and together we guess: "The Medal of Honor." And we are right. Bent over a task, preoccupied, I am startled by the peculiar speech of the nerve-worn, its sudden clarity. Up here, each day is the same, a refrain, and nothing can be taken for granted, nothing.

I know how many people hate nursing homes — hate the word, the notion, the possibility. A friend of mine lives next door to a local nursing home, and she tells me she hears people screaming in the evening, their voices leaping the tall fence between. She assumes the worst, my friend: that they scream from neglect, from abuse, from terror. She says it is a "terrible place," never having been inside. (I am similarly fearful around big machinery, in boiler rooms and factories. I am out of place, adrift, and fear the worst: Is that shower of sparks routine, or does it signal disaster? What is that loud noise?) I tell her that in every nursing home, there are people who scream; that they scream without warning, at private phantoms. I ask her where such people should go; I ask her how she would stop them from screaming. She listens, and I know she is unconvinced. Nursing homes are terrible places, she says, and it is because what happens there is terrible.

I enjoy my work, but I enjoy it in moments that are separated from each other by long stretches of fatigue and concentration. I enjoy it best when it's over. I catch myself, hot and worn at the end of the day, hoping the man I keep expecting to die will live until the next shift. I get irritable, and the clock creeps past 11:30, past midnight, and I'm still sitting with my feet propped up, trying to decipher my scribbled notes. The undone tasks, the unexplained events that want explaining badger and chafe. And everybody dies.

My ideals twitch on occasion, like a dog's leg in dream. I want no one to lie in urine a moment; I want every ice pitcher filled at every moment. For many years, I disliked the use of sedatives and antipsychotics to knock out the wound-up chatter of restless, disoriented souls. The orders read: "P.R.N. agitation" — as necessary — and this is the nurse's power to ignore, and the power to mute. So easy to misuse, so simple. But just as a shot of morphine can break the spiraling cycle of pain, so can a spiral of panic be broken — not for my comfort, but for the comfort of the panicked.

Sadie screams at me from far down the hall: "Help! It's an emergency!" And she screams over and over, rocking back and forth, till I come to see; she leans over and points at the blazing fire under her bed, a fire she sees and hears and smells, raging out of control. I see no fire. I coo to her, hushing; she babbles on. Finally I lie beside her on her bed; she is stiff and yearns to leap up. And, at last, I go to prepare the syringe: "From the doctor," I tell her, because Sadie loves her medicine, and she falls asleep.

The responsibility is mine, the consequences are mine. I have to be sure about choices no one can be sure about. I call for nurse's aides to come and hold the flailing arms and legs of Charlie, more than six feet tall, furious at the world that confounds him so. He squirms and tries to bite me when I hit his hip with the needle. We all fall across the bed together, grunting. And I know that the visitor, passing by, sees only the force, the convenience, the terrible thing we have done to this person: the abduction.

The same is true of the smell, just barely there, acrid in the heat. It's true of the drooling, the patter of nonsense in the dining room. Visitors tremble, knowing Grandma is here, and wish they had the courage to bring her home, out of this awful place.

Could this inadvertent audience, my patients' families, see these scenes and believe me when I say it is a labor of love? Some do; they bake blueberry pastries and bring doughnuts for the nurses, pat us on the arm, cluck their tongues. "I don't know how you can stand it," the niece says, after an afternoon at Aunt Louise's bedside.

The difference here is in what we each call love, the gap between their definition and mine. Their burden — and they seem to really want to know — is a burden of despair, a personal burden, bred of fear and impotence in equal amounts. This personal despair imagines as its opposite, its anima, a personal love and a personal sorrow. The visitor sees May defecating helplessly on the rug before anyone can stop her, and it strikes his identity, his self. It is as though the observer himself stood there, revealed. I have the advantage of knowing May will forget it in an hour; he does not. In breathless confession, waiting for the elevator, the visitor says: "I pray to God I die before this happens to me." I am told this again and again. "I pray to God." A kind of ego-terror is born, and with unbordered empathy comes flight. Suddenly the sounds and smells oppress, overwhelm. Suddenly it's time, more than time, to go.

A labor of love, love for fading people who dwell in shadows. I am saved from the need for flight — I am uninjured — because I let them do their own suffering. This is a cold-sounding excuse, I know. Call it compassion instead of love. (I am surprised, and pleased, just now, to find that *Roget's Thesaurus* lists as a synonym for *compassion* the less lofty *disinterestedness*.) I have learned not to make personal what I see. Just as the witness imagines himself, complete, transported to this place and trapped, so does he grant full awareness to those who are. He assumes Maud is aware of her plight, ruminates on her fate. I grant Maud plenty, without granting full cognizance to her withered brain. Down the hall from Maud is a man in his forties, paralyzed from polio, limited to a respirator, and he *is* fully cognizant: no pity for him, either. Pity makes distance, creates a separation of witness and participant; by assuming a person is absorbed in suffering, the witness prohibits him or her from participation in anything else.

I close the curtains, keep my voice down, as a point of etiquette as much as sympathy. I have a spring in my step; I can see and hear; I can eat and digest and control my urine, and I know these for the blessings they are. I am young enough, still, to take care of the old. But these are the most transient of graces, these graces of health, and I might lose them all tomorrow if the brakes fail. Old and sick comes later — but it comes.

Here, everybody dies. We tell black jokes. (I laugh and laugh at a cartoon of an old man sitting up in bed, surrounded by impatient doctors: "These are my last words," the old man says. "No, *these* are my last words. No, no, wait. . . .") We have a three-part mythos of death here, and first is that no one dies when we think they will, always later. Second, if a person long ill and silent suddenly comes to life, he or she will die soon. And last, people die in threes. Within a day or week of one death will follow two more. Just last week, Monte died, days after we'd predicted, and now Mr. H. down the hall is talking again, after months of sleep.

Death is anticipated, waited on in suspense. It is like waiting in a very long line that snakes around a corner so you can't see the end. When the last breath is drawn, it is startling; here is a breath, and another, and another. Death is the breath after the last one. Always fresh, always solemn, and not unlike a child-birth: the living let their own held breaths go, and smile, and in the solemnity is an affirmation. Here it is. I stroke the skin so suddenly and mysteriously waxen. I pull out tubes and patch holes. I like dead bodies: at no other time am I so aware of my own animation. This isn't because I am lucky and this poor fool is not, but because here before me is the mute, incontrovertible evidence. Some force drives these shells, and it drives me still. I am a witness, an attestant, to a forsworn truth.

Still I have my own despair. For me it is the things undone that break my back sometimes, the harried rush with people calling, and all those unexplained events. I wish we could ferret out the meaning in all this chaos, talk it out. No time — sometimes the ice pitchers are dry all night. Last week I had a shift like this, split in the middle by an impatient doctor who snapped his fingers at me and tapped his toe in frustration at my slowness. An hour later another doctor dropped by, and I asked her to see a new patient with a minor but uncomfortable problem. She refused, and then explained. "Medicine is the kind of job where you have to be really careful not to let people take advantage of you," she said. "Somebody always wants something." And all I could do was look at her, and get back to work.

I have to remember to temper my criticism of the aides, who work at least as hard as I do in a job of numbing repetition and labor. Hardest to remember, when so much is left unfinished, is what I have managed to do. I think I've been of no help at all, and then I realize how little help I'd be if I got discouraged and quit. Every task, no matter how late, every kind word, no matter how brief, makes a difference.

In my first job as an aide, I cared for a Swedish woman named Florence, who had only one leg. She was happy and confused and didn't know she'd lost her limb. Time and again she would try to walk and fall. I tied her in her chair, in her bed, and over and over she managed to untie herself and fall, thud, to the hard tile floor. She was always surprised. Exasperated at last, I stood over her and asked, "What am I going to *do* with you?" And she looked up from where she sprawled and said, "Don't stop trying, dear."

Don't stop trying. This is far from the best nursing home. It isn't the worst. I rant, jump to complain, go home frustrated. It should be better. But the sheets are changed, people are fed, for the most part each one is treated with kindness — a clumsy, patronizing kindness at times, but many of them don't discriminate on these fine points. Kindness is enough. Thousands, no, hundreds of thousands of people have joined these ranks, saved. There is no place to go but on, and on.

I like dead people and all their apprenticed fellows like Maud, who, slowly, is learning to die. And I like this place, with its cockeyed logic. I will feed Maud her squirts of purée, and a few minutes later Sadie will announce she is the queen of Germany and requires royal treatment. Celia will cough up blood, and, as I consider my options, I will hear the distant bedrails shake, the curses, the rhythmic, pattering singsong. Sometimes the borders shift even further. I sprawl across a bed, fiddling with Roberta's leaking catheter, trying to disentangle her fingers from my hair.

The feeding tube drips on my leg. Who is keeper, who is kept? This is the Marx Brothers all grown up, slapstick matured, life imitating art imitating life. Down the hall the Greek chorus begins, explaining the meaning and the mystery as the melodramatic story limps along.

November 1985

The Evidence Of Miracles

Jaimes Alsop
from a line by Robert Bly

I don't know why the snow line stops a few feet short of every
   house.
No doubt there is a simple explanation. Something to do with
   the warmth inside
melting the snow before we notice it. But there is still the
   surprise that it does.
As though it obeyed an order given long ago; a little
   forbearance shown us
by an otherwise indifferent God, caught on a good day and in
   a better mood.

How a man sitting at home alone one evening might take a
   familiar book
from the shelf and, opening it at an accustomed place, see for
   the first time
something he never saw before. And believe he has misread it
   or projected his own
interpretation onto the printed page, but look again and see he
   read correctly
what the writer always meant to say and realize in that instant
   he was never alone,
through all of it, and never feel that solitude again.

Or see his own mortality written on his children's foreheads
   and not mind
the years creeping up on him and accept the minor role from
   then on
with gratitude and a certain humility, feeling somehow
   honored by it.
How even grief may teach us something, wisdom, if it has to,
   and leave us cleaner

and the better for it. Able to wonder at incongruities or the
  merely insignificant.
To look for the evidence of miracles in the most ordinary
  events of our lives; something holy
in snow lines stopping short of every house. Even the empty,
  the apparently abandoned ones.

February 1993

Instrument Of The Immortals

Jake Gaskins

Play Chopin only on an excellent piano.
— André Gide

I understand that later you come to an age of hope, or at least resignation. I suspect it takes a long time to get there.
— Jane Smiley

It's a Steinway vertical, the "professional" model, taller, with longer strings than a spinet, and three pedals, like a grand. It has a satin finish, each of its five coats of black lacquer applied and then rubbed in turn with fine steel wool, producing a sheen rather than a hard shine, the elegance of a top hat. It's an aristocrat, born with a patina and a name. It bears, in fact, the autograph of John Steinway — now seventy-two, great-grandson of the founder — on the rear right-hand corner under the lid, in magic marker.

I hadn't played for years when I went shopping for it, so I was shy about trying it out in the store. I played a few notes as quietly as I could, felt the touch, heard the tone. I couldn't find anything to criticize, but I knew I hadn't given the keyboard a real workout. When I heard someone playing one of the grands in another part of the store — which was a warehouse, pianos lined up like desks in an insurance company, row upon row, grands in one section, uprights in another — I asked the salesman to ask the performer to play something on the piano I'd selected. The performer, the manager, was glad to comply. But first he shook my hand and congratulated me on my purchase. "Is this your first Steinway?" (Is this your first child? Your first Oscar?) A man about my age, forty-four, dressed conservatively in a dark blue suit, the manager sat and began playing Chopin's Waltz in F Minor, op. 70, no. 2. That's when I felt my upper lip tremble, and I held my hand over my mouth. I was a fool, but I thought I had never heard anything so beautiful.

Miss Eva Hodges, my piano teacher for eight years, now deceased, would be gratified to learn that I bought the Steinway. She'd be *proud* of me. She used to shout down from the window of her studio, between the Ionic columns above the front steps of the high school where I went twice a week for lessons, "Hold your head up!"

She looked like George Washington on the dollar bill: the same Roman nose, same thin lips, same chin. She had frizzy red hair and a redhead's pale complexion, pale blue eyes behind thick, rimless spectacles. She was tall and slender. She wore sensible shoes with low, thick heels; in summer she wore printed organdy dresses. She never married. She rented out the upstairs of her house; once, the tenants let the tub run over, and water seeped through the ceiling and damaged her Baldwin grand below. She drove a '52 Chevy that her brother, an orthopedic surgeon, had given her. She was the organist at St. Paul's Episcopal Church.

And she taught piano to talentless and recalcitrant children.

At first she glued stars in my assignment book each week, gold if I'd practiced well, silver and blue for B and C. But she soon stopped that, because I was too smart for her. I knew that stars weren't real grades, and music (I took "music," not piano, lessons) didn't really matter. I was going to be a doctor. I didn't have a lot of time to practice, anyway. And I had never been much of a sight reader. I could play better by ear. I didn't practice scales or Bartók's *Mikrokosmos*, although Miss Hodges assigned them. I'd play the show tunes and popular ballads I liked: "Old Man River," "Someone to Watch Over Me," "But Not for Me" — all in the key of C, but with feeling. I played for fun. I stopped the lessons altogether when I was sixteen because I knew I would never be *really* good.

She kept in touch over the years. One night when I was still in high school, she phoned to let me know that there would be an all-Gershwin program on the *Bell Telephone Hour*. When I graduated and was heading off to Johns Hopkins, my brilliant career as a physician assured, she gave me a plaid canvas suitcase. Every Christmas, I received a card from Miss Hodges. The summer of my sophomore year, she invited me and another young man who'd taken lessons from her to lunch. She served chicken salad (I mistakenly complimented her on the tuna fish) and said she was proud of us both. The other young man attended Yale and was majoring in physics. I had switched from pre-med to English, having failed chemistry and managed only a C in cell biology. But she was proud of me.

I told her I'd developed a taste for Brahms. We all loved Chopin, of course. But I couldn't understand Beethoven, I admitted. I said he seemed so noisy, all those kettledrums, all that pounding. Miss Hodges had taught me only a couple of pieces by Beethoven: the *Turkish March* and, later, the slow movement of the *Funeral March* Sonata, which is comparatively easy. But Beethoven in general was what I didn't "get," I said. I was hoping Miss Hodges would leap to Beethoven's defense and justify his reputation. But she didn't. Maybe she thought that attempting to explain Beethoven to a twenty-year-old would be pointless.

Or maybe she didn't know herself why Beethoven was so great. Unlike me, Miss Hodges lacked all pretension. I doubt that she was bothered by the thought that she was not a great artist herself, that she didn't compose, that she was even something of a type: a spinster with a rigid expression, tight-lipped, frizzy-haired, bespectacled, smelling of bath powder.

The only thing she did that went against type was to join a fundamentalist congregation out in the country after she was diagnosed with liver cancer in her seventies. She started attending revivals, apparently convinced that she would be healed. She wasn't.

I'VE HAD THE piano for six months now, and I play every day, probably more than I have time for, more than I should. I'll play while I'm waiting for French toast to heat in the microwave (two and a half minutes) in the morning. I'll play between the time I finish the breakfast dishes and the time I head out for class. (I'm a teacher, not a physician.) I'll play after supper, before I start grading papers, and I'll play before going to bed, from 10:30 on, until I'm scared the neighbors will call the police. If there is something desperate about my playing, it's only natural; there's something desperate about my having bought the piano in the first place — the result of a typical midlife realization that I am not going to live forever. Some men have affairs at this age.

In six months I've learned two waltzes, four mazurkas, and a nocturne by Chopin. I've learned two "easy" (that's a laugh) sonatas by Beethoven, the nineteenth and twentieth, and the (relatively) slow movement of the eighteenth. I am wrestling with the other movements of that sonata. I've mastered two impromptus by Schubert, nos. 3 and 4 of op. 90. And I've learned Gershwin's Second Prelude, which I couldn't play as a teenager because of the wide intervals.

I used to struggle just to read the notes. For each new piece I attempted, Miss Hodges would have me read the left- and right-hand parts separately, then put them together later. She would assign a few measures each week, marking them off with her number-two pencil and dating them. Now I forge ahead on my own, playing both hands at once. I am reading music faster than

I can memorize it. Amazing what the fear of death will do for one's concentration.

I am not, however, a real musician. Sometimes I'll be tired, and the fourth finger of my right hand, which seems longer than it ought to be, gets in the way. I'll be negotiating a rapid scale, and that finger will trip me up. It'll scrape the edge of a key, and that minor mishap will be enough to transform my Steinway, this "instrument of the immortals," into a mere contraption, a row of wooden blocks, a box of Tinkertoys, fuzzy hammers striking tinny steel, wound wire. It doesn't sing. It plinks and plunks, twangs and buzzes. I think, *What for?* I know I'll never be any good.

MY FIRST PUBLIC performance was a disaster. I was six or seven; it was shortly after I'd begun taking lessons. Miss Hodges gave me "The White Knight" to learn, a piece composed for a child. I felt it was beneath me. But I agreed to play it anyway in a recital to be given by a few select pupils of local piano teachers in an auditorium at the local college. I was second on the program, after a little girl who played her simple piece without a hitch and walked off the stage to polite applause and adoring smiles. (And where, may I ask, is that little girl today?)

My turn came, and I wasn't nervous. The piece was in my fingers. I'd played it at home countless times without a mistake. It was easy. I slid off my seat, marched down the aisle, climbed the steps to the stage, and approached the piano, a Steinway concert grand. I turned, faced the footlights, and bowed as I'd been told to, from the waist, with a smile. Then I sat down. I waited a moment until I felt comfortable on the stool, not too close, not too far away from the keyboard, per Miss Hodges's instructions. Then I lifted my little hands, lowered my little fingers, and began playing the wrong notes.

Startled, I stopped and started over. Wrong again. Where was I? I couldn't find my place, the right keys to begin on. The keyboard appeared strange, intractable, a friend turned traitor. I got up and fled the stage. I went straight for my grandmother's lap, where I curled up and endured the rest of the recital, whimpering sporadically amid a long program of others' musical triumphs. I lived to play again, of course, even in public. But the experience left an impression.

ARE YOU GOING to play for us, Jake?" This from my nephew's bride at the dinner table on Christmas Day. I've not met her previously, but I can tell she's been cued. The whole family knows I bought the Steinway. "Jake can really tear up a piano," says my sister's husband, a hardware contractor. Is he merely being kind? Or does he actually believe I play well? One does not say of Murray Perahia, for example, that he can tear up a piano.

Later in the evening, after dessert, after the dishes are cleared away, after we've finished opening presents and I'm on my way to the kitchen for another cup of coffee, my sister grabs me in the hall. "Please play some carols, Jake," she begs. "I miss your music." I decline because I'm not going to play some carol in the key of C that nobody can sing. If I attempt something more ambitious — Chopin's *Barcarolle*, for instance — I'll fall on my face. No one will listen, anyway. It's too long. When my brother-in-law says I tear up a piano, when my sister says she misses my music, when they all say they want me to play, they don't mean it, or else they don't know what they're talking about. I have my pride.

My pride isolates me. I am forty-four, as I've said, single, unattached, bald. As the years pass and I watch my nephews marry, have children, solidify their positions in the community, I feel more and more alone and defensive about it. I live across the street from a park. April through September, I hear the gleeful shouts (and sometimes the profanity) of children playing baseball, flying kites, throwing frisbees. They hear me practicing the piano, if they hear me at all, dimly. If they were told that the pianist is an English professor, middle-aged, unattached, what would they think, if anything?

I used to pity Miss Hodges. With the arrogance of youth, I pitied her. Had she never been in love? Why hadn't she married? Was she normal? Was she human? Now, as I approach the same age, and the same fate, I am contemptuous of the questioner, the child I once was. Now I answer my own questions: we're *all* normal; we're *all* human in our loneliness.

On April 12, 1956 (I can see the inscription now, her distinctively neat handwriting), Miss Hodges gave me a copy of *The Music Lover's Encyclopedia* inscribed: "With best wishes always to my little pupil." I was eleven, and I wasn't particularly interested in the chapter on counterpoint, for instance, or the collection of brief biographical sketches of famous composers, including Deems Taylor's once-familiar essay on Wagner, titled "The Monster." But I kept the book on the piano at home, beside the metronome she also gave me. And as I grew older, I'd open the book and read only the inscription: "With best wishes always to my little pupil." At first it embarrassed me, like a declaration of love that I was in no position to reciprocate. It seemed sentimental, the words of an old lady to a boy, sweetly inappropriate. Later, the inscription would serve to remind me that ten, then twenty, then thirty years had passed since I was anybody's little anything.

My mother tells me I began playing the piano when I was four, standing at the keyboard and picking out tunes I'd heard on the radio. I inherited the ability from her. She worked for a time in a music store, demonstrating sheet music. I remember her playing "Glow Worm." When I was older, my father bought me an accordion, a manlier instrument. But I preferred the piano, our upright Hamilton that Mother had bought used. It was smaller than the Steinway, less resonant, tinklier, with a flaccid touch and real-wood sharps and flats. It was also infested with moths. Mother didn't find out until it was too late, and the felts were nibbled ragged. But before the piano reached its eventual state of dilapidation, some notes sounding *clunk*, it stood up to my practicing; it bore the brunt of my youthful enthusiasms, my Gershwinisms, my Broadway melodies. When I was eight, I composed "Ducks on a Pond" and "Halloween Night" on that piano, and, when I was fifteen, "Metropolitan Midnight," my biggest hit. There's a tape somewhere of me playing it at a high-school assembly to thunderous applause. Miss Hodges transcribed that one and sent it off to a music publisher, which rejected it.

I played after school to relax, and I played after finishing my homework to wind down. I played while my parents fixed supper, and I played as quietly as I could after my father went to bed early because he was tired. I played through the years that saw my father lose his grain-storage business to an unscrupulous partner, that saw my mother return to work as a salesclerk in a women's-clothing store, that saw my father struggle with alcoholism and depression. After I left for college, I played only when I was home on vacation. By the time I got to graduate school, I'd stopped playing altogether. I received my doctorate in English in 1979 from the University of Iowa and have been teaching here at Southeast Missouri State ever since.

My father died five years ago, at seventy-eight, of Alzheimer's. My ability to grieve, like my musicianship, is not remarkable. I'm an amateur. God forbid that I should become an expert. There are millions of us out there, in any case, who are grieving. Nor am I the first man to feel, on the death of his father, that he, too, is mortal. But what my father's death — as compared to, say, Miss Hodges's passing — made clear to me was not only that life ends but that life can disappoint, dreams can fail. His death taught me that I am no better than anyone else, not only because I, too, will die, but because I will probably leave the world as he seemed to, defeated first by circumstances and finally by disease.

AGAINST THIS BACKGROUND of loss, then, and feeling my middle age, I have bought a Steinway to realize at least one dream. On the practical side, I think it makes me healthier — along with walking and lifting weights and eating bran flakes — to practice every morning, to learn new pieces, to discover that I am capable of improvement, that my brain is not dead. I'm learning the *Goldberg* Variations, and I'm on the sixth one now. Never having studied any Bach before, I feel a real sense of accomplishment. I may not play them well, but I play the right notes, in tempo, and my playing of Chopin has improved as a result — perhaps because Chopin himself played Bach, or because playing Bach strengthens the fingers.

At my most sanguine, when I am able to sit down and play a

variation through without faltering, as a piece of music and not a series of notes, with my heart and not just my fingers, then the music sings. My sense of accomplishment derives not only from knowing that my practicing has paid off, but from the feeling that my spirit has prevailed: muscles and nerves, bones and ligaments, wood and steel — all in service of the Ideal. In this sense, buying the Steinway, taking up "music" again, was an act of courage, of defiance appropriate to Beethoven, a pounding of kettledrums against God's sullen silence, a beating down of despair.

E.M. Forster defended his own playing, bad though it was (said he), because it helped him listen, taught him something about construction, about the importance of keys and their relationships. But that's if you like music in the first place. The philosopher Susanne Langer wrote that music expresses the form of feeling, and that feeling, expressed as form, is more than raw emotion: it is thought. It is thought that is beyond words. And since music is not words, I can't repeat what it says.

With Beethoven, there is evidence of a brilliant mind at work in what he does with a key, a theme. You don't have to be a musicologist to admire that. But the beauty speaks of something beyond logic and words and even feeling. It's not what music "says"; it's what music intimates, the existence of spirit, something shared but ineffable, something within us that we can only acknowledge without knowing fully, something we share with Beethoven, genius though he was. Because music is an argument for the spirit, I have reason to hope.

Would Miss Hodges confirm my speculations? No matter. The important thing now is her abiding, incontestable presence — an argument in itself for the immortality of the spirit. "Best wishes always to my little pupil," she wrote. Apparently she meant it. I want to believe in the truth and efficacy of that benediction. I want to be redeemed. I want to emerge whole and intact from this phase in my life, this humiliatingly ordinary midlife crisis, with its fears and regrets. I want to be saved.

The religious analogy is apt; I have taken up playing the piano again the way people return to their faith after abandoning it in their youth. They didn't realize what they had until they'd lost it. But now they know that their faith is a part of them they can't shake, ingrained, like their temperament. Returning to their faith is returning to themselves, to the selves they neglected in order to pursue degrees, careers. Humbled by experience, what they once rejected as childishness they now embrace as wisdom.

March 1991

THE PRAYER
OF THE BODY

AN INTERVIEW
WITH STEPHEN R. SCHWARTZ

Sy Safransky

I'd been melancholy for weeks, dogged by feelings I couldn't name. Then my wife went out of town; I didn't want her to go.

You might say I was ready for a good cry. Yet how tempting to ignore sorrow, as if it were a beggar. Those dark, accusing eyes.

I almost rushed past Stephen Schwartz, too. A booklet describing his workshops sat on my desk, unread, along with dozens of other brochures promising to unfurl my petals. Who has time for workshops? Life is so busy rearranging us already, and truth such a flirt. Read the words of the master, spend an evening with the master: there's no telling whether you'll get enlightened, or herpes. No, I didn't want another teacher.

But when I finally picked up the booklet, I was intrigued. Here was a spiritual thinker who shunned spiritual dictums; who suggested that the body doesn't need to be transcended, or the personality fixed; who insisted that self-knowledge has more to do with feelings than with philosophy — feelings, *not psychological insights, not our thoughts and stories about our feelings. I was intrigued, too, that the booklet had been sent by one of Schwartz's admirers. Schwartz doesn't advertise, and his books aren't available in most stores; you need to stumble upon his work. I ordered a tape.*

I listened to it on my way to the airport, but I wasn't really listening. I was thinking about my wife, eager to see her but determined not to show it, embarrassed I'd been so sullen about her leaving, as if she weren't coming back. I was sick of myself, the scratchy soundtrack of my days, the desperate longing. What was Schwartz saying? Sadness isn't wrong. Pain isn't shameful. *I listened more closely.* We condemn and deny our loneliness. We take our longing for love and turn it into shame. *His voice so ardent, so sincere.* Yet loneliness is a prayer, a deep longing to know and feel God's presence. *His words like the sea, rocking me; like a big wave, whacking me in the chest.* There is great strength in wanting. There is dignity, not shame, in loneliness. *My mind was a hanging judge who showed no mercy. I resented my wife for leaving, but I resented myself even more for feeling abandoned.* We can't keep measuring ourselves against some enlightened ideal, as if self-hatred could be a path to love. There's no disgrace in any experience we've ever had. There's nothing we need to run from. *No, I thought, sorrow isn't the enemy. In the mirror I caught a glimpse of myself, tears streaming down my cheeks.*

Since then, *I've spent many hours reading Schwartz's writings and listening to his tapes. The beauty of his language, the passion and lucidity of his message, continue to move me.*

Schwartz is more than just a good listener and a psychologically astute writer with a flair for the right phrase. At times, he inspires comparison with the great Indian philosopher Krishnamurti, who insisted that truth is a "pathless land" that cannot be approached through beliefs. Also, those familiar with A Course in Miracles — *the radical rendering of Christian thought with which Schwartz was once identified — will recognize his debt to the Course, as well* as his independence from it. Instead of parroting the Course, he embodies its teachings on love and forgiveness, threading the Course's lofty ideals into the frayed fabric of ordinary lives.

At his workshops, Schwartz asks people to sit in a circle with their eyes closed. Then he talks them through a process similar to meditation, with its focus on the breath, and to therapy, with its emphasis on feelings — but dramatically unlike either. He gently but persistently encourages them to turn to the pain, not the ideology about the pain; to the truth of the body, not clichés about the truth; to the actual feelings, not the words that wrap the feelings in too many layers, like a mother nervously bundling her infant against a gentle breeze.

Usually, Schwartz explains, thoughts are so clustered around a feeling that it seems as if the definition is the feeling. But the feeling is different from the thought, different from the interpretation we've always given it. The body feels. When we feel loneliness, when we feel anger, when we feel love, we feel it in the body. Our fears and mental turmoil, he says, are the result of trying to place limiting labels on the innocent feelings of life. Therefore, turning to the body with compassionate attention is the first step in really caring for ourselves.

There's a more esoteric aspect to Schwartz's work that's difficult to describe and may leave skeptical readers shaking their heads. To him, the physical body is only the visible portion of "an invisible field of radiant energy." This field, he says, is made up of exceedingly subtle filaments, which function as conduits or passageways through which energy is given and received. Feelings, Schwartz suggests, are the movement of energy at these subtle levels.

Although he doesn't call himself a psychic, Schwartz says he can sense these energies in others. He believes that, given the right conditions, he can merge with another person on a feeling level; he speaks of entering into "their silence, their presence, their depth." As a consequence, he's able to speak to them about their feelings in an intimate, helpful way.

His dialogues with workshop participants bear this out. He gently encourages people to move their attention from the tangle of thought to the energy of the body. By persistently asking where a feeling is being experienced, he helps distinguish between what is actually occurring in the body and the conditioning, the descriptions, the self-defeating ideas carried by the mind.

Even those who have experienced the process have difficulty explaining it, he acknowledges; it's easier for them to quote him than to describe the intuitive scaffolding for the work. I'm aware that my words, too, reduce the rich complexity of his work. It's like reading aloud the lyrics to a song, minus the tune.

Ironically, the very act of writing this — hunched before a glowing computer screen, jazzed on coffee and deadline anxiety — distances me from my body. But under just about any circumstances, it's hard for me to experience feelings as pure energy, rather than as something already encased in meaning, mummified, cataloged. I wander anxiously through the museum of myself, rules on every wall. Even Stephen Schwartz can become just another exhibit as he reminds me not to pretend I'm less damaged than I really am, or less holy; reminds me, as well, that there are doors up and down these hallways, and they're not locked.

SCHWARTZ, FORTY-THREE, LIVES in upstate New York, on a mountain-top overlooking the Hudson River, with his wife, Donna, and their two sons.

While studying literature at Brandeis University during the sixties, he got involved in antiwar and civil-rights activism but eventually became disenchanted by all "the divisiveness, the anger, the blame." He went through a "big turnaround," giving up psychedelics and committing himself to meditation, which he's practiced regularly since 1966.

During the next two decades, he worked as a shoe salesman, a deliveryman, a (vegetarian) butcher, an insurance agent, a teacher; ran a secondhand bookstore; sold exotic plants with his father; and wrote two theatrical productions, one based on the work of William Blake and the other on James Joyce.

For a while, Schwartz enjoyed a modest reputation as a teacher of A Course in Miracles, though his maverick approach to the Course got him in trouble. He says people would come to his talks carrying copies of A Course under their arms, then storm away angrily, offended by his insistence that salvation has nothing to do with be-

liefs; that we need to free ourselves of all ideologies, even those as sublime as the Course.

His teaching has continued to evolve, his language less overtly spiritual now, more accessible. I respect his willingness to re-create himself regularly, to risk offending his admirers. He seems more interested in discovering what's real, what's next, than in enshrining a system; more dedicated to honoring mystery than to footnoting it.

When I met Schwartz for the first time last year [1991], I expected to feel intimidated. But he was warm and relaxed, not the kind of person who engages in spiritual grandstanding. He's willing to kid around, to be honest about his doubts, even to parody his own vocabulary.

His books include Problems Are the Doors through Which We Walk to Peace *and* The Compassionate Presence. *The books, as well as information on his tapes and workshops, are available from Compassionate Self-Care Publications, P.O. Box 28214, Seattle, WA 98118, (206) 725-6920.*

A seemingly robust, indefatigable man, Schwartz nearly died earlier this year after he collapsed at home and was rushed to the hospital, where he was diagnosed with cancer. When I heard the news, I was stunned. We spoke about this during our interview, and Schwartz has just written a book about the experience. True to form, he intends to call it I Accept in All Gratitude: Cancer, Crisis, and Compassionate Self-Care.

Safransky: I've heard you talk about loneliness in a very moving way — as a prayer to God. What do you mean by that?

Schwartz: Inside our loneliness is a longing to be released from the pain of separation and the confusion it entails. We've all been taught that there is something wrong with or even dangerous about being lonely. But such an assumption is based on a misunderstanding of what loneliness is and how it relates to our life here. Loneliness is a kind of wisdom, a recognition of something, an urge toward genuine transformation.

There is nothing to fear about loneliness. There's no reason to run from it or to tighten down when it comes. If we allow ourselves the chance to attend to the loneliness, to be with it at a feeling level — physically — then the harsh overtones dissolve. What we called "loneliness" turns out to be something else entirely.

Each of us is longing for something. This longing runs deep. Sometimes it manifests as loneliness, sometimes as grief, anger, or something else. Whatever way it comes, we can be with it respectfully and openly, allowing it to exist. This so changes our relationship to it that we never need fear it or run from it again.

There are times when the body is calling for attentive care. There are times when the signal is there, but our response is self-hatred or dislike, and the body's call gets ignored. Loneliness is such a call. We need to turn to ourselves as a mother to a child and wait, without judgment.

Safransky: I deal with loneliness by either wanting to blame someone — say, my wife for being away — or explaining the loneliness to myself in terms of the past.

Schwartz: You do miss your wife, and that is real. It's important not to create a moral attitude about this. We can say that on level A, you miss your wife, and on level B, you are longing for a deeper relationship with her and with life itself. One layer of truth does not contradict another. At both levels, the call is for gentle respect, not moral evaluations.

Disliking feelings or deciding they're wrong never solves problems. The reason we dislike them in the first place is that we've been taught to. There is nothing in the feeling to dislike. It is a movement in the body, a flow of something, maybe a hurt or a woundedness that we assume to be weak, neurotic, or wrong. All those labels are made up.

Feelings come and feelings go. The interruption of this flow comes from conditioning, from habitually imposed responses. When a person turns toward whatever is being felt in the body, it is always different from what they initially thought.

Something is happening to us as human beings that can't be explained by surface events or by the psychological dogmas that parade as truth. Deeper than the content of life dwells a mysterious force — a presence, if you will — that is guiding us toward an unknown end.

As far as explaining loneliness in terms of the past, it's important to see the past not as the cause of a problem but rather as a formative, evolutionary force. It's something that we let work on us, not something from which we try to release ourselves. The only piece of the past that one would try to release oneself from is the hypnotic conditioning that prevents us from having a rich experience in the present.

Safransky: I want to be braver. I want to be someone who isn't hung up on his own fears, but who extends himself compassionately toward others. Yet here I am, with my shameful dependency on my wife.

Schwartz: The idea that you have to be something else, to prove something, is disheartening. Heroism is not overcoming what we perceive to be negative about ourselves or anything else, but rather facing right into those things — finding the core. A heroic act is a naked encounter with what we've judged to be dangerous and then, perhaps, the discovery that it is something else entirely.

Safransky: Let's take another example. I become jealous because of something that's said. I'm angry, but also ashamed. I'm full of self-righteousness, and at the same time I know I'm being possessive and unfair. One minute I'm justifying my feelings, and the next minute I'm condemning them.

Schwartz: The justification and the condemnation are exactly the same. They represent a moving away from the actuality of experience. Jealousy is a good example. On a certain level, human beings are fragile and vulnerable, at least in terms of our physical existence here. The forces of nature are more powerful than we are. The force of a disease, for instance, may be more powerful than our will or our desire to have it healed. We live in a vulnerable place, on an edge between what we think is going on and a great mystery. Rather than allowing ourselves to enter into the truth of our vulnerability, we create a stance to protect ourselves from it.

It's very beautiful, if you think about it. We are trying to protect and take care of ourselves, but in a backward way. It seems dangerous to encounter our vulnerability directly, so we create a fantasy scenario that keeps us away from the feelings we assume to be dangerous.

Safransky: What do you mean by a "fantasy scenario"?

Schwartz: I mean the way in which we substitute a kind of cartoon made up of personal beliefs, images, assumptions, and associations for the mystery and spaciousness of what exists.

The fantasy grips us because the body responds as if it were real. First we interpret what is occurring "out there" as dangerous; then the body tightens, as if to protect itself. The tightening of the body seems justified, even helpful. But when we really get down to it, this tension is an attempt to avoid a deep encounter with an unknown, an edge, a mysterious moment in which we are suddenly aware of how little we know and how delicate this life is. It is an attempt to run from discovery. But we can't.

As odd as it may sound, just being with a feeling — not trying to change it or get over it — leads to a radically different sense of what that feeling is. I wish I could express the beauty and power of this, but it can come only through practice and patience.

In this work we turn an experience like jealousy upside down. Maybe the reason you have a low threshold for jealousy is that you are a person with a natural inclination toward something mysterious, delicate, and vulnerable, and there has been in your life, for one reason or another, such abuse of that vulnerable place that you will do almost anything to avoid it, including turning your body against your body.

Maybe your low threshold for jealousy represents a kind of giftedness — an openness to life rather than a problem. I would be willing to bank on that. It seems to me that what we struggle with most, what torments us, represents something profoundly important. When we face right into what we've been calling our weakness, we can find our gift, our unique energetic expression.

Safransky: You are saying, in effect, that jealousy is not a psychological problem.

Schwartz: Our so-called psychological problems take place in a fantasy realm that has almost nothing to do with the actual space in which human life is unfolding. These problems seem so real because our attention is consumed by thinking, and the real space is sacrificed to a substitute reality.

So much of what we call psychology is actually a mystification of experience. The question of where an experience actually takes place is rarely addressed. When someone comes to a psychologist and says, "I'm lonely," how often does the psychologist ask, "Where is the loneliness?" Or, "What is it like to be you having this loneliness? Is it in the upper chest, the stomach? Is it vertical, horizontal? Does it permeate the whole body?" These questions change the nature of our feeling experience. The loneliness becomes felt in the body instead of remaining dangerous and abstract.

The work of compassionate self-care is not about trying to find out why a particular psychological problem exists, but dropping out of the problem altogether and merging into the natural spaciousness that exists in and around the body.

Safransky: This is why people meditate, isn't it? But there's the temptation, when we try to be more detached, of pushing away who we are psychologically.

Schwartz: In a meditative approach to life, the emphasis is entirely on the disengagement. In self-care work, the emphasis is on using the problem and the pain as a bridge between the tight place that we find ourselves caught in and spaciousness itself.

In this work, the mental struggle is not ignored. It is attended to and then dissolved through a practice in which a rhythmic interplay is developed between feeling the pain in the body and expressing the dilemma as it appears in the mind. We come back to the body, feel the body, attend to the body, and alternately speak about the problem, even in great detail. Over time, the mind's grip on its particular point of view is increasingly defused of power, and something else becomes available instead. It may be pain. But no matter what it seems to be at first, we have entered another side of our experience as human beings. We have found the gateway to mystery.

We don't enter into this process to gain an insight that will allow us to see through the problem. Our goal is simply to recognize that on some level the problem is a fabricated story, a substitution for a sublime truth about ourselves and each other.

Safransky: There's the story I tell myself about jealousy — and there's something the story leaves out. There are my beliefs about intimacy — and there's something more mysterious.

Schwartz: Absolutely. For instance, the word *intimacy*, even in the most radical psychology, doesn't mean the same thing in the context of the self-care process. Generally speaking, intimacy tends to refer to sharing something personal — perhaps in a felt way, but primarily through information. Intimacy in the context of self-care means opening energetically to another human being — not knowing them as anything other than a radiant mystery.

For me, the greatest fulfillment in my work arises from experiencing another person as a great opening, a huge space. We sit together with the eyes closed, and I can feel who they are. This body only appears to be an enclosure. It is actually a passageway — like an entry to a cave or a cathedral. It is quite the opposite of the way we've been taught to perceive it.

Safransky: I value intimacy, yet I'm afraid of it. I fear that loving people may lead to *making* love with them.

Schwartz: That's an interesting statement, but there is something more radical and scary in this than what you're talking about directly. The encounter with another person on the edge I'm describing is much sexier than sex. It's highly erotic, but deeper than any physical act.

It all could lead to some sexual act, but the greater challenge and the greater risk is not having it lead to that but letting it go *beyond* that. Sitting with another person and discovering them as an opening, as opposed to an enclosure — finding that space and entering into it on a subtle energetic level — is as intimate and erotic as any experience could be.

The physical act of sex is a kind of metaphor for the deeper encounter we speak about here, and yet when there's no sexual encounter, we have little intellectual justification for jealousy. This is the heart of what attracts us and frightens us about human relationships. We want to have a deep and open exchange with other human beings, and yet we spend a lot of time running away from it. We don't know what that exchange really is.

Safransky: It's easy to be confused about this, to mistake one thing for another.

Schwartz: Especially with sex. It seduces you toward it; then whatever it is that enticed you isn't actually there when you get to it. Isn't that true with sex, in the long run? The way it's been built up and the aura that surrounds it make it seem like an answer to something, but there is always a longing for some greater contact.

Safransky: That longing becomes suspect, like so much else.

Schwartz: Often we treat certain aspects of ourselves as junk, having no value. We try to throw parts of ourselves in the garbage. But a human being is an ecosystem, and everything in that system is of value to the whole.

The community in which a person lives — the sphere of friends and family — is another kind of ecosystem. It is necessary as a social service to be respectful and open to the qualities within ourselves that we can bring to the whole. Disparaging and attacking certain aspects of ourselves is not a service to those who are involved with us in the greater ecosphere.

If we look at ourselves from a slightly different angle — a little more compassionately, perhaps — we can see that there are no negative emotions, and no positive ones, either. At the heart of every emotion is an innocent wave of energy that is inherently free from psychological and moral dichotomies.

Feelings are energy, and by their very nature ascending or expansive. They are trying to create, expand, grow, and move in various ways. Thoughts about feelings exert a downward pressure, creating a stranglehold on that ascending force. The purpose of the self-care process is to touch the ascending force and transcend the downward force.

It is vital that we look at the human being as an exotic life form, rather than as some familiar thing that we've grown so accustomed to, something that bores us. We are a mobile life form, moving about on a planet, in a universe that we know almost nothing about. But we carry a bizarre unconscious assumption that we know almost everything about it.

Safransky: Yet how little I really know about my own feelings, about jealousy, about loneliness.

Schwartz: There is something beautifully paradoxical about what you're saying. Not knowing leads to another kind of knowledge that is filled with possibilities — a knowledge of the heart. The kind of knowledge we've been taught to trust is imposed on a mystery and is different from the knowledge that arises from our willingness to be ignorant.

It's like waking up in the middle of the night on the edge of some incredible dream and not remembering where you are or what your life is about. Suddenly, there is that incredible line between terror and profound attraction. It is so much more interesting to wake up in that dream space than to wake up in a space that has been made familiar by an arbitrary system of beliefs.

Safransky: Still, it seems that understanding oneself requires understanding the past.

Schwartz: Insights about the past come from thought and not from the experience itself. For the most part, we live our lives without any experience of a sudden moment when the past is no more — when there isn't something that happened, there isn't a good belief or a bad belief, there isn't anything at all except something very mysterious and alive. We as human beings aren't those memories; we aren't those traumas. Certainly they exist and have an impact, but ultimately there's something underlying all of that — something that can be experienced. And when a person does, it's liberating, even though it might not be comfortable at first. It's liberating because it's alive. It's not *about* anything.

Safransky: So you don't ask questions about the past?

Schwartz: I'm not saying that if somebody started to talk about the past I wouldn't follow them, but we'd never stop there. I'd never say, "Oh, there it is. It's the way your mother treated you."

When I began this work, I saw that most of us were carrying around a sense of wrongness or inadequacy that was actually an ideology. In the dialogues, I would just follow someone until the very net of their own conflict caught them. There wasn't any place to take it on a conceptual level, and something would split open. Then, and only then, could I feel that person, could they feel me, could the people in the room suddenly be aware of the real human connection. From this I realized that communication can't take place through beliefs — no matter how great the beliefs, no matter how powerful or sublime or spiritual. If you take the lid off the belief, you can see that it was an idea protecting you from something you didn't want to feel, something very raw and vulnerable.

What I'm focusing on in any so-called neurotic situation is the disruption of a bodily process at a very subtle level. Mundane thought, by its very nature, is confused; you can't solve anything at that level. What we're working toward in these settings is an experience that takes place in the rhythmic opening and closing of the subtle, physical membrane of the body.

Safransky: You talk frequently about the front of the body.

Schwartz: This developed very naturally. When a person spoke about emotional difficulty, we'd start moving to the body to feel that pain. Over and over again, it was experienced in the front of the body. That interested me. In the work I do, I can literally feel another person's space. I wish I could describe that a little better. It's not psychic. It's not intrusive. But I have a tremendous sense of that person in a spatial way. I discovered that, in sitting face to face with someone, I would have a much greater sense of that person than I had ever had before.

Safransky: Do you see colors? Patterns?

Schwartz: No, I don't see colors. Patterns, yes. But this experience of emanations is not like an aura reading. These emanations are a unique space that we carry around us and in us, and I'm speaking very physically here. I can feel when someone's space is being enlivened and when that space is closed off.

Safransky: Can't compassionate self-care turn into a kind of self-absorption, into an unhealthy preoccupation with feelings?

Schwartz: I understand what you're saying, and that's why I try to put this work into an expanded context. We do not own our feelings. They become ours only when the mind entraps them in highly personalized definitions. Prior to that entrapment, they are energies given to us in much the same way that the air we breathe is given to us.

I always go back to the saying "Man does not live by bread alone." There is another source of sustenance, another energy through which life is sustained.

When we turn toward ourselves in a certain way, we end up turning toward the universe. As we deconstruct emotions, we are taking apart an intensely introverted sense of identity, which is always caught in its own confines, and coming to something much broader, much more encompassing. We can give love to the world only when we know ourselves to be much bigger than a complex string of memories, ideas, and beliefs.

Safransky: Would you say that practicing compassionate self-care inclines one toward a greater sense of social responsibility? If people are more sensitive to themselves and others, will they act in a more responsible way?

Schwartz: I think so. I think they act with more kindness, which results from disengaging from one's own obsessiveness. The natural outpouring of this would be a very spontaneous ability to appreciate someone and to be sensitive to them in a way that one wasn't before.

Safransky: Would someone be more inclined to fight injustice in the world?

Schwartz: There's certainly no contradiction between the work I do and an indignation about injustice. But I would say that implicit in the work is the understanding that the creation of enemies has never solved a problem. We can look at history: after every enemy is vanquished, the central problem still remains.

Safransky: The psychologist James Hillman describes someone going into a therapist's office angry about the fact that there's too much pollution, and the therapist saying, "How does this remind you of when you were a child?" Hillman says this results in one less person who's going to go out and fight pollution, because the passion has been placated.

Schwartz: I think there's validity in that. I wouldn't want anyone to think of what I'm doing as placating. But it's also true that anger like that doesn't necessarily lead to heroic acts of political courage. With or without therapy, people end up conflicted, confused, and helpless. There is something more than therapy, and something more than political action. A greater possibility exists for all of us.

Rather than finding the causes of anger — your mother or your father or your past, or even pollution — I would say we can find a message in it. That is, there's an energy in the anger that is a message — a hieroglyph from the universe. This may sound poetic, but it's very real.

Most of what we call anger is mainly fantasy to start with. There's also a physical tightening against it. And underneath that there's a kind of force, a very powerful movement, even a direction. If we can lift the fantasy and lift the tightening, we can come to the force.

All of our conflicts — whether about pollution or money or anything else — are an interplay between downward, restricting, limiting conditioning, and upward, ascending, creative urges. That is, between expansion and contraction.

Safransky: How would this apply to someone who's being persecuted? Not just to someone contemplating the existence of injustice, but to someone actually suffering?

Schwartz: I've worked with people who have committed horrible crimes. I've worked with people who are the victims of crime — rich and poor. These understandings do not change. When self-acceptance and self-care form the basis of our response to what's going on, there is a much greater opportunity to create and to move forward.

So many of us assume that hating our lives the way they are is a way out. Quite the opposite is true. Only more pain and more oppression come from such an internal stance. Deep and direct acceptance is not the opposite of creative action. It is the beginning of a new life.

We are most powerless when we believe what we've been told about ourselves, when we don't find out what we really are and what life is about. The greatest power comes from approaching our own experience honestly and directly, not through the dictates of conditioning. It's as possible for someone who's poor as for someone who's rich, for someone who is not educated as for someone who is. It's fundamental. There's a feeling. There's an interpretation of that feeling. You locate the feeling in the body, breathe with it, listen to it as if it were a kind of message. The feeling becomes a guide. But its guidance is far different from the guidance that comes from the mind.

I've worked with people in prison, and I respect their anger. The anger — without the conditioning about what the anger is about, and without the physical close-down — is their key, their power, their strength. I would never seek passivity over anger.

I've always been very moved by stories of religious conversion. I'm particularly moved by *The Autobiography of Malcolm X*. Malcolm X went from a culturally induced state of humiliation — believing that it was better to be white, that he was trash, that there was no life for him but in crime — to an unbelievably clean discipline and an orientation toward a higher goal than mere survival. In that ferocious discipline he imposed on himself, he was able to be a moving force for other people, highly inspirational.

Safransky: What if someone in one of your workshops were expressing racist sentiments? Would you still be able to be respectful?

Schwartz: Absolutely. Until you respect it, there's no room for change. Until you actually come to that person, acknowledging their stance, acknowledging why it arose and where it came from, there's no way that person is going to trust you enough to get into that more marginal, vulnerable space.

Safransky: How would you define evil?

Schwartz: You can make a spiritually irresponsible statement about evil. For example, I've heard people say that Hitler was crying out for love. That may be true. But there was a massive anti-ecological, anti-evolutionary act that he brought into existence, and one could call that evil. There's a danger in using the deepest level to excuse the other levels. So it seems to me that there is evil. There are acts that oppose the flow of life and growth and human dignity. They must be dealt with courageously. They must be dealt with by warriors.

Safransky: Do you think there are occasions when killing is justified?

Schwartz: I think there are occasions when almost anything is justified.

Safransky: So your stance is not nonviolent?

Schwartz: My stance is totally nonviolent, but that doesn't mean that one can't raise the sword. This work is an exercise in nonviolence toward oneself — approaching everything we've called an enemy with absolute, undeviating love. It is an exercise in personal nonviolence, in treating oneself with the most unbelievable compassion. Yet one can raise the sword with compassion.

Safransky: Can one bring it down with compassion?

Schwartz: I think so. I think that's rare, though, and one has to be very cautious about making that into a position. Gandhi said that if it's fear that keeps you from raising the sword, then it's better to raise the sword. Nonviolence is total fearlessness.

There are people who would try to bring nonviolence into the world, even though they can't approach their own loneliness with complete nonviolence. How could they deal with somebody else in a nonviolent way? To know nonviolence means that you have no enemies inside yourself; that your loneliness, your grief, your anger are not your enemies.

Safransky: You were diagnosed with cancer recently and went through a harrowing ordeal. Were the principles of compassionate self-care available to you then?

Schwartz: I came into the hospital with a distinct and identifiable prejudice against the whole system. I didn't care for the way most doctors practice medicine, or the way hospitals are currently set up. I still don't. Part of the reason it took me so long to do something about my illness was that I knew instinctively it was going to lead me into that system. Finding myself on chemotherapy in the intensive-care unit of a major New York City hospital was a direct violation of my beliefs.

When a belief system is violated by an experience, we can choose torment and conflict, or the return to something very simple. That choice, for me, involved caring for my feelings and my body at a level beyond meanings and ideas.

I had nothing to hang on to. I was being cured by what I hated, thrust into what I profoundly disagreed with. Only with the process of self-care could I carry myself through that encounter. When I felt pain, fear, or loss, I turned my attention to it and took care. Even when I was groggy from anesthesia, even when chemotherapy was being injected into my veins, that's all I could do.

But to say that the self-care process "worked" is a kind of absurdity. It didn't get me out of anything. It brought me into direct and compassionate contact with what was there — no filters, no distractions, no abstract ideas. What I found within myself wasn't bad: My fear wasn't bad. My loneliness wasn't bad. My sadness wasn't bad. My brush with death wasn't bad. It all just was.

When I was in the hospital during those ten days and became frightened by thoughts, I realized I couldn't afford to indulge them; it was weakening me to do so. The fear in my body wasn't weakening me, but the obsessive descriptions, speculations, and fantasies about what might be happening to me were indeed destructive. I had to return my attention over and over again to my body; that's the only way I could take care of myself.

Safransky: Can you give me an example?

Schwartz: I'm told that I have a tumor larger than a basketball in my chest. I go into the hospital for a simple biopsy. I end up in the surgical intensive-care unit unable to speak, eat, or drink. I realize, from overhearing conversations, that I almost died on the operating table because my windpipe was twisted so tight. Chemotherapy is administered to me in the middle of the night, but I'm not even aware of it because I'm so out of it. Suddenly, I begin to feel my death is inevitable.

I'm confused and ill. I realize that my wife and children really need me, that I might very well die and leave them not taken care of properly. I think about my parents, who live on the same property with us. They are going to have to bear the burden of my passing and, in a sense, of decisions I have previously made. All this becomes a roaring fantasy, a round of torment. Having that fantasy does not assist in recovery, nor does it make the situation better in any way; it actually worsens it. It's an indulgence. It may appear to be realistic, but it's not at all.

I know I'm speaking about an extreme circumstance, but it can serve as a metaphor for the way so much of our life is lived. I was completely isolated, unable to communicate what was going on within me. At such a moment, obsessing about the situation is a complete turning against oneself and a denial of the real need at hand. My only resource was to breathe, to come to the body; to feel the fear and the pain, and not engage in that kind of sabotage.

Safransky: Not everyone has had such a close call with death. But we all live with the knowledge — which we usually repress — that one day we're going to die. How do we live more consciously with that fact?

Schwartz: I don't think this can be done in an intellectual or philosophical way. People have developed some pretty mighty beliefs about dying, but you don't go to death with your beliefs. Even your belief about whether there's an afterlife doesn't amount to anything when you're touching up against the fact that the body is about to go. Any belief that is held out of fear or as a mental construct is wrenched away.

What you have really taken into yourself, absorbed in a true way, is carried with you. But if you're holding on to a mere mental structure, it is dissolved. It's no longer available.

It is not useful to hold on to beliefs. We can only keep coming to the mystery and doing the best we can. When you really need to take care of yourself, beliefs get in the way. They topple easily and leave you with something that can't be explained — and that is what you must turn to in the end, anyway.

Safransky: You came close to dying just before entering the hospital.

Schwartz: I had fallen asleep at home one evening, when my windpipe became almost totally constricted. It was impossible to breathe. My wife, Donna, heard me moan just a little bit. She leaned over and touched me and realized that my body was stiff; I was leaving.

My identity — who I assumed myself to be — had become quite diffuse and was expanding out into a field, a radiance surrounding my body. I could actually see my body behind me on the bed, gasping for air. But it had little to do with who I was anymore. I felt no particular attachment to it.

As I faded from my earthbound identity and entered into something much bigger, I felt sorrow, but no fear. Then, in a way that doesn't resemble our experience here on earth, I was given a choice (I didn't know who was giving it to me, or why): I could go back or continue on. It was as simple as that.

I knew suddenly that I had to go back. There was no need for deliberation or the complexities of thought. I had to return. It was then that Donna put her hands on my back and I saw myself being sucked into the body, being re-formed and, in a way, restrained by my physical contours.

At other times during my illness, I was acutely aware of something lying on the other side of this physical realm. It was like an invitation. There was also, side by side with this invitation, a deep awareness of how precious life in this body really is. There was also fear. The recognition of the sacredness of this life and the fear of leaving it were two different things, however. My sense of life's sacred value was not a fearful holding-on to anything. It was soft.

Allowing ourselves to be without the familiar labels we use to define our lives, and especially our feelings, frees us from a reality that is largely made up. We can then begin to experience ourselves as a living phenomenon, a mystery, something that can't be understood along the usual lines. This is exactly what happens at the moment of death, and this is why even brief encounters with ourselves in an open way are so transformative. They offer the opportunity to be reborn.

As long as we keep looking at ourselves in the same old way, we lose sight of the ever-changing process of life. We are not static, framed, or caught. We can't be boxed in for long. Insight and conceptual analysis relate to still frames. But this life is all change. Nothing ever remains the way it was even a few minutes before. Here lie the beauty and the fear, the adventure and the desire to hold back.

Safransky: There are so many myths about illness and healing. How has your recent experience changed your ideas about what it means to be sick?

Schwartz: Somewhere inside, I must have thought that I should have been exempt from a disease like cancer because of the way I had eaten and lived for so long. I could feel sometimes a kind of humiliation, which arose from the contradiction between an image of myself and what was actually going on. I don't feel that anymore at all. I am grateful that cancer came into my life. Nothing has had more power to transform; nothing else could bring such relief from the burden of unnecessary self-images. Cancer has been an ordeal of fire, and a kind of grace as well.

The message of my work from the beginning has been that there is no relationship between the circumstances of one's life and the level of spiritual development one has achieved. We can't measure spirituality in terms of what happens to us. In fact, there is no way at all to judge where we are in spiritual terms. All we can do is treat our lives, both inner and outer, with deep compassion and acceptance. Nothing else really matters.

Life can throw very difficult challenges our way. Illnesses may come, great losses, changes of all sorts, but those situations are no indication that we're holding negative beliefs or thinking negative thoughts. Such an approach to life is moralistic and abstract. It doesn't really support the unity and mysterious purpose of all things.

Safransky: Eating a certain diet or living a certain way doesn't make us invulnerable. There's no protective shield.

Schwartz: After my stay at the hospital, I would return there for tests and see people on the street living in cardboard boxes, scavenging food from garbage cans, sitting around a fire drinking from dirty bottles. If this illness comes as a result of bad diet or negative thinking, I would wonder, why don't those people have it? There is no human logic to any of these things. All we can do is participate, cooperate, and in so doing find the transcendent silence that dwells in the heart of all things.

I am grateful, and I don't mean this lightly, for having had to face this experience, but I would not want to go through it again, nor would I wish it on anybody, enemy or friend. It was a walk through fire. Nothing else has ever brought me this kind of purification and change.

Safransky: Can you be more specific about how it has been a purification?

Schwartz: I want to be cautious in my response so as not to be misunderstood. There is something about suffering, taken in a certain spirit, that can be tremendously purifying. A more accurate statement might be that there was something purifying for me as I went through it. And it was not the illness alone that caused the suffering; it was the illness in combination with the chemotherapy.

I didn't want to be in bed for six months. I wanted to feel the fullness of my life during this period of time. To stand up and continue my work — even with the toxicity in my body — has been strengthening, not weakening. I had to find a force within, the will that made it possible to go on.

There were also certain things in my life that I had to clean up, but I couldn't quite identify what they were. I just felt there was something that wasn't entirely clear — a subtle heaviness, perhaps — and this has lifted it for me. Because I came so close to death in a conscious way, my priorities have gotten straightened out. It can happen very quickly. Lying on a bed with a respirator, unable to talk, with my family surrounding me, I was offered a sense of what is really important on this earth. You can feel the danger people put themselves in by making unimportant things seem important.

Safransky: With all the attention you give to the body, I can't help but wonder why you didn't know such a tumor existed.

Schwartz: I knew something was wrong. A part of me thought that I could deal with it myself. Interestingly, when I sat in meditation or with other people in the circles, the feeling of imbalance would go away. In my deepest internal connection, I would transcend it.

I would facilitate a weekend meeting, for instance, with no sign of the physical problem, and then, when the meeting ended,

I'd get a fever. The fever would continue for about four days, then go away. This pattern went on for a year.

During that time the fevers never felt like an illness. It was like being at the edge of something, an extreme delicacy. I've felt much sicker from having the flu.

Also, as I explained earlier, I had a profound philosophical disagreement with the medical system. I didn't want to end up in the hospital.

In some peculiar way, going all the way with this was important to me — I don't know why, but it was. I wonder sometimes whether, had I discovered the tumor earlier when it was relatively small, I would feel the kind of expansiveness and lightness that I do today. There's something powerful about going all the way down and coming up the other side.

At any rate, it's impossible to make sense of this in any conceptual way. In trying, it becomes twisted and unrelated to the actuality of the experience.

Donna and I used to walk at night together along River Road, just at the edge of the Hudson River. We would stop to look at the Tappanzee Bridge and watch the golden lights spanning the water. We would pretend for a while that we didn't know anything about where we were or what we were doing there, and suddenly the bridge would become a jeweled necklace, a thing of exquisite natural beauty. It was no longer the concrete structure we had seen so many times before. A new meaning permeated everything, as if we were in a dream.

I contemplate the last few months in this way sometimes: What if it had been a dream? What would it all mean? How could I tell it to others so that it could be felt and understood deeply? I can't answer these questions in a concrete way, but during those periods of contemplation, the memories have a different tone than they do when I dwell on the "whys" and question all the decisions I've made.

You know, Sy, when you ask me about the sensitivity to my body and why this came about in the way it did, I have to say that I truly don't know how I do the work I do. I don't know where it comes from — it's just there. But I don't feel removed or different from anyone else. My role is different, perhaps, but the needs, the dreams, the hopes are exactly the same. I am not necessarily going to handle what happens to me in the world any better than somebody else. I don't even know that there *is* a better way to handle anything. I am out there doing what I can against the backdrop of an incredible mystery. And that's the way it should be.

October 1992

The Body Knew

Tim Seibles

Long before there were words
long before there was *patience*
the body was twiddling its thumbs

Long before this haze of lies this
swirl of stupid things
said and done
the body knew

Long before the animals ran
from men before the lands
were named before the clouds
rose up and flew
the body knew

The body knew the tongue
would come up with something to say
that the ears would listen that
the words would come like ants
that soon the brain would be
infested and the head would grow
hard and heavy
The body knew the body

would be forgotten
The body knew the body
would be used to take the brain
there and there to make
money to make *relationships*
to assume the countless postures
of idiocy — to sign the contracts
and treaties to stock the stores

the homes the schools the offices
the streets the prisons the battlefields
the body bags   The body knew

it would be lost
under fabrics that soon the belly would
hang and the back would be stiff
that the days would pass the months would
pass the years would pass
The body knew

it would be *rated x*
because the body knew words
would be used to deceive to
decorate to pack the space between bodies until
reaching out meant climbing the mountains
of things said
The body knew

the brain would be a bully
that the face would be a canvas forever
painted with words that *love* could never be
what they said it was that a word
was always a mask
The body knew the body

would dream of headlessness the way
a breast dreams of bralessness of blouselessness
of sunlight and weightlessness
The body knew that someday
it would have to move to forget to
dance to forget that it knew
what it knew
that it knew

May 1986

Tully

D. Patrick Miller

The tree is the regenerative, unconscious life which stands eternally
when human consciousness is snuffed out.
— C.G. Jung

Now, in the long evenings after dinner, she often found herself standing before the bathroom mirror, trying hard to glimpse some of the prettiness her husband had always championed. The glass portrait started well above her head, and just enclosed her perfectly circular, perfectly average breasts, but left out the last three inches of her fine and golden hair. Sometimes she would gather it all and bring it round one shoulder, holding it in her hands so that she could see the ends. Her hair

---

---

was her only feature that satisfied her, but she feared it also made her vain. With so much time and aimlessness on her hands, she washed and brushed it sometimes twice a day. Could she ruin her one claim to excellence?

The thought made her eyes shift nervously to her face, which she almost never liked. Her eyes were certainly too large — something Andy had thought was wonderful! — and she had never fixed their changeable greenish color exactly, which always gave her pause on forms that asked for that detail. Her thin lips and the size of her mouth were OK, she guessed, but just OK — she had never understood why Andy would so often trace her lips with one finger, quite kindly, before kissing her, as if he had to use a spell to open a door always accessible to him. . . .

She blinked and sank deeply into another memory she found inescapable of late. Every time it had been her turn to undress him, she would pull off his socks last of all and then climb him with a growl or a giggle, and grab the ever-ready standing part of him like a branch on a tree. Then she would drop her head over him and mouth him with graceless enthusiasm, always feeling she must make a good show to compensate for not knowing exactly what to do. She knew she had no technique, and couldn't control his arousal as she should — but nonetheless he flailed about and whimpered like a puppy. Sometimes this made her forget what she was about and begin to laugh, which always caused him to pull her head up to his and her mouth to his, and then he always wanted her too fast, when she wasn't quite warm enough —

The familiar cramp below her stomach seized and unfolded, releasing nausea, and she found her face in the mirror again. Why couldn't she stop this? How many months would it take to forget that kind of thing? It was only when snapping out of these reveries that she thought she might really need help, for sometimes they were even worse than the dreams. But she could make it: she simply had to concentrate. She concentrated on the weakness below her lips. *No chin*, she thought sadly. *No chin at all.*

She reached up to the hook on the door and drew from it her thin white gown, which when brought round her did little more than veil the dark texture of her nipples. She knotted the frail sash just under her breasts, and looked up into the mirror to find a strange, demonic grin on her face: a demon inside her remembering the glint of desire Andy had always shown when she'd come to bed wearing that suggestive garment. Once, she had sat on the edge of the bed and, pushing away his large hand with both her own, asked innocently, "Why does this *particular* item of clothing make you crazier than usual, mister? It's practically the same as naked."

His reply had stolen her breath, like all his best surprises. "Because it makes you a secret," he'd whispered dramatically, raising his long forearm to push her back to the bed while nuzzling her stomach through the gown.

She decisively clicked off the bathroom light just in time to short-circuit the nausea — *There*! she thought. *I did it* — and strode forcefully down the hall to the living room. She took her usual cross-legged seat on the floor by the beat-up orange couch and, before reaching for the TV's remote control, tried to retrace her thoughts. She was trying to learn how to think continuously, without lapsing into memories and mournful fantasies, and sometimes it was helpful to back up to the point before her mental lapses. *Chin*, she thought, and rubbed her palm wistfully over the chin that wasn't there.

Andy's sister Andrea — now, there was a woman with a chin! Of course, she had Andy's square chin, but smaller and somehow stronger. You noticed that long, defiant jaw first, and then the high cheekbones that made Andrea — or "Andrew the First," as she often wryly introduced herself — a real beauty. Brother and sister had the same dark, almost predatory eyes, but Andrea's were fuller and wiser — Andy had usually chosen not to look into things too deeply. His wife marveled at his ability to sidetrack his own intelligence so deliberately, but perhaps it had kept Andrea from abandoning him. What a rivalry those two had had! Andy had told her many times of the fifteen years of long silences and short skirmishes with his sister, from the time Andrea had left for college in California until their parents had both been lost in an air disaster. It was then Andrea mysteriously left her university position and her life on the West Coast to live near the younger brother she supposedly despised. Andy had silently closed the deal by surrendering the verbal sparring, becoming habitually clownlike in Andrea's presence.

Why, wondered his widow, had he chosen to act so strangely? Why, in fact, had his ambition seemed to founder after losing his folks? His mourning had mutated him somehow, cutting off his sense of direction, so that he floated from job to job, mostly manual labor, in the three years of their marriage, and took up more passing hobbies than she could remember. She had felt helpless to counter or calm his thrashing about — probably because she was no mental giant herself. So they both became charges of Andrea, who would call at least weekly, inquiring after everything from Andy's current job dissatisfaction to whether they were eating well. It irked Andy, but he would not challenge the precarious peace within the family he had left, and his wife actually liked the attention. *Andrea was always smarter than both of us*, she thought, finally reaching for the remote control to push a button, but seeing a flash of Andrea's anger instead.

They had been eating out, she and Andy and Andrea and her friend from the Richmond Institute of Women's Studies. Andrea was explaining why their names were only one letter different. "You see, I was really Andrew the First. Our young man here was the boy my father wanted, but he futzed around and got born second." Her fork clattered to her plate, interrupting her tone of forced affection, and she finished her story, her fiery gaze traveling to a far corner of the restaurant. "That's why I had to be so goddamned smart." Later, she had commandeered the check and charged it, forcefully signing the slip, "Andrea Waycross, Ph.D."

Andy had always laughed off these bad moments, his wife recalled, with a high, nervous titter that made her uneasy. It was nothing like the clear and easy laughter that had seduced her instantly on their first meeting. She'd been a silly sorority girl, floating through an average Virginia college with only the dullest of aspirations — she shook her head at the chilling realization of just how pointless her former life seemed today. She'd been giggling and nervously trying to act her way through a mixer with boys from a slightly classier school when she spied a long-legged, unfashionably long-haired odd bird looking lost by the corner of a table. Feeling suddenly responsible for playing hostess, she boldly approached the stranger as he pulled a pipe from a deep coat pocket and began fooling around with it in the manner of a total novice. She was fond of his ungainliness already. "So," she said in the sweetest, blondest voice she could muster, "are you from Jameson?" — a question so dumb she still winced at the thought of it seven years later.

But the boy gave her his first surprise. "Sort of," he choked, darting his eyes around like a comic fugitive from a silent movie.

She was not yet accustomed to his unexpected behavior. "Oh," she answered vacantly. "That sounds interesting."

He smiled tightly and gradually brought his dark eyes to focus on her. A too-long moment passed in silence. "Well, what's your name, I guess?" he said in a rush, a dark red flush rising from his neck.

"Tully," she said carefully, practicing her new, clipped non-drawl — she badly wanted to be from *northern* Virginia, and not almost North Carolina. But then the truth spilled out before she could stop herself. "My name is the most interesting thing about me."

The boy started, and then his mouth opened wide in the uninhibited, ringing laughter Tully would always wait to hear in their subsequent life together. It came in four or five overlapping peals, as if from a deep, rich bell at a great height. He dropped the charade with his pipe and shook his mane of dark brown hair as if to refocus on her. For an instant, he seemed incredibly adult and wise. "Really?" he said, grinning, seeming thoroughly pleased. "Well, at least you're a damn sight more interesting than all these other *fas-ci-nating* ladies," he declaimed grandly, making a broad gesture that somehow ennobled the tackily decorated hall she had felt abandoned in just minutes before.

Tully squeezed the remote control hard and watched a blurry image of a fatherly Bill Cosby materialize on the TV screen. Andy was *so* charming, she remembered bitterly, with that Georgia grandiloquence and complementary lack of know-how. Ten minutes later, they'd been out on the humid balcony of the ancient building, his teeth clicking around her ears because he seemed not to know exactly where to start. She didn't care; she was hugging him as fiercely as she had the fireman who'd saved her life when she was five. She could feel his strong erection against her lower belly.

So young, she thought, squeezing her eyes shut to push out the tears. She marveled at how much older she felt now — "But you're still a young woman!" her mother's voice barked inside her head. It was true, she realized, but for now she much preferred to remain a hermit. It had been nine months since Andy's death, but her so-called girlfriends — most of them divorced, she thought meanly — already wanted her to go out with them. "At least you can *try* to have a good time," her mother had cajoled yesterday, taking her friends' side. Tully turned up the volume on the television, but not quite to where she could make out the dialogue. Only Andrea seemed to understand; she would walk into the apartment, undisturbed by the lingering presence of Andy's possessions, and deliver an answer before any question was asked. "It's OK," she always said. "It's OK, for now."

Tully wished Andrea would come to visit her tonight, and almost every night, but the proud, handsome woman came on Tuesdays and Fridays only, always punctually, always matter-of-fact. She would give the usual excuse: "The light is *so* much better at your place," she would say in mock envy, dumping her tower of books on the couch. But Tully knew that Andrea studied at home, too, inferior light or no. She was always studying.

Andrea would read, pencil clenched in her teeth and papers rustling in her lap, while Tully watched the quiet television blankly, with the disinterest of a drunk. And she was a little drunk, intoxicated by the beauty of this woman with the firm, forward breasts and the lean, worked-out legs and the unmistakable resemblance to Tully's lost husband. Tully felt guilty for the sensation, as if she were cheating or tempting fate. But she didn't care to do anything else and had no reason to send Andrea away. She would just as soon give up her life, and it shocked her when she realized this was so.

The evenings with Andrea passed so much faster than those she spent alone; that was one reason Tully wanted her there. They talked very little, yet a mutual sense of comfort settled over them. Late — usually about midnight — they would go into the kitchen and eat together while making small talk, and then return to watch whatever was left on television. Tully would go back to her seat on the floor, but Andrea would push her books aside and sit forward on the couch, her legs lightly embracing the widow by the shoulders. And for an hour or so Andrea would play with Tully's natural prize, braiding, unbraiding, swirling, and stroking her long, wonderful hair. Sooner or later, something would rise in Tully's throat, and she'd have to clench her jaws to keep from crying or turning to embrace one of Andrea's protecting legs. Because this she could not do — envisioning it, as she often had, led her to an edge of mystery full of anxious murmurings.

She was certain Andrea knew about it all, and yet she never did anything. Once the stroking stopped, Andrea would soon release a full, collapsing sigh, and then Tully would excuse herself for bed, as if on cue. She had to be up at seven to reach the reception desk at Dr. Morgan's office at eight; she was proud that she had never been late and took only a week off following Andy's death, though the gray-haired patriarch had begged her to "give yourself more time" and take some extra money, too.

Andrea usually made some joke about "the life of an intellectual" and feigned interest in an old, late-night movie, or even David Letterman, whom Tully despised. Tully always slipped easily into sleep on these nights — Andrea's hands in her hair worked like a narcotic — and she never knew when Andrea left, though she suspected it might be very late.

Tully blinked and noted with minor alarm that the *Cosby Show* was ending — she had really meant to watch it, but her mind was running loose again. *What would Andy think of this . . . this friendliness with his sister?* Tully wondered suddenly. Andrea's rare interest in men had always been a subject that Andy pretended innocence about, and he studiously avoided socializing with Andrea's crowd from the Institute. Last January, just six months before his death, Tully had collared him into going to dinner to celebrate Andrea's single-handed procurement of a survival grant — she was *so* smart! — on the grounds that he would really hurt his sister by not going, and besides, there had to be at least a couple of other men there.

She was wrong and gulped when the two of them approached the eight women seated at a long table in the Italian restaurant Andrea haunted. None of the women were like her — basically passive, Tully thought defiantly — and Andy could hardly deal with his sister when she was alone and sober, much less drunk and amplified by her friends and their accolades. Tully suffered through a long night, wincing every time Andy made a lame remark, which was often, since his charm failed utterly among women with quicker tongues.

Tully also remembered that night for Andrea's behavior, which turned moody near the end of the party and included several long, unfocused stares into Tully's eyes. Tully would then slip her hand to Andy's bony thigh, squeezing it softly to reassure herself and somehow ward off Andrea's attention. After one such silent episode, Tully looked to Andy only to see him staring open-mouthed at Clarissa, a huge bear of a woman who was vociferously recounting her dressing-down of a professor. "And then I said," Clarissa boomed with mounting enthusiasm, " 'Dr. Roundhouse, you are one *stupid* son of a bitch!' " Then she banged the table too hard, eliciting less laughter than she obviously expected. Tully caught Andrea rolling her eyes heavenward, as if to say, "This never happened," but instead she told Clarissa in a measured, maternal tone, "Good for you, Clarissa, good for you." Clarissa fell silent. Andrea's solemn gaze returned to Tully and brought a hot flush to her cheeks.

After the dinner Andy regained the playful mood that always delighted Tully. As they walked home together on the unusually warm winter night, he was actually skipping, ecstatic that he had simply escaped the dinner intact. Abruptly, he did an about-face and grasped his wife gently by the shoulders, his mock-serious face highlighted by the soft glow from a streetlamp above. "Tell me something, Tully — do you think some of Andrea's pals might actually be lesbians?" His face flickered both with conscious, high good humor and an animal brooding underneath, and Tully held her arms out in front of her as if grasping a barrel and boomed as gruffly as she could, "Andrew Waycross, you are one *stoo-pid* son of a bitch!" And he literally slapped his knee in delight, laughing a little too hard, but making Tully feel safe again.

She placed the remote on the floor carefully and leaned back against the couch, not smiling despite the warmth of the memory. He *was* a stupid SOB, she thought; he fell out of a tree and died instantly on a green suburban lawn. Dr. Morgan assured her that he died instantly, falling so flatly on his back, his head whipping violently backward over a limb that had fallen just seconds before him. It was the first job he'd kept for more than six months since the death of his parents, and he had begun to take it seriously, bringing home thick manuals on tree surgery Tully would never have guessed existed. In six months, he had gleefully told her on the bright August day before the fall, he was certain they could "work on that baby of yours."

Now Tully closed her eyes in deep reverence, the word *baby* bringing slightly less pain than in previous times. *At least I still have Andrea*, she told herself — *and the wooden man*.

She leaned over to the coffee table and grabbed a six-inch-high Buddha crafted from solid green glass. She hadn't told even Andrea about the wooden man, though as a psychologist, she might be expected to understand — no, she had told no one. The secret gave her a bizarre delight, as if she had the perfect right to tempt insanity on her long evenings alone. Yet she doled out the craziness to herself in careful portions: she thought she should not dream of the wooden man more than once a week. Tonight he was due again, so she ritualistically placed the small statue on the floor facing her and stroked his static braids much the way Andrea petted hers on their nights together. *It's the Buddha*, she told herself girlishly, *who brings the wooden man*. Then she recoiled slightly, the fear of madness hissing again along the back of her neck like a serpent. *What am I doing?* she thought. She didn't even know what a Buddha was supposed to be.

Andy had found him at a flea market, lying on his side between a corroded toaster and an old radio with vacuum tubes. Tully watched in wonderment as he hefted the thing like a trophy, then lofted it into the air, where it spun rapidly before returning to his open palm with a loud smack. "I kinda like this guy," he said breezily, and held it out toward the crafty-looking old fellow sitting on a worn pickup's running board, who seemed to have an eye for softies: "Twelve doller," he wheezed, " 'cuz it's real jade."

Tully shook her head instinctively at a sham even she could see through, but Andy only smiled like he was somehow getting the upper hand. He took twelve of their precious dollars from his wallet and handed them over, and Tully felt a flash of anger that made her wish she knew how to discipline her husband sometimes. Instead she thought, *Just wait till Andrea hears about this* — and Tully would make sure she did.

But despite Andrea's vitriolic lecture about budgeting, the Buddha became a kind of pet that even Tully began to regard as having a personality. That winter Andy announced that "the Buddha does love his basketball," and turned the statue to face the games that absorbed Andy's attention so completely. "Charlie," Tully once heard him say to the green man after his beloved Cavaliers dropped a very close one to North Carolina State, "what do you think of these Wahoos?" Tully paused in her dinner preparations to listen for what might come next — Andy was drunk on a six-pack, and he could be very funny in that condition. "Honey," he finally called, exaggerating his own slurred accent, "will you bring Charlie here a beer? He's absolutely *disgusted*."

Tully smiled to herself and cupped the Buddha's head in one hand. She touched it only to dust it or on the nights when she wanted to dream of the wooden man — who always hurt her, and whom she longed for in the depths where he hurt her.

The wooden man was Andy, of course, but she preferred not to think of him by name, and she had that choice now. He had Andy's features, although chiseled crudely out of the dark mahogany-like wood that composed his entire body. He would always be standing naked by her bed when she looked up from her lonely sleep, and she no more than had to think, *Yes*, before he climbed upon her with a terrible, awkward weight and pushed himself inside her, and then he would poke and prod her in clumsy desperation, the unforgiving straightness of his passion causing jabs of pain in every corner of her womb. From his throat came a strange, ghastly crying like the rubbing together of high limbs in a windstorm, and she was nearly consumed by fear of him — yet she clung to him, knowing he was all she had left. Sometimes she even felt something like an orgasm, a partial clutching and starry release nearly lost among her cries of pain and the wooden man's stormy groaning. But he never seemed satisfied — he would only stop inexplicably and stand, murmuring something low, then retreat into the closet Tully knew to be a forest at night.

The last time, the wooden man had ceased his murmuring after their lovemaking and looked at her tenderly — she had never noticed his carved black eyes before — and said clearly, "You. . . ," as though he meant to speak a sentence but forgot its meaning. Then he began his usual retreat, but Tully lunged after him, grabbing at the always-erect branch between his legs. She wanted it to be wet and slippery with herself, for then she would rub him and cajole him back to bed, no matter what the pain. But her hand closed on a dry wooden knob, and she awoke, her throat parched and her cheeks tracked with salt.

The next day, Tully was riding with Andrea in Andy's old green Volvo — Andrea had wordlessly claimed it several days after her brother's funeral — when she noticed the ball on the old gearshift vibrating near her knee. She tingled with the vision of herself straddling the stick, wondering if she could accept the cold shape inside herself, and then suddenly Andrea's hand covered the ball, downshifting the car decisively into third with the same strong-arm movement Andy would have used. Tully gasped so loudly at the merging of the visions that Andrea turned her head and said with great concern, "Tully? Are you all right?"

The widow felt tears sting her eyes — would she ever stop crying? — and looked into the beautiful dark face of her husband's sister, lit with Andy's fire. Tully felt it impossible to hold herself back. "I really do love you, Andrea," she said softly, lowering her eyes in a habit ancient and commanding.

Andrea pushed the ball and stick into neutral and the car drifted to the side of the road, rattling slightly as she braked to a stop. Tully felt a cold fear rise in her bosom — what would happen now? She imagined Andrea moving toward her, and wondered what she would have to do if the older woman tried to kiss her. Then she felt a warm familiar hand at the side of her neck, tugging softly at her hair as usual. Tully looked up. When Andrea spoke, the words came in Andy's rolling Georgia lilt — the accent she claimed to have lost in California years before.

"You're all the family I've got now, sweet Tully," Andrea said calmly. "I don't think I could live without you."

They seemed to shudder together, and soon both were crying freely in their unchanged positions. Andrea halted herself with a harsh half sob, drawing away from Tully to rub her nose vigorously with the back of one hand, forcing a laugh. "What would Andy do now?" she said with difficulty. "Both his women weeping at once would scare that boy to death."

Tully nodded, laughing and crying, noticing but not caring about Andrea's slip.

Now she lifted the Buddha back to its place on the table and made her way to bed. She swallowed one of the sedatives Dr. Morgan had given her, hoping that it would not black out her dreams as it sometimes did. She slipped out of her gown and into bed, absent-mindedly pulling and brushing the hair between her legs with one hand as she pictured the wooden man, preparing herself for the assault and the delirious intimacy. In a short time, her hand fell still on her thigh.

When he appeared, the wooden man looked different — he was partially covered in a growth of new leaves and twigs. She thought he looked like Robin Hood, and told him so silently; he only smiled broadly and reached out a hand, unexpectedly warm, to massage her breast, stealing her breath for a moment. Then he moved with uncustomary gentleness upon her and filled her perfectly, slipping his two broad, sinewy hands beneath her in a way he never had before. His largeness and care made her feel warm and liquid, as if her inner heat was melting everything solid in her body. "Oh, my Andy," she murmured, but she felt him begin to shake his head, and she opened her eyes to see what he meant.

"I am growing away," he said clearly, too slowly for her to have made a mistake in hearing, his half-wild face even showing a mischievous delight in his pun. "You . . ." — and then his speech stopped again, as before.

Tully shifted her hips slightly, the ease of their union washing her heart with pride, and grabbed the tree-man's face, traversed by two vines, with both her hands. "You can't leave without telling me!" she said with sudden gaiety.

"You. . . ," her faithful lover sighed, straightening his spine; a cool breeze blew between their bodies. "*You are the light of the world*," he sang inexplicably, driving hard into her with that part of him that was wooden, but growing, too, and it struck electrically into her like a root, branches, fibers, and tendrils burning all the way from her opening out to her toes and fingertips. Her head swam as the wooden man disappeared, and she felt a flash of lonesomeness, a farewell, before she moaned and rolled on her side to look toward the window she saw first thing every morning. But it was spinning, detached from everything, high in an early blue sky. As its rotation slowed she saw that its curtain was actually her filmy white gown, and when it had swirled to a stop around the window frame, a soft puff of air blew the gown open, allowing just an instant of absolute golden light to blast through the open window, stunning the young woman into wakefulness, which she entered gasping, clutching at her bedsheets, laughing with a long-lost gladness, forgiving her husband.

March 1986

THE WARD

Bruce Mitchell

L'Hôpital de Caen was an ancient building riddled with bullet holes from the Second World War and covered with eighty-six years of pigeon droppings. The hospital ward reflected this history, for it was filled mostly with wounded men who had long been forgotten, monuments to self-sacrifice or chance or poor judgment. The man next to my bed, a Monsieur Girard, had a severed spine from an American bomb dropped to liberate Normandy. He never mentioned the war nor expressed any bitterness over having been in a bed since 1945. Girard, like the other long-term residents of the ward, no longer cared about how he'd gotten there. The only thing that really mattered was the ward itself, the routine.

Each day began with the entrance of the nurse, Madame

Charoing, at exactly six, turning on the lights with a quiet but determined "Bonjour, messieurs," which meant "All right, gentlemen, it's time to get those bowels moving." Those of us who were immobile had ten minutes before she returned to collect the bedpans and urinals. If they were not full, they would not be collected until the next shift, which meant they would sit, stinking up the ward, until four. None of us wanted to compromise the hygiene of the ward, so there was immediate action. Each man, tested by some degree of immobility, pulled the bedpan or gooseneck urinal from under his bed, pushed it into its proper position, contorted himself into an often bizarre posture, and, looking straight ahead, strained to do his best. No matter what his disability, it was expected that each man perform. For many, this was the most demanding task of the day, for it often required monumental individual — and sometimes communal — effort. Monsieur Girard was lifted by two other patients and held as the pan was placed under him. I, who had a broken leg from a rugby match and was in traction, had to do a reverse push-up, arching my back, arms wrenched backward. We all had ten minutes.

After the ward's waste was collected by two subalterns, a peace settled over the room. Everyone had done his job, and now we could wait until the smells of success dissipated and breakfast was served. Breakfast was each man's reward, and he savored it, the victors slowly putting just the right amount of jam on a croissant or sipping just the perfect quantity of café au lait so that croissant and café would be finished at the same moment. There was very little talk during this time, each man's enjoyment honored and respected.

Once the breakfast trays had been collected, the ward was enlivened by the arrival of the newspapers. We all read, commented, and even argued across the room as the papers rustled and seemed to speak — crisp, vertical sheets emitting outrage or laughter or disgust.

In my second month in the ward, this daily ritual was interrupted by the arrival of another patient, a young man whose entire face, except for the hole of his mouth, was covered with bandages. He was carefully placed on the bed, quietly moaning, obviously in great pain. To my surprise, the papers continued to crackle and banter.

Lunch was served, and as usual we ate slowly, appreciating any little change in the menu and often continuing our discussion of issues raised by the morning papers. Lunch was a pulling back, a time to savor how well the morning had gone, to slow everything until a nap just sort of happened as a natural extension of our well-being. Even the moans and muted crying of the young man (who, we learned, had been in a motorcycle accident) did not change this well-rehearsed process. He was ignored, his agony merely an unpleasant background noise, until it began to interfere with the afternoon nap.

It was understood, but never stated, that conversations would diminish, sharp noises would cease to exist, and we would doze off without notice, one by one. Even I, half the age of most of the men, began to like this part of the day, letting my body find the most comfortable position possible (this itself a challenge), losing myself in the sounds of a distant lawn mower on the hospital grounds or the nurses' muted talking far down the hall. Even if one did not sleep, it was time to be quiet. The young man, blind and in excruciating pain, could not have known that he was inadvertently making enemies, that with every cry or violent inhalation, he was interrupting the established routine.

On the third day of his terrible recuperation, a small contingent of the ward's older men circled his bed to have a talk with him. His bed was across the room from mine, and I could catch only the gist, which was hard and unforgiving — they understood his pain, but he must bear it without disturbing the others. Incredibly, he responded with silence. Even in the wrenching horror of his eternal darkness, he sought to be accepted, to reach beyond his own suffering. He held out until that night.

The lights went out at exactly eleven, and for half an hour, the room was completely black. At 11:30, the first match flared, illuminating a face with a cigarette, then another and another. It was against hospital rules to smoke, but in the ward the rule was ignored if broken discreetly at night. The butts were always carefully extinguished, wrapped in toilet paper, and put into the bedpans for the morning pickup. For me, it was a wonderfully introspective time, smoking my meerschaum pipe, reflecting over my life and possible future, watching the orange, breathing fireflies of light, enjoying the communal darkness.

The first noise was hardly audible, like the whimper of a hurt child, a primal crying that went far deeper than language. That hurt had lost all anger and selfishness; it spoke only of its existence, incapable of any control, gurgling its rawness. The Gauloise cigarettes continued to glow, and the animal wound continued to bleed, and a tension began to develop between the ward and the pain. No one rang for a nurse, who could have been induced to give yet another morphine injection, that magic balm that lifted one out of the body, beyond the pain. Somehow we knew that this profound whimper was not asking for a placating flow of drugs, but that the hurt itself cease to exist.

One by one, the orange tips of light began to disappear, and the men of the ward began to move inexorably toward the boy. Monsieur Girard was carried, with difficulty and perhaps even at great risk, to a bed vacated for him, next to the boy. The same two "porters" removed me from my traction and carried me to an empty chair at the foot of his bed. The entire ward, eighteen men in various contortions, surrounded the blind, softly crying young man and began to touch him, tentatively, gently, wherever we could, feet, hands, legs, arms, shoulders, massaging the torn face through the bandages. No words were spoken, but we each, through our hands, tried to talk to the pain, to tell it that it was not alone, that we understood, that we felt it. Each hand spoke from its own experience, some firm and steady, some fluctuating between compassion and resolve, some light and afraid.

Our identities and feelings must have become well-known to the young man, for he became part of our routine. He knew that each night the first acrid smell of a Gauloise signaled that one of us would sit with him, hold him. The nurses never knew of this secret therapy, but when I was finally released, I knew that the young fellow would survive, that his darkness had become part of the ward's attending to itself.

March 1992

CONTRIBUTORS

Jaimes Alsop is the founder and publisher of the online magazine the *Alsop Review* (www.alsopreview.com). He is also the author of a poetry chapbook, *Small Lies* (Monkshood Press), and is currently at work on a longer manuscript titled *Sarah's Gate*. He lives in the San Francisco Bay Area.

When not walking in the wilderness or traveling to perform his poems, **Antler** lives in Milwaukee, Wisconsin. The author of *Factory* (City Lights) and *Last Words* (Ballantine), he's trying to find a publisher for his new manuscript, *Ever-Expanding Wilderness*. His work appears in many anthologies, including *Wild Song: Poems from Wilderness Magazine* (University of Georgia Press) and *American Poets Say Goodbye to the 20th Century* (Four Walls Eight Windows). "Somewhere along the Line" won a Pushcart Prize in 1993.

Dan Barker is executive director of the Home Gardening Project Foundation in Portland, Oregon — "giving away free raised-bed vegetable gardens to people in need since 1984." The foundation's current goal is to help start two hundred other garden-building projects around the country. The *Utne Reader* has named Barker as one of America's 100 Visionaries.

Yaël Bethiem and her husband, Doug, operate a spiritual center called Circle of Light out of their home in the Ozark Mountains outside Eureka Springs, Arkansas. She remains housebound, but with better medication she is able to be up more. She hopes to publish a book of meditations this year.

James Carlos Blake was born in Mexico and raised in Texas and Florida. He is the author of *In the Rogue Blood* (Avon Books), which won the 1997 *Los Angeles Times* Book Prize for fiction. His novel *Red Grass River* (Avon Books) was awarded the inaugural Chautauqua South Book Award in 1999, and his collection of short stories *Borderlands* (Avon Books) received the 1999 Southwest Book Award. He lives in a different town every year.

Dana Branscum currently writes in the margins of a life crowded with mothering and spiritual exploration. She has a spouse (male) and a guru (female). "Obviously I have some gender issues to work through in this life," she writes. "I haven't yet become the writer 'What It's Like' (my first publication) promised I could be. But more than a decade later, I'm grateful I'm no longer the damaged creature who wrote that beautiful piece of squalor." She lives in the town of Hope in the miracle of Maine.

Chris Bursk lives in Langhorne Manor, Pennsylvania, and teaches at Bucks County Community College. He is the author of six books, most recently *Cell Count* (Texas Tech University Press).

He is immensely proud of two things: being a grandfather, and being a contributor to *The Sun*.

Stephen T. Butterfield lived in Shrewsbury, Vermont, and taught English at Castleton State College for twenty-six years. Author of *The Double Mirror: A Skeptical Journey into Buddhist Tantra* (North Atlantic Books), he also sang traditional Celtic tunes and played the guitar and the bouzouki in the band When the Wind Shakes the Barley. He died in 1996.

David C. Childers grew up in Mount Holly, North Carolina, and attended the University of North Carolina at Chapel Hill, where he studied creative writing and other unmentionable subjects. His first two books of poetry, *American Dusk* and *The Monster* (both out of print), taught him not to expect much from being a poet. Since 1991, he has concentrated mainly on songwriting and has recorded three albums: *Godzilla! He Done Broke Out!*, *Time Machine*, and *Hard Time County*, all currently available on Rank Records. These days, he travels widely throughout the Southeast performing original songs with a four-piece rock-and-roll band. For more details, visit his website at www.davidchilders.com.

Gloria Dyc lives in Gallup, New Mexico, and has been published in numerous small journals and a number of anthologies, including *Southwestern Women: New Voices* (Javelina Press) and *Seven Hundred Kisses: A Yellow Silk Book of Erotic Writing* (HarperSanFrancisco). She is currently in search of a publisher for her manuscript *The Actress*, an erotic comic novel set in Detroit. She taught in Native American communities for eighteen years and was fortunate enough to learn the ceremonial ways of the Lakota and Dine. A student of Tibetan Buddhism, she has practiced under lamas in exile.

Jake Gaskins is a native of Greenville, North Carolina, and director of the writing center at Southeast Missouri State University in Cape Girardeau.

Since he left teaching in 1991, **John Taylor Gatto** has traveled 1.5 million miles on the lecture circuit, encouraging an end to forced schooling. In lighter moments, he likes to hunt for mushrooms, fire a pistol at water-filled beer cans, and raise money for a film about the history of education. His books include *Dumbing Us Down: The Hidden Curriculum of Compulsory Schooling* (New Society Publishers) and *A Different Kind of Teacher: Reflections on the Bitter Lessons of American Schooling* (Berkeley Hills Books). He divides his time between Manhattan and upstate New York.

Eleanor Glaze is a Tennessee fiction writer and playwright. Her books include *Fear and Tenderness* and the feminist science-fiction novel *Jaiyavara* (both out of print).

David Grant is coordinator of the Nonviolence Education and Training Program for the International Fellowship of Reconciliation. He currently lives in Europe and has been writing, off and on for fifteen years, a novel and a screenplay that will change the world. He writes, "Reading my writing of a decade ago forces me to reckon my failures and successes by my own yardstick. I've been to world capitals and to the dumping grounds of human rights, adding my two cents wherever I go. Do I have 'hope'? I've found it's not a requirement."

David Guy lives in Durham, North Carolina. The themes that he explored in essays for *The Sun* in the late eighties eventually culminated in a novel, *The Autobiography of My Body* (Plume). He is also the author of *The Red Thread of Passion: Spirituality and the Paradox of Sex* (Shambhala Publications) and has co-written two books with Buddhist teacher Larry Rosenberg. Since writing "What's Eating Me," he has: (1) resolved to quit weighing himself all the time, because it was making him too anxious; (2) gained eight pounds in a single year (which he probably never would have done if he'd been weighing himself); and (3) gone to Weight Watchers and lost eighteen pounds, which was probably overdoing it a little. He still has no idea how to eat.

Fred Hill has followed his dreams from a small cotton farm in the Arizona desert to a blacksmith shop in Idaho to managing a corporate website. For the past twenty years, he has lived on six acres in Meridian, Idaho, with his wife and assorted dogs, cats, and horses. Writing has been an occasional pastime, along with reading, landscaping, welding, traveling, and music. His articles have appeared in *Earth Watch*, *Llama* magazine, and *Western Horseman*.

John Hodgen won the 1993 Bluestem Award for his first collection of poems, *In My Father's House* (Emporia State University Press). A former gravedigger, now a creative-writing teacher, he reports that the jobs have their similarities.

Richard Hoffman is a poet, essayist, and fiction writer whose work has appeared in the *Hudson Review*, *Shenandoah*, *Bostonia*, *New Age*, the *Boston Globe*, and elsewhere. He has been awarded fellowships from the New Jersey Council on the Arts for poetry, the Massachusetts Artists' Foundation for nonfiction prose, and the Massachusetts Cultural Council for fiction. His memoir *Half the House* was published in 1995 by Harcourt, Brace and is available in paperback from Harvest Books. He lives in Massachusetts.

Dan Howell's collection of poems *Lost Country* (University of Massachusetts Press) was runner-up for the 1994 Norma Farber First Book Award and shortlisted for the *Los Angeles Times* Book Prize in Poetry. His writing awards include the Tom McAfee Discovery Prize (*Missouri Review*) and a citation for Notable Essay in *Best American Essays 1993*. His poems have recently appeared in *Another Chicago Magazine*, *Green Mountains Review*, *New American Writing*, and *Rhino*. He lives in Chicago and teaches at Northwestern University.

Ivor S. Irwin teaches writing at Roosevelt University in Chicago. He has recently published a book of nonfiction, *A Peacock or a Crow* (Willes E-Press), and placed work in *Playboy*, *Actos de*

Inocencia, *Sonora Review*, and *Sycamore Review*. He is a fanatical supporter of Manchester United Football Club, the only thing he misses about his hometown in Great Britain.

Gillian Kendall believes that consumption and creativity are opposites: if you spend enough time writing stories, dancing, making up songs, cooking, and so on, you won't have time to go to the mall. (And vice versa.) She lives in San Martin, California, and misses her friend Mark O'Brien, master of creativity, simplicity, and many struggles.

Carl-Michal Krawczyk is an associate professor of history at Washington State Community College. He is still very much in love with America and has been writing a biweekly column called "Images of America" for his local newspaper. Since the original publication of "A Soccer Hooligan in America," he has expanded the story into a novel. *The Sun* also published his story "He Wears Black." He lives in Marietta, Ohio.

When we last heard from **Kathleen Lake**, she was living in Mount Desert, Maine, and trying to get some money in the bank so she could go back to being a poet for good.

Lou Lipsitz has been sleeping a lot lately, often in the afternoons. He is planning a memoir about listening to jazz while awake during the night. His new poetry manuscript is called *In the Huge Surf of the Dark* and includes yet more autobiographical poems. He and his son were surprised to catch some northern pike while fishing in Lake Champlain. The pike were surprised, too. He continues practicing psychotherapy in Chapel Hill, North Carolina.

Pat LittleDog lives in rural Texas, where she tends her grandkids, ducks, and gardens. When *The Sun* published "A Clouded Visit with Rolling Thunder," she was known as Pat Ellis Taylor. She writes, "Reading the piece over, I enjoyed the memory of the great old man and his tremendous generosity of nature. His strong desire for peace and understanding put him on the road, traveling through many foreign and hostile environments, when he was well into his seventh decade. He was truly one of the great elders of our time."

Alison Luterman is a poet and writer living in Oakland, California, where she teaches, writes, dances, and plants garlic. Her book of poems *The Largest Possible Life* won the Cleveland State University Poetry Contest in 2000 and is available from Cleveland State University Poetry Center, (888) 278-6473.

Deena Metzger lives at the end of the road in Topanga, California, with her husband, Michael Ortiz Hill, and their two wolves, Akasha and Isis. Her books include *Tree: Essays and Pieces* (North Atlantic Books), *Writing for Your Life: A Guide and Companion to the Inner Worlds* (HarperSanFrancisco), the novel *What Dinah Thought* (Viking/Penguin), and the poetry collections *Looking for the Faces of God* and *A Sabbath among the Ruins* (both Parallax Press). Her latest book is *The Other Hand* (Red Hen Press), a novel about a woman astronomer meditating on the nature of darkness and light.

Lorenzo W. Milam was born at a very early age. When he was five, his parents ran away from home. He is founder of a dozen or so community radio stations and author of *Sex & Broadcasting* and *The Radio Papers* (both Mho & Mho Works). He is also Carlos Amantea, author of *The Lourdes of Arizona*, a book on family therapy. At times, too, he is Lolita Lark, radical-feminist editor of the online magazine *RALPH: The Review of Arts, Literature, Philosophy, and the Humanities* (www.ralphmag.org). When he is in a sour mood, he becomes Pastor A.W. Allworthy, author of *The Petition against God*.

D. Patrick Miller wrote books for three major publishers before seeing the light and founding his own press, Fearless Books of Berkeley, California. He is the author of *A Little Book of Forgiveness*, *The Book of Practical Faith*, and a novel, *Love after Life*. He publishes an online magazine, *Fearless Reader*, which presents "information with inspiration," at www.fearlessbooks.com, and his *Fearless Reviews* showcases the best works of other independent publishers.

Bruce Mitchell was born in the Midwest, where he has spent most of his life. For the past thirty-two years, he has been an English teacher at Evanston Township High School in Evanston, Illinois. In 1962, he broke his leg in a rugby match in Caen, France, and spent three months in L'Hôpital de Caen. He continued to play and coach rugby until 1980. During the summers, he has worked in a fishing camp in Canada, a brick factory in France, and a landscaping company in Australia. Recently, he and his wife have helped in the rebuilding of medieval fortifications in St. Victor laCoste, France. They have two grown children and one grandson.

Mark O'Brien was a poet, essayist, and journalist who lived in Berkeley, California. His life was the subject of Jessica Wu's film *Breathing Lessons*, which won the 1997 Oscar for Best Documentary. His books *Love and Baseball, The Man in the Iron Lung,* and *Five at the Center: Essays on the Disability Experience* are available from Lemonade Factory. He died in 1999.

Veronica Patterson is the author of three poetry collections: *How to Make a Terrarium* (Cleveland State University Press); *The Bones Remember* (Stone Graphics Press), with photographer Ronda Stone; and *Swan, What Shores?* (New York University Press). Her poems have appeared in numerous publications, including the *Southern Poetry Review,* the *Louisville Review,* the *Bloomsbury Review, Willow Springs,* and the *Colorado Review.* She lives in Loveland, Colorado, and has been an artist-in-residence in Rocky Mountain National Park. She loves loons and all their names, especially "cry-in-a-necklace."

Candace Perry's earliest writings appeared on small scraps of paper. "Candy Perry lived here," she wrote, added the date, and then tucked the folded paper into a crack in the wall or behind the window molding every time the air force told her family to move. She lives in Wellfleet, Massachusetts, and is currently at work on *Troopers*, a collection of short stories about the lives of military brats.

John C. Richards won the award for Best Screenplay at the 2000 Cannes International Film Festival for *Nurse Betty*, which he co-wrote with James Flamberg. (The script was based on an unpublished short story.) In May 2000, the West Coast Ensemble of Los Angeles produced Richards's play *The Picnic Basket*. He currently has two scripts mired in the swamp of Hollywood development; whether they'll ever emerge is anyone's guess. He continues to write short stories and has just completed his first novel, *Nolan's Gift*. "Heart Too Big" was inspired by a previous incarnation as a musician in southern Louisiana. He now lives in Los Angeles with his wife and three children.

Edwin Romond is the author of two books of poetry, *Home Fire* (Belle Mead Press) and *Macaroons*. He lives with his wife, Mary, and their son, Liam, in Wind Gap, Pennsylvania.

John Rosenthal is a photographer and writer who lives in Chapel Hill, North Carolina. The essays in "Amazing Conversations" were originally broadcast on public radio. Safe Harbor Books recently published a collection of his photographs, *Regarding Manhattan*.

Howard Jay Rubin is a full-time professional magician and part-time teacher of Jewish mysticism who lives in Hollywood, California. He was *The Sun*'s contributing editor from 1981 to 1985. Of the forty interviews he conducted for *The Sun* — including encounters with Allen Ginsberg, Pete Seeger, and Abbie Hoffman — his conversation with Rabbi Dovid Din had the greatest personal impact on him. Rubin credits their brief meeting with inspiring his own passionate inquiry into his Jewish spiritual roots. Rabbi Din died not long after the interview was conducted.

Sy Safransky is the founder and editor of *The Sun*. He left New York City and the world of professional journalism to start *The Sun* in 1974 and is the author of a book of essays titled *Four in the Morning*.

Tim Seibles has written five books of poetry, including *Hammerlock*, published in 1999 by the Cleveland State University Poetry Center. His work has appeared in several anthologies, among them *In Search of Color Everywhere: A Collection of African-American Poetry* (Stewart, Tabori, and Chang); *Outsiders: Poems about Rebels, Exiles, and Renegades* (Milkweed Editions); *A Way Out of No Way: Writings about Growing Up Black in America* (Henry Holt and Company); and *Dark Eros: Black Erotic Writings* (St. Martin's Press). He is a member of the English faculty at Old Dominion University in Norfolk, Virginia, where he teaches in the M.F.A. writing program.

Ona Siporin is the author of two poetry chapbooks, a book of essays, and a book of fiction and history with Anne M. Butler. She is also a regular contributor to Utah Public Radio. She recently returned from an extended stay in Italy and is currently working on a piece about the architectural furniture of Venice.

R.T. Smith lives in Lexington, Virginia, where he is the editor of *Shenandoah: The Washington and Lee Review*. His most recent book is *Split the Lark: Selected Poems* (Salmon Poetry).

Sparrow actually has an M.F.A. in creative writing, although he tries to write like an illiterate barbarian. He lives above a children's-novelties store in tiny Phoenicia, New York, and descends on Manhattan occasionally to watch plotless Asian films. Though widely read — he has finished Boswell's *Life of Johnson* and a thousand-page history of the Korean War — Sparrow is also deeply pious, seeing the benign face of God on every acorn and feather. His unique journey has taken him through two presidential campaigns and to the banks of the Ganges, where he contemplated the inner geometry. He lives with Violet Snow, the herbalist and huntress. Their daughter Sylvia can sing *My Fair Lady* in its entirety.

"Night of Dying" was **Maureen Stanton**'s first published essay. Since then, her work has been included in several anthologies, translated into Dutch, and nominated for the Pushcart Prize. Her essays have been published in *Creative Nonfiction*, *American Literary Review*, *Fourth Genre*, *Grand Tour*, and other magazines. She lives in Bath, Maine, and holds an M.F.A. from Ohio State University.

Michael Thurman is a practicing astrologer and therapist who incorporates fifteen years of study in Buddhism, yoga, and Jungian psychology into his astrological counseling. He lives in Lexington, Kentucky, and has taught workshops on transpersonal astrology at the Open Center in New York and the Oasis Center in Chicago.

Sallie Tisdale is the author of six books, most recently *The Best Thing I Ever Tasted: The Secret of Food* (Riverhead). She no longer practices nursing. Today she is writing, continuing her longtime Zen Buddhist practice, and getting old.

T.L. Toma lives in San Antonio, Texas. His novel *Border Dance* was published by Southern Methodist University Press.

Ashley Walker lives in Dallas, Texas, with her husband, Lynn Harris, and a shifting population of cats. For the past eighteen years, she's worked at various high-tech corporations as a marketing-communications writer. Prior to that, she was a painter, photographer, and faculty member at the Universities of Iowa and North Texas. She is currently working on a collection of short stories and a novel.

A NOTE ON THE TYPE

The text is set in a typeface known as Garamond, designed by the sixteenth-century printer, publisher, and type designer Claude Garamond. The Garamond typeface and its variations have been a standard among book designers and printers for four centuries. It was adapted for the computer by Adobe in 1989.

The titles are set in Carlton, designed during the early 1900s for the Stephenson Blake Typefoundry.

ALSO AVAILABLE

Four In The Morning: Essays By Sy Safransky
$13.95
Since Sy Safransky founded *The Sun* more than thirty years ago, his essays have been an integral part of the magazine's identity. This collection brings together thirty of his best.

A Bell Ringing In The Empty Sky: The Best Of The Sun, Volumes I & II
Each volume, $15.95
This two-volume paperback, totaling more than a thousand pages, is filled with the best interviews, short stories, essays, poems, and photographs from the first ten years of *The Sun*.

Sunbeams: A Book Of Quotations
$15.95
Unusual for a book of quotations, *Sunbeams* is designed to be read: the selections aren't organized into subjects but appear just as they originally did on the back page of *The Sun*, with all the startling juxtapositions that honor the humor and pathos of our lives.

Subscriptions to *The Sun:* Subscriptions are the blood in the vein, the meat on the bone, the smile on the face of a healthy magazine. Six months, $17. One year, $34. Two years, $60. Five years, $150. Ten years, $300. Lifetime (yours or ours), $1,000. We send a free set of back issues with every lifetime subscription. Canada and Mexico residents, add $15 a year. All other countries, add $20 a year.

Back Issue Sets
$350. More than 120 back issues of *The Sun* are still available. Individual back issues are $5 each. Orders of ten or more back issues receive a 10 percent discount.

Order by mail, phone, fax, or online.
The Sun
107 North Roberson Street
Chapel Hill, North Carolina 27516
(919) 942-5282 ▪ Fax (919) 932-3101
www.thesunmagazine.org